Environmental Discourses in Science Education

Volume 5

More information about this series at http://www.springer.com/series/11800

Arthur J. Stewart • Michael P. Mueller
Deborah J. Tippins
Editors

Converting STEM into STEAM Programs

Methods and Examples from and for Education

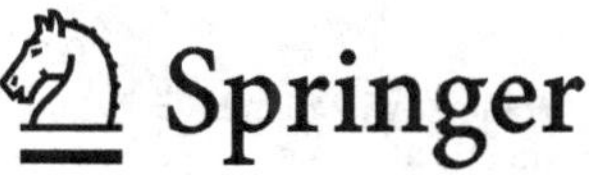
Springer

Editors
Arthur J. Stewart (Retired)
Oak Ridge Associated Universities
Oak Ridge, TN, USA

Michael P. Mueller
University of Alaska Anchorage
Anchorage, AK, USA

Deborah J. Tippins
University of Georgia
Athens, GA, USA

ISSN 2352-7307   ISSN 2352-7315   (electronic)
Environmental Discourses in Science Education
ISBN 978-3-030-25100-0   ISBN 978-3-030-25101-7   (eBook)
https://doi.org/10.1007/978-3-030-25101-7

This Springer imprint is published by the registered company Springer Nature Switzerland AG.
The registered company address is: Gewerbestrasse 11, 6330 Cham, Switzerland

Prologue

When the Soviet Union successfully launched the first artificial satellite, Sputnik I, in October 1957, things changed: that event started the space age and triggered a space race. Not long thereafter, multiple education initiatives in the USA began emphasizing disciplines related to Science, Technology, Engineering, and Mathematics (STEM). As one might expect, the emerging emphasis on STEM education manifested differently in different countries, and differently among agencies within countries, and within the USA, differently from state to state.

As changes occurred in STEM-related science education, science education research began changing, too. STEM education investigators began conducting studies and publishing papers on various aspects of STEM education. The number of STEM education-related papers published annually has increased substantially since 1990 (Fig. 1).

Of course, ideas about science education – philosophy, objectives, research methods, and the like – all change with time, too. So even as teaching began focusing more on STEM-related disciplines, with a rationale of educating students for future jobs in a technologically complex world (see, e.g., http://www.sciencepioneers.org/parents/why-stem-is-important-to-everyone), some education practitioners and investigators began questioning the emphasis on STEM. Perhaps (some discussion went), the focus on STEM was too narrow and too exclusive. Perhaps (other discussion went), STEM should be made more accessible to students who were marginalized by factors such as race, gender, or interests; and perhaps (some discussion went), STEM-focused students were at risk of losing touch with knowledge outside of STEM.

Education practitioners in non-STEM disciplines took some exception to the growing focus on STEM, because such focus could depreciate or marginalize teaching Arts or other non-STEM courses. The Arts (so the discussion went) are worthy of teaching on their own merits: the Arts do not need to be introduced *with* STEM disciplines, because they are sufficiently self-worthy even when taught alone.

Now, truth is, the business of splitting seems to be fundamental to human nature. Our sensory systems are very good at discriminating among inputs – sounds, colors, motion, odors, heat, or cold. Based on these sensory inputs, we classify, we

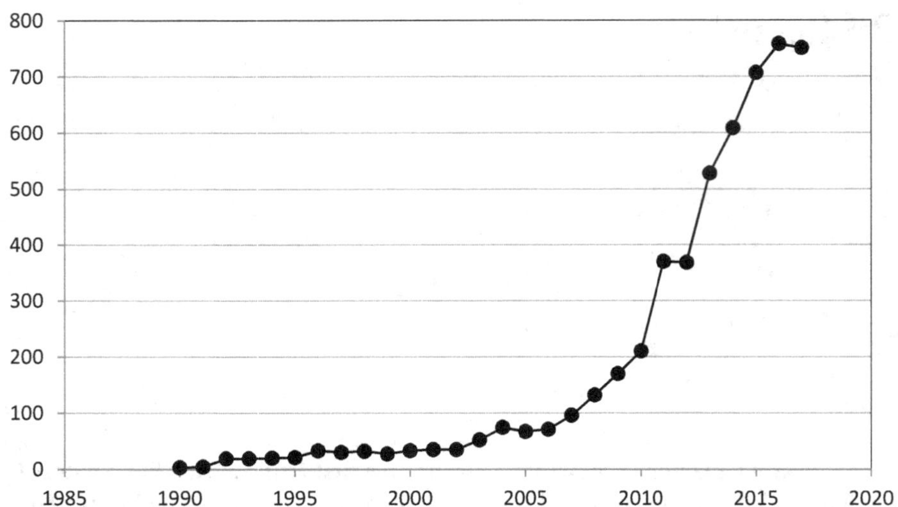

Fig. 1 Number of papers published annually (Y-axis), from 1990 through 2017 (X-axis), dealing with STEM education. Data are from a search on topic, keywords "STEM education," conducted using the University of Tennessee, Knoxville's version of Web of Science (Reuters) on February 5, 2018

discriminate, and we sort things out, moment by moment, this from that. We pose unstated questions about differences we detect, and we seek explanations based on causality, consciously or subconsciously, so as to better predict what might happen next. These abilities have survival value. They also fundamentally establish the roots of science. Instant by instant, we push sensory inputs into packets, arrange the packets, and rearrange them again, in association with inputs from other sensory structures: sounds with motion, estimates of size with estimates of speed, and odors with visual data and auditory data (are the burgers on the grill done yet?). Then, we compare the results of such efforts to our prior experience to make sense of the world. These constant artistic and individual creative fusions of sensory data allow interpretive assessments of conditions and things in our environments. By nature, we are all deep in the business splitting, just as we are all deep in the business of putting together. The matter largely devolves to what and how finely we split and how we go about putting the pieces back together.

Long ago, the Greeks considered the Arts and the Sciences to be the same thing. As science was formalized, the process of science was deliberately rendered "stand-offish," because the idea then was that science should be objective: different viewpoints, if tolerated at all, were to be constrained narrowly to data-driven interpretations, with minor speculation sometimes permitted near the end of the discussion section. An entire scientific language developed, focusing on details – passive voice, past tense, sentences built often on Germanic structure, with long, gnarly sentences and not infrequently with the verb, like this, near the end of the sentence hanging. That's all part of our STEM heritage (cf. Bohm 1998).

So, on first blush, the current trend of incorporating the Arts into STEM teaching, thereby establishing STEAM programs, is merely a return to an earlier state – an arc and a swoop from here towards where we were. The chief difference is that now, we know a lot more about the bits and pieces of things that go into STEM and the Arts, and we know a lot more now about what works, and what does not work so well, in terms of education effectiveness and educational policies. We have a lot more data, and we have much better access to these data. We have a deeper and broader context for analyzing and understanding these data, so we can establish more effective pathways for making changes in how we teach, and we can develop stronger and more effective arguments for why policies should or should not be changed for better educational outcomes.

Many chapters in this book testify to some of these ideas. Some of these chapters advance the idea that STEAM helps students become more creative or become better at problem-solving, for example. We appreciate such arguments. But to be fair, we should mention several other points that we should not forget about as we again start constructing and trundling out new teaching methods and revise curricula to be more inclusive. Let's start with the idea that an important education objective – and one reason for converting towards STEAM education – is that of helping students develop better problem-solving skills. Well and good. This seems like a worthy objective. But why wait to start this effort until after children are in school?

A study by Leonard et al. (2017) reports this: "…adult models causally affect infants' persistence and that infants can generalize the value of persistence to novel tasks." Let's restate this important idea. Leonard et al.'s (2017) study shows that adults can model persistence and effort in solving problems and that infants who watch these adults do these things learn value in time and effort spent in problem-solving. They – the infants – then apply these behaviors (i.e., they spend more time and effort) to new problems they encounter. So, this is a strong point supporting Michael Roth's chapter on the value of lifelong learning but on the younger side of the age spectrum! One take-home message here could be this: teachers, don't make things too easy for your students. Rather, show how, with work and perseverance, you *can* solve the thing. Solving wicked problems will require lots of perseverance in addition to creative problem-solving skills. STEAM alone might not be enough to save us. Modeling effort that leads to success – even incremental success – might help.

But what about this? Other studies show that infants are, in part, hardwired to understand certain basic aspects of cause and effect. Experiments by Waismeyer et al. (2015), for example, show that toddlers can learn about cause and effect without trial-and-error or linguistic instruction, simply by observing the probabilistic patterns of evidence resulting from the imperfect actions of other social agents. Furthermore, Gopnik (2012) notes that "preschoolers test hypotheses against data and make causal inferences; they learn from statistics and informal experimentation, and from watching and listening to others. … (these discoveries) suggest both that early childhood experience is extremely important and that the trend toward more structured and academic early childhood programs is misguided." Such arguments suggest early development of critical thinking and creative problem-solving skills and that early intervention expanding this ability might pay big dividends. Similarly, early-grade intensive intervention programs show long-term education outcome benefits (c.f., Reynolds et al. 2018). Jaschke et al. (2018) report a positive effect of long-term music education on cognitive abilities, such as inhibition and planning, and a positive effect of visual arts education on a visuospatial memory task. So once again, let's consider how, and when, we might most advantageously teach STEAM, or STEM, and how to go about it. We need more creative, passionate scientists with cross-disciplinary interests and expertise if we're to solve the wicked problems that now plague us.

Later deployment of STEAM alone might be helpful, but the benefits of this approach may be smaller. Or perhaps, the benefits of later interventions are large, relative to the way we generally teach now. Or perhaps, STEM is better than STEAM, or perhaps, starting STEAM earlier is much better than if started much later. The point here is that there's a great deal about optimally implementing STEAM that we don't know yet – and yet, even as we puzzle on it, we're still going forward. That's why this book is so timely and important.

The authors of the 18 chapters offered here offer multiple perspectives on STEAM at various functional, educational, and spatial scales. They provide ideas, data, and contextual information and recommendations about STEAM efforts, ranging from elementary/middle school-level STEAM education through high school and into the university level and beyond. Spatially, the chapter authors address STEAM education considerations relevant to single teachers in individual schools, through the school system level, to whole-country level STEAM implementation efforts (cf. Jeong et al., Chap. 16). Also varying, by intent, are different philosophical perspectives among chapter authors: What should we be doing with respect to STEAM, and why? The chapters offer, too, a satisfying range of nuts-and-bolts information for policy-makers, superintendents, and prospective teacher-implementers as they contemplate injecting STEAM ideas into STEM education or develop new ways to make STEAM work.

But things always take on deeper meaning with deeper context, and we want to keep pushing our knowledge forward and to use the best possible evidence for making good decisions as we learn more about STEAM. So, let's take a moment here to consider how our understandings of STEM education, STEM education research, and STEAM education relate to one another.

One way to consider these three things is to plot the number of papers published annually for topic searches conducted on each of the three terms. We did this, for searches conducted on February 2, 2018, using Web of Science (University of Tennessee's subscription), with the following search parameters: years, 1990–2017, and search terms (topic), enclosed in quotes to ensure exact phrase. The three search terms were "STEM education," "STEM education research," and "STEAM education" (Fig. 2). Specifying exact search parameters is important because various subscriptions are available for accessing the Web of Science system and because the Web of Science core collection can change later if new journals are added.

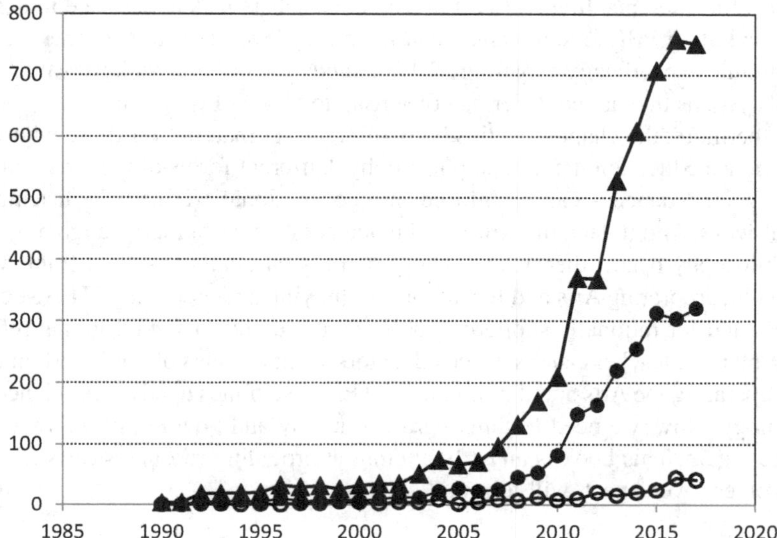

Fig. 2 Y-axis, number of papers published per year; upper curve (triangles), Web of Science (WoS) search term; topic, "STEM education." Middle curve (solid circles), number of papers published per year (X-axis, years) for WoS search term; topic, "STEM education research." Lower curve (open circles), number of papers published per year for WoS search term; topic, "STEAM education"

As one might guess, research on STEM education lags STEM education papers by about 3–5 years – approximately the life-span of typical grant. The number of papers published annually on STEAM education currently appears to lag STEM education research by 7 or 8 years (based on 50 publications per year as a comparison point). We might then anticipate that publications classified as focusing on STEAM education will begin increasing rapidly within the next 2–3 years, and they should then continue to increase rapidly thereafter for about a decade more, presuming similarity with the historical increases in number of publication per year in the STEM education and STEM education research categories. So, STEAM investigators, some advice: buckle your seatbelts and begin drafting your proposals.

The chapters in this book set the stage for multiple excellent opportunities for future STEAM education researchers. For example, several chapters describe potentially effective STEAM interventions and describe outcomes of such interventions, whereas other chapters describe rigorous reasonable assessment pathways, largely unencumbered with data. STEAM research clearly could benefit from uniting these two aspects, because in the world of STEAM, we have more anecdotal information than we have compelling, evidence-based data on what actually works best.

The chapters in this book also offer clear guidance for STEM- or STEAM-related policy-makers. Problem-based learning (PBL) methods (see Chaps. 6 and 7, by Ubben) are identified by the authors of several of the chapters as the preferred platform for implementing STEAM. Yet many teachers lack experience in designing or operating PBL lesson plans. So at the policy level, effective teacher training opportunities specifically targeting PBL methods could be especially valuable in school systems that are considering converting to STEAM programs.

Furthermore, the chapters in this book offer strong material for those who work with science education theory and philosophy. Different forms of the Arts – visual, music, poetry, dance – can be infused into conventional STEM-style teaching in various ways. And in fact, the Arts and the Sciences were originally taught together. But philosophy begets theory, and theory permits rationally designed frameworks for best incorporating Arts and for rationally studying and assessing STEAM effectiveness and for rationally supporting policies needed to drive educational reform. Policy-makers, teachers, and science educators spent decades of hard work in carefully separating the Arts and the so-called STEM disciplines to advance science and technology. Now, we need to think hard about why and how to put the two areas together again. This book is an early attempt at providing resources for such task. Many others, we expect, will follow.

Oak Ridge Associated Universities   Arthur J. Stewart
Oak Ridge, TN, USA
University of Alaska Anchorage, College of Education   Michael P. Mueller
Anchorage, AK, USA
University of Georgia   Deborah J. Tippins
Athens, GA, USA

References

Bohm, D. (1998). *On creativity* (L. Nichol, Ed.). London: Routledge, 125 p.
Gopnik, A. (2012). Scientific thinking in young children: theoretical advances, empirical research, and policy implications. *Science, 337*(6102), 1623–1627.
Jaschke, A. C., Honing, H., & Scherder, E. J. A. (2018). Longitudinal analysis of music education on executive functions in primary school children. *Frontiers in Neuroscience, 12*, Article 103, 11 p.
Leonard, J. A., Lee, Y., & Schulz, L. E. (2017). Infants make more attempts to achieve a goal when they see adults persist. *Science, 357*(6357), 1290–1294.
Reynolds, A. J., Ou, S. R., & Temple, J.A. (2018). A multicomponent, preschool to third grade preventive intervention and educational attainment at 35 years of age. *JAMA Pediatrics* https://doi.org/10.1001/jamapediatrics.2017.4673. Published online January 29, 2018.
Waismeyer, A., Meltzoff, A. N., & Gopnik, A. (2015). Causal learning from probabilistic events in 24-month-olds: An action measure. *Developmental Science, 18*(1), 175–182.

Book Abstract

For centuries, education usually incorporated aspects of the Arts and the Sciences into a single cohesive education framework. Some separation of these two discipline areas began in the 1950s with the onset of the space race, which was triggered by Sputnik I. And now, 50–60 years after Sputnik, there's an emerging effort to reblend the Arts and the Sciences. We refer to the results of these latter efforts as STEM-to-STEAM conversions: that is, some educators now are beginning to add the Arts back into educational programs focusing on Science, Technology, Engineering, and Mathematics or beginning to infuse science content into the Arts. The authors of this book provide a diverse medley of views on the nature of these efforts at various spatial scales. The STEM-to-STEAM information provided by these authors runs the gamut from workshop room and classroom scale efforts to whole school efforts, to school district efforts, to whole country efforts. The authors offer tools, perspectives, examples of methods, and STEM-to-STEAM conversion results data to help readers understand this fascinating science education trend.

The range of information provided by the authors of this book is not just broad with respect to scale. The information also provides readers with diverse windows into philosophical rationales associated with the efforts to push education from STEM to STEAM. Changing from one educational strategy to another at a whole school, school district, or whole country scale requires considerable effort by many people, exerted over years. This book is among the first to attempt to capture a significant change-in-scope educational thrust early in change history. So yes: it will be heterogeneous and incomplete. But it also will be interesting, and we think it will offer many readers a multitude of new ideas, which hopefully will motivate teachers, policy-makers, and educational program directors to begin devising and encouraging new and more effective ways to teach science. We need broader-thinking, courageous, highly interdisciplinary students – many of them! – if we're to solve the multiplicity of "wicked problems" we now face.

Oak Ridge Associated Universities   Arthur J. Stewart,
Oak Ridge, TN, USA
University of Alaska Anchorage, College of Education   Michael P. Mueller,
Anchorage, AK, USA
University of Georgia   Deborah J. Tippins,
Athens, GA, USA

Contents

About the Editors

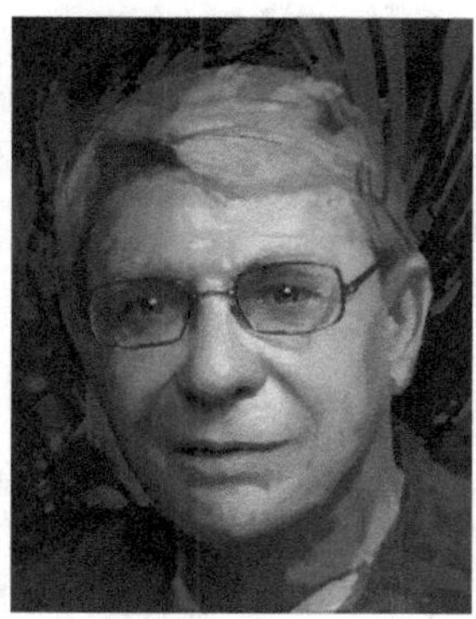

Arthur J. Stewart of Lenoir City, Tennessee, is a scientist, science educator, and poet. He earned his Ph.D. at Michigan State University in Aquatic Ecology and worked at the Department of Energy's Oak Ridge National Laboratory for 17 years as an ecologist and ecotoxicologist before becoming a science education project manager for Oak Ridge Associated Universities. In addition to publishing over a hundred scientific articles, book chapters, and technical reports, his science-inspired poetry has been published in many literary magazines and in more than a dozen anthologies.

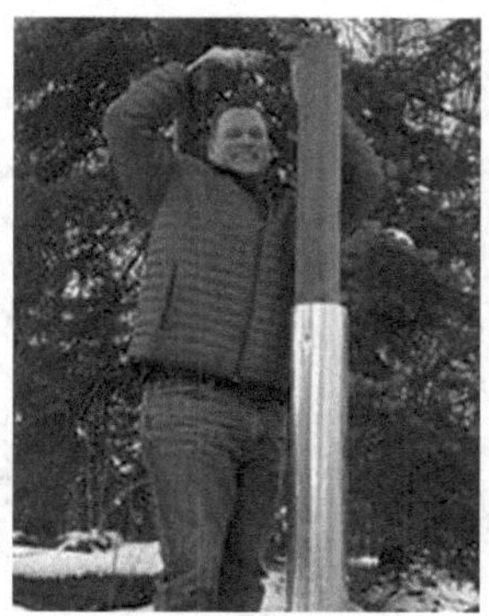

Michael P. Mueller is a professor of secondary education with expertise in environmental and science education in the College of Education at the University of Alaska Anchorage. His philosophy now focuses on how privileged cultural thinking frames our relationships with others, including nonhuman species and physical environments. He works with teachers to understand the significance of cultural diversity, biodiversity, and nature's harmony. He is the coeditor in chief of *Cultural Studies of Science Education*.

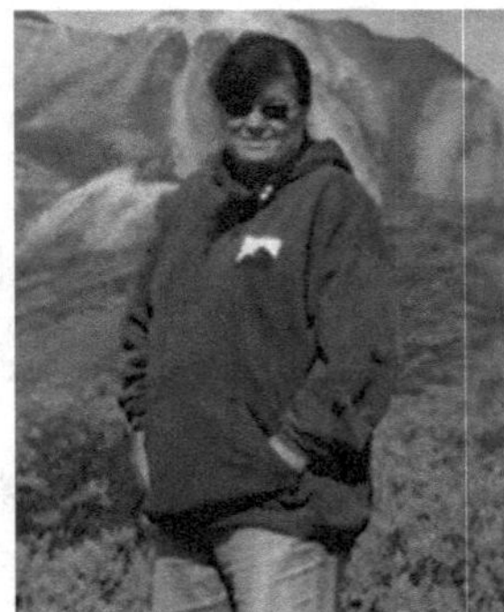

Deborah J. Tippins is currently a professor in the Department of Mathematics and Science Education at the University of Georgia. Her scholarly work focuses on encouraging meaningful discourses around environmental justice and sociocultural issues in science education.

The original version of this book was revised: The chapter author's, "Sophia (Sun Kyung) Jeong", first and last names has been corrected. The correction to this book is available at https://doi.org/10.1007/978-3-030-25101-7_19

From STEM to STEAM: How Can Educators Meet the Challenge?

Mariale M. Hardiman and Ranjini M. JohnBull

1 Introduction

The global competitiveness for a workforce capable of innovation and creative problem-solving takes us back to the late 1960s, with the launch of the Russian satellite Sputnik, and it leads us now to focus on STEAM—the integration of the arts into the study of Science, Technology, Engineering, and Mathematics (STEM). The journey of integrating the creative forces of the arts with scientific inquiry began with a national spirit that was more than just a race to space. The wake-up call that Sputnik triggered led to an urgent focus on preparing a pipeline of innovative thinkers able to propel our nation to greatness in technological advances and innovations across all sectors of industry and intellectual endeavors. The answer coming from policy-makers at that time was to reform our education system by focusing on a "back to basics" approach by setting standards for teaching core subjects, and holding elementary and secondary schools accountable for student performance.

Years later, the 1980s' publication of *A Nation at Risk: The Imperative for Educational Reform* (National Commission on Excellence in Education 1983) moved school reform to a heightened level, by focusing public discourse on comparisons of international test scores of US students with those of other countries. The weak performance of US students compared with those of other countries created fear that our nation will lose economic strength. The fear propelled a renewed call for reforming of our nation's education systems.

The conflict still wages. Indeed, the achievement gap between US and foreign students on international tests such as the Programme for International Student Assessment (PISA)—an assessment focused on higher-order thinking skills in mathematics and science—places the performance of our nation's students 31st

M. M. Hardiman (✉) · R. M. JohnBull
School of Education, Johns Hopkins University, Baltimore, MD, USA
e-mail: mmhardiman@jhu.edu; Rmjohnbull@jhu.edu

© Springer Nature Switzerland AG 2019
A. J. Stewart et al. (eds.), *Converting STEM into STEAM Programs,*
Environmental Discourses in Science Education 5,
https://doi.org/10.1007/978-3-030-25101-7_1

among 35 industrialized nations in mathematics, and near the middle of the pack in science (Barshay 2016). Scores on the National Assessment of Educational Progress (NAEP) have been largely flat over time. In response, federal legislation, including the 2001 No Child Left Behind Act and now the Every Student Succeeds Act, accelerated the focus on standards in English/language arts and mathematics as well as high- stakes accountability metrics for schools (National Assessment of Educational Progress 2015).

Enacted federal legislation and work by national and state entities resulted in several clear shifts in educational practices and policies over the last two decades. First, a well-documented "narrowing of the curriculum" effort was implemented to make more room in the school day for tested subjects. This effort often resulted in schools eliminating or reducing nontested subjects. Schools with tight budgets were particularly affected, and a disproportionate share of those schools service children in low-income communities. Walker (2014) found that 81% of elementary teachers reported that time devoted to mathematics and language arts instruction resulted in less time for other subjects. Among the nontested subjects that lost instructional time in schools, the arts, in particular, suffered significantly (Mishook and Kornhaber 2006). In 2010, the US Department of Education reported that 40% of high schools did not require any coursework in the arts for graduation (Bryant 2015, April 27). A national survey conducted in 2011 revealed that half of all teachers reported that art and music classes were being eliminated in their schools (Walker 2014, September 2).

At the same time, another significant change in education policies and practices focused on industries' need to employ workers with highly refined skills in STEM disciplines. This change brought to light the critical shortage of workers in STEM fields and suggested a need for schools to build a pipeline of students pursuing STEM careers. Thus, while the arts were taking a back seat in school curricula, relegated to the category of "fringe subjects," STEM-centric initiatives proliferated.

After Sputnik, Presidents D. W. Eisenhower and J. F. Kennedy directed the country's focus to the sciences and technology by establishing the National Aeronautics and Space Administration (NASA), and this attention on the sciences and technology continued over the next several decades (Marick Group 2016). The term STEM was first established in the 1990s by the National Science Foundation, as the National Science Education Standards were launched and curriculum began reflecting the country's STEM focus more strongly.

In 2005, STEM began to draw greater attention in education policy and reform when an alarming report from the US National Academies of Science, Engineering, and Medicine showed that the USA was falling behind other countries with respect to mathematics and science preparedness. In 2009, President B. H. Obama announced the *Educate to Innovate* initiative with the goal of moving US students to the top of the pack in science and mathematics achievement over the subsequent 10 years (Marick Group 2016). This initiative included increasing federal investment in STEM and preparing 100,000 new STEM teachers by 2021. However, despite these initiatives, the USA is still lagging behind other developed countries and is not producing enough STEM-proficient workers to meet market demands (American Youth Policy Forum 2012).

2   STEM to STEAM

As schools focused on how to increase the number of students entering the STEM pipeline, many arts advocates pointed to a growing body of research that showed the power of the arts to influence broad domains of learning. While many agree that the arts are important "for their own sake," strong evidence also suggests that arts can improve learning in other disciplines and help to engage students in subject matter more efficiently than traditional instructional techniques. The arts, when used as a pedagogical tool for teaching non-arts subjects—known as arts integration—show promise for learning and retaining academic content, transferring knowledge to other domains of learning, and developing creative thinking and problem-solving skills. For example, Burnaford et al. (2007) describe arts integration as a way of promoting the transfer of knowledge and skills from arts to non-arts domains by helping students draw connections among different disciplines. Anderson (2015) found, through qualitative analyses, that integrating dance arts into mathematics lessons for students with disabilities, over 1 month periods, improved mathematics achievement, dance arts skills, and social emotional learning. Additionally, Scripp and Gilbert (2016) developed a "multiple literacy" tool to assess shared skills between music, literacy, mathematics, and the arts after engaging in arts-integrated lessons. Their findings (op. cit.) indicated that arts literacies, mathematics and reading learning, and achievement all increased over time and that these gains were statistically greater than for students in control schools, where arts-integrated lessons did not occur. Further, Scripp and Gilbert (2016) found greater levels of arts skills, competencies, and insights versus the control schools.

A recent quasi-experimental study by Teske and Pittman (2017) explored the effects of arts-integrated learning on memory for science content within a middle school in Iowa. This study took place over a 4-day period in 1 week, and included two lessons in which students explored the form and function of modern or fossil organisms in both arts-integrated and conventional teaching conditions. Six course sections included 128 students who engaged in one lesson via arts-integrated instruction; the second lesson occurred through conventional science instruction. Content-based assessments were then applied: the investigators reviewed the responses students constructed to questions, and analyzed student drawings to evaluate memory for content. Post-test results for the two groups of students did not differ statistically, but the delayed post-test results indicated large effects and a statistically significant gain in learning for students who had the arts-integrated lessons. These findings affirm Hardiman et al.'s (2014, 2019) findings, that arts-integrated science instruction benefits long-term memory for science content.

Csikszentmihalyi (1997) described the importance of the arts in cognition by underscoring the emotional responses that the arts can produce, constructing novel ways of thinking that "…break through the gray affectless daily routines and expand the range of what it means to be alive" (p. 36). He reports how the arts can engender a state of deep concentration that leads to the "aha" of creative thinking.

Adding the arts to STEM subjects may be essential to students effectively learning the critical twenty-first-century skills important to the workforce of today. No longer is knowledge of a content area alone enough. According to the Partnership for Twenty-First Century Skills, "many of the fastest-growing jobs and emerging industries rely on workers' creative capacity—the ability to think unconventionally, question the herd, imagine new scenarios and produce astonishing work" (2008, p.10). A workforce capable of creative and collaborative problem-solving requires techniques not necessarily supported by conventional teaching. The arts may serve as a springboard for imbuing traditionally taught STEM subjects with the types of creative thinking that encourages innovation.

Recent research seems to support this view. For example, Kong et al. (2014) and Kong and Huo (2014) found that infusing STEAM activities into elementary schools resulted in statistically significant increases in positive attitudes toward science education, higher levels of self-efficacy for STEAM subjects, and increased interest in scientific learning. Similarly, Lim et al. (2015) found that design-based STEAM programs for 5th- and 6th-grade students resulted in a statistically significant increase in students' awareness about careers in biotechnology and medical engineering. Another study investigated the effects of STEAM approaches to teaching computing coding and programming (Yee-King et al. 2017): in this study, students who learned programming through an arts-integrated approach earned higher grades and developed more sophisticated programming skills, compared to students who learned programming in conventional classes. Oner et al. (2016) found that middle and high school students who engaged in STEAM-project-based lessons believed that they used their creativity within the instructional activities: this result indicates an increase in self-efficacy for these subjects. Similarly, Rule et al. (2016) explored the effects of a project on creating dioramas of the lives of successful women mathematicians. Qualitative analyses of 24 5th-grade girls indicated that students were more motivated, had higher levels of creativity, and increased arts and cognitive skills. Taken together, such studies suggest that arts-integrated learning within STEM not only improves academic achievement: it also affects attitudes toward STEM careers, interest in these subjects, and awareness of STEM-related career trajectories.

3   Proving Some Causal Connections: Science via the Arts

While research on the effectiveness of arts integration in general, and within STEM subjects in particular, is encouraging, most studies are correlational and do not claim causal connections. To add to the body of research on the potential effectiveness of STEAM and to begin providing evidence for causation, our research team at the Johns Hopkins University School of Education conducted pilot randomized control trials to test the effectiveness of science units taught through traditional STEM approach compared with STEAM units that incorporated various arts-based activities.

We developed science units for 5th-grade students in astronomy and ecology, each covering approximately 15 days of instruction. For each content area, one unit was designed as STEM, using conventional instruction (control condition). The other version embedded various art forms into teaching activities to create arts-integrated units (treatment condition). The unit pairs were closely matched to provide the same content; all activities were allotted the same amount of time; and each unit used similar modalities (visual, auditory, and kinesthetic). For example, when students were singing a song to reinforce content in the arts-integrated unit, the students in the control unit were engaged in choral reading of the same content. Four randomized groups of students received one body of content through an arts-integrated unit, and a second body of content in a control unit that used a traditional presentation. The design was counterbalanced so that half of the students received arts-integrated units in the first session while the other half received the control units first.

Our team designed curriculum-based assessments that we administered at the end of each unit. From pretesting to posttesting, we found that students learned about the same amount of information, regardless of the way they were taught. As expected, students did not initially learn more or less when taught through arts integration. However, 10 weeks later, delayed test scores showed significantly better student performance for the arts-integrated condition. This result indicates that infusing the arts into science disciplines may help students remember content better, compared to traditional teaching. The study revealed a differential benefit when comparing students according to levels of proficiency in reading. Interestingly, compared to higher-achieving students, students at the lower levels of achievement were more likely to retain significantly more science content when given arts-integrated lessons, compared to traditional science instruction.

In an expanded study with 16 randomly assigned groups, we tested four sets of arts-integrated treatment units matched to control units, using four science-content areas: astronomy, chemistry, life science, and environmental science. Similar to the first study, students performed better in delayed testing when taught using arts-integrated instruction, and students at the average and lower levels of achievement benefited the most (Hardiman et al. 2019). These studies provided evidence for the efficacy of using the arts to teach science in a STEAM approach, compared to teaching science using traditional STEM methods.

Our findings also raised some interesting questions about whether or not learning through the arts resulted in residual benefits. Findings from the second study suggested the possibility that once taught using arts-integrated instruction, students may later apply the strategies they learned, even during subsequent instruction through conventional methods. Data from the second study shows that students who experienced the arts-integrated units first performed significantly better in subsequent non-arts-infused units, compared to students who had never experienced the arts-integrated approach.

While other factors, such as familiarity with the unit structures, may account for better performance of all students the second time around, our findings nevertheless raise questions that warrant further investigation. In particular, emerging research on creative thinking and problem-solving might connect learning with and through the arts as a fruitful alternative to conventional methods.

4 Creative Thinking and Problem Solving

We suggest that the arts may influence not only memory for content and engagement in learning but also enhance creative thinking abilities and problem-solving skills—and these two aspects have become paramount in the call for teaching twenty-first-century skills, because they are deemed essential for the much-needed workforce in STEM careers. Rostan (2010) supports this view, positing that engaging in high-quality arts learning demonstrably cultivates creativity and provides an advantage for related forms of critical thinking.

While traditional instruction focuses on convergent thinking (in which students are seeking the right response), a hallmark of creativity is divergent thinking (in which multiple solutions and ways of thinking are encouraged). Within traditionally taught STEM instruction, teachers might plan a lesson in which a problem is presented, a solution is sought, and processes for finding the solution are clearly defined. Lessons that involve more divergent types of learning allows students to define problems, and then encourages them to find as many different ways to solve the problem as possible (Beghetto 2017). The arts can drive this divergent process by providing students with multiple ways to think about, and demonstrate, content and concepts.

For example, students can learn about the theoretical perspectives of different scientists by embodying the scientists in dramatic play, and even enact debates between different scientists with opposing views. They can explore different chemical processes that take place during the creation of art, such as learning how silk paintings can be made by dissolving Sharpie ink in water or how a water-soaked popsicle stick can be formed into a bracelet. Movement and dance can be used when exploring different states of matter, with students using "tableau"—a group of motionless performers depicting a scene or story—to demonstrate the properties of each state. Students can be asked to draw, in various ways, in science lessons: they could paint the different components of cells or make sketches to visually demonstrate vocabulary concepts. They can use simple tunes to recite the phases of the moon or to remember algebraic formulas.

Studying works of art also can provide valuable learning opportunities for students. When studying chemistry, students can examine works by artists such as Etsuko Ichikawa and Andy Goldsworthy, who use chemical processes such as burning paper and the formation of ice in their work. During astronomy lessons, historical drawings of the universe can be used to show how humans' understanding of the universe has changed throughout history.

Providing opportunities for students to create their own designs can also be beneficial, especially during project-based learning. When studying the environmental impact of humans on Earth, students can take on the role of landscape architect to design an environmentally friendly city neighborhood. In lessons about sustainable resources, T-shirt design using renewable and nonrenewable fabrics can help to bring this concept to life.

5 STEAM Instructional Framework

The examples above show that activities that embed the arts into STEM instruction can be accomplished through simple but potent arts-based activities. Such activities can help students acquire and retain content and encourage them to make interdisciplinary connections linked to required content standards. Arts-integrated teaching, however, requires a shift from traditional pedagogical approaches. In our experience, several important features are essential for successful arts-integrated STEAM instruction. First, professional development and ongoing coaching support is necessary to help teachers plan lessons in which arts-based activities support and enhance content instruction that can be accomplished within a reasonable timeframe.

We have observed that some teachers intuitively know how various art forms can support the content they are teaching. However, we also see that most teachers benefit from training, collaborative planning, and ongoing coaching support. Further, school-based visual and performing arts educators and artists-in-residency programs are important resources for teachers moving from STEM to STEAM instructional approaches, particularly when they all discuss and incorporate the required content standards.

Another critical tool for teachers is ready access to an instructional framework that guides them in understanding how arts-integrated instruction is part of a total instructional program, not just an activity "pasted on" at the end of the lesson. We have successfully used the Brain-Targeted Teaching (BTT) Model® (Hardiman 2003, 2012) as a framework for arts-integrated instructional unit planning. This model, informed by research from the learning sciences, delineates six components of the teaching and learning process. Fundamental to the model is using the arts to support learning across six domains: (a) establishing the emotional climate for learning by creating a positive learning environment and mitigating factors that produce stress; (b) designing the physical learning environment to foster student engagement through the use of novelty, displaying students' artistic work, and attending to environmental features such as visual and acoustic inputs; (c) providing students with "big picture" concepts of a unit's content through visual concept mapping; (d) teaching for mastery of content, skills, and concepts by using the arts as a tool to enhance retention; (e) teaching for application of knowledge through real-world problem-solving tasks; and (f) evaluating learning through the use of techniques such as the "artist's portfolio" and performance-based assessments. We found that an effective way to design arts-integrated units is for classroom teachers and arts educators to work together to plan an arts-integrated unit using the BTT framework. Observation checklists associated with the model provide tools for administrators and academic and arts-integration coaches to provide meaningful feedback and support.

We believe that the most effective implementation of STEAM teaching requires robust professional development accompanied by ongoing individual and group coaching. It also requires collaborative planning among classroom teachers, consultation with arts educators and artists-in-residence, and an instructional framework within which content instruction is planned in a holistic manner, which involves multiple instructional strategies and meaningfully infuses arts-based activities into scientific-content areas.

6 STEAM and Educational Reform

As we reflect on the proliferation of the STEAM movement as a relatively new educational initiative, it is important to acknowledge that the arts have informed scientific discoveries and innovations for centuries. Some of the most creative inventors of all times were also artists, musicians, and sculptors. For example, Louis Pasteur drew on his experience of creating lithographs to understand how some chemical crystals present as mirror images (Klein 2017).

In short, we agree with arts advocates who insist that the arts should no longer be viewed as the *victim* of public policy but instead become the *driver* of reform (O'Brien 2017). We believe that the STEAM movement will help advance STEM instruction and even enhance pedagogy in general, moving our students toward more creative and innovative ways of thinking by experiencing the "aha moments" that are associated with artistic insights. We believe that the arts can and should be a core driver of education reform.

References

American Youth Policy Forum. (2012, January 27). *American youth policy forum.* Accessed at http://www.aypf.org/wpcontent/uploads/2012/05/STEM%20Facts%20and%20Resources%20 Handout.pdf

Anderson, A. (2015, July). Dance/movement therapy's influence on Adolescents' mathematics, social-emotional, and dance skills. In *The educational forum* (Vol. 79, No. 3, pp. 230–247). Routledge: Taylor & Francis.

Barshay, J. (2016, December 6). *U.S. now ranks near the bottom among 35 industrialized nations in math.* Accessed at http://hechingerreport. org/u-s-now-ranks-near-bottom-among-35-industrialized-nations-math/

Beghetto, R. (2017, October). Inviting uncertainty into the classroom. *Educational Leadership, 75*(2), 20–25.

Bryant, S. (2015, April 27). New NAMM foundation study shows parents and teachers in harmony about students learning music. *National Association of Music Merchants Foundation.* Accessed at https://www.nammfoundation.org/

Burnaford, G., Brown, S., Doherty, J., & McLaughlin, J. (2007). *Arts integration frameworks, research & practice: A literature review*. Washington, DC: Arts Education Partnership. Accessed at http://209.59.135.52/files/publications/arts_integration_book_final.pdf

Csikszentmihalyi, M. (1997). Assessing aesthetic education: Measuring the ability to "ward off chaos". *Arts Education Policy Review, 99*(1), 33–38. https://doi.org/10.1080/10632919709600763.

Hardiman, M. (2003). *Connecting brain research with effective teaching: The brain targeted teaching model*. Lanham: Rowan & Littlefield Education.

Hardiman, M. (2012). *The brain-targeted teaching model for 21st-century schools*. Thousand Oaks: Corwin Press.

Hardiman, M., Rinne, L., & Yarmolinskaya, J. (2014). The effects of arts integration on long-term retention of academic content. *Mind, Brain, and Education, 8*, 144–148. https://doi.org/10.1111/mbe.12053.

Hardiman, M. M., JohnBull, R. M., Carran, D. T., & Shelton, A. (2019). The effects of arts-integrated instruction on memory for science content. *Trends in Neuroscience and Education, 14*, 25–32. https://doi.org/10.1016/j.tine.2019.02.002

Klein, J. (2017, June 14). How pasteur's artistic insight changed chemistry. *The New York Times*. Accessed at https://www.nytimes.com/2017/06/14/science/louis-pasteur-chirality-chemistry.html?emc=eta1

Kong, Y. T., & Huo, S. C. (2014). An effect of STEAM activity programs on science learning interest. *Advanced Science and Technology Letters, 59*, 41–45.

Kong, Y. T., Huh, S. C., & Hwang, H. J. (2014). The effect of theme based STEAM activity programs on self efficacy, scientific attitude, and interest in scientific learning. *International Information Institute (Tokyo). Information, 17*(10 (B)), 5153.

Lim, Y. N., Min, B. J., & Hong, H. J. (2015). Development and application effect of design-based STEAM program for boosting the career consciousness of 5–6th grade elementary school students for natural sciences and engineering. *Journal of the Korean Association for Science Education, 35*(1), 73–84.

Marick Group. (2016, May 16). *A look at the history of STEM (and why we love it)*. Accessed at http://marickgroup.com/news/2016/a-look-at-the-history-of-stem-and-why-we-love-it

Mishook, J. J., & Kornhaber, M. L. (2006). Arts integration in an era of accountability. *Arts Education Policy Review, 107*(4), 3–11. Retrieved from http://eric.ed.gov

National Assessment of Educational Progress. (2015). *2015 mathematics & reading assessments*. The Nation's Report Card. Accessed 08 Mar 2016, at http://www.nationsreportcard.gov/reading_math_2015

National Commission on Excellence in Education. (1983). A nation at risk: The imperative for educational reform (p. 65). *The Elementary School Journal, 84*(2), 112–130. Accessed at http://www.jstor.org/stable/1001303

O'Brien, W. (2017). *How disruptive neurotechnologies are changing science, arts and innovation: Federal Brains on Art*. Panel participant at the 2017 International Conference on Mobile Brain-Model Imaging and the Neuroscience of Art, Innovation, and Creativity, Valencia, Spain.

Oner, A. T., Nite, S. B., Capraro, R. M., & Capraro, M. M. (2016). From STEM to STEAM: Students' beliefs about the use of their creativity. *The STEAM Journal, 2*(2), 6.

Partnership for 21st Century Skills. (2008). *21st century skills, education, and competitiveness: A resource and policy guide*. Accessed at http://www.p21.org/storage/documents/21st_century_skills_education_and_competitiveness_guide.pdf

Rostan, S. M. (2010). Studio learning: Motivation, competence, and the development of young art students' talent and creativity. *Creativity Research Journal, 22*(3), 261–271. https://doi.org/10.1080/10400419.2010.503533.

Rule, A. C., Atwood-Blaine, D. L., Edwards, C. M., & Gordon, M. M. (2016). Arts-integration through making dioramas of women mathematicians' lives enhances creativity and motivation. *Journal of STEM Arts, Crafts, and Constructions, 1*(2), 8.

Scripp, L., & Gilbert, J. (2016). Music plus music integration: A model for music education policy reform that reflects the evolution and success of arts integration practices in 21st century American public schools. *Arts Education Policy Review, 117*(4), 186–202.

Teske, J., & Pittman, P. (2017). Eight graders explore form and function of modern and fossil organisms. *Journal of STEM Arts, Crafts, and Constructions, 2*(1), 79–84.

Walker, T. (2014, September 2). The testing obsession and the disappearing curriculum. *The National Education Association Today*. Accessed at http://neatoday.org/

Yee-King, M., Grierson, M., & d'Inverno, M. (2017, April). STEAM WORKS: Student coders experiment more and experimenters gain higher grades. In *Global Engineering Education Conference (EDUCON), 2017 IEEE* (pp. 359–366). IEEE.

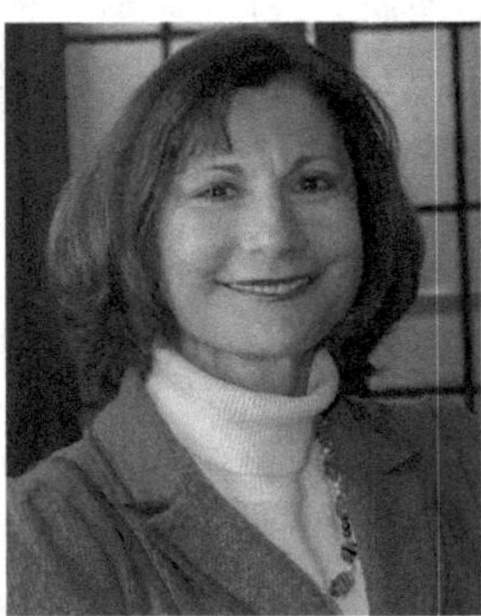

Mariale M. Hardiman is Professor of Education and the Co-founder/Director of the Neuro-Education Initiative, a cross-disciplinary program that brings to educators relevant research from the learning sciences. Her research focuses on the effects of arts integration on long-term retention of content in the STEAM content areas, and studies teacher professional development on topics related to neuroeducation.

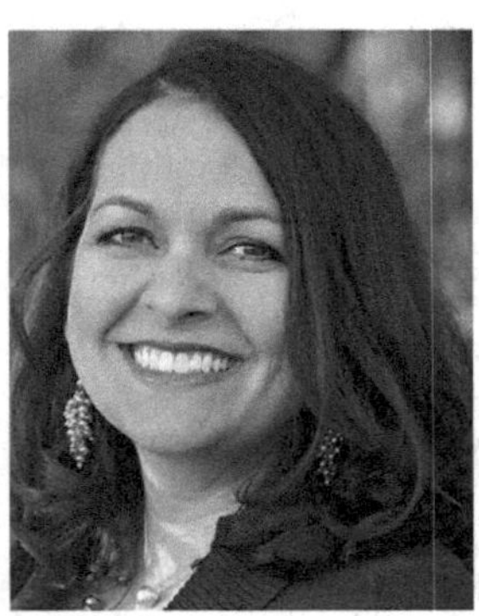

Ranjini M. JohnBull is Assistant Professor in the Johns Hopkins University School of Education, and she serves as the faculty lead for the Mind, Brain, and Teaching master's and doctoral specializations. Her research focuses on teacher efficacy beliefs, cultural competence, arts integration, and professional development on neuroeducation.

The Importance of Integrating the Arts into STEM Curriculum

Michelle Land

1 STEM Education in the USA

1.1 Investing in STEM

The USA's education system is composed of both public and private sectors, separate from the federal government. In public education, the state and local governments are primarily involved with establishing educational standards for their respective districts. Public school educators develop curriculum based on these educational standards. Although state and local governments are responsible for developing the standards, the federal government still wields much influence within the education system, primarily through funding. A major initiative currently in place is the STEM (Science, Technology, Engineering, and Mathematics) movement. Many professionals hold the belief that by focusing on these key areas, current students will later propel the global competitiveness of the USA.

Historically, the USA was considered the leading country in innovation. However, in 2016, the World Intellectual Property Organization (WIPO) ranked the USA fourth, behind Switzerland, Sweden, and the United Kingdom. Even though the USA continues to be one of the world's most innovative nations, the WIPO scores the country lower in STEM related areas, such as educational expenditures and number of STEM graduates (Cornell 2016). "As the economic activity of our nation and the world continues to rapidly transform, the need to invest in education that promotes innovation and creativity has become primary to the central themes in this on-going public dialogue" (Immerman 2011).

Progress does not come from technology alone, but rather from the melding of technology and creative thinking through art and design. According to professionals

M. Land (✉)
Riverside Elementary School, Alexandria, VA, USA
e-mail: mhland@fcps.edu

© Springer Nature Switzerland AG 2019

A. J. Stewart et al. (eds.), *Converting STEM into STEAM Programs*,
Environmental Discourses in Science Education 5,
https://doi.org/10.1007/978-3-030-25101-7_2

such as Dr. Michael Nance (personal communication, September 22, 2012), a Chief Information Security Officer at Lockheed Martin, the USA is riding on the coattails of technologies such as personal electronics and personal computing creations instead of putting brain trust into new scientific and technological breakthroughs. If the USA wants to remain a global competitor, we must find new or better ways to foster creative thinking and practice. As long as an individual is pushing personal boundaries and developing his or her own conceptual methods innovatively, a person can have a creative practice in any field. The arts can help develop STEM skills because of the more divergent approach. For example, Robert Root-Bernstein's study of scientific Nobel laureates demonstrated that most all of the scientific "geniuses" between 1902 and 2005 were proficient not just in science but also in the arts (Root-Bernstein et al. 2008).

2 STEAM: The What and Why

2.1 The Need for Creativity

During the Industrial Revolution, education in the USA became accessible to children of all socioeconomic backgrounds. Before the government began allotting funds for teaching and learning, only students from wealthy families could afford a formal education. Curriculum was developed to teach students how to be a part of an obedient and industrialized society. Today, technologies have replaced factory workforces and many information-age jobs. For example, the Whiteley (2015) educational documentary highlights a company called Narrative Science that used an algorithm to write a *Forbes* article without any human involvement. This documentary, *Most Likely to Succeed*, challenges viewers to think about what will happen when people feel like their muscle and brainpower are no longer valued and have been replaced instead by a tiny box.

The push for the STEAM (Science, Technology, Engineering, Art, and Mathematics) platform derives from a perceived shortage in creativity and innovation in recent college graduates in the USA. Currently, our education system teaches students how to execute given tasks fluidly, but it rarely fosters curiosity and self-motivation. Given the strong demand for highly creative STEM workers, the education system and government entities need to collaborate to not only invest in the arts, but also to change the way we view education as a whole. Our antiquated system should be restructured to foster different essential skills. Ultimately, what do we want our students of tomorrow to know when graduating high school or college? When history facts and mathematic equations can be accessed through the click of a button, students need be taught communication and collaboration. Educators, like myself, hope to instil empathy and curiosity in students through critical and creative thinking. We can help students become more goal-directed and resilient individuals through our instructional delivery methods and cross-curricular collaboration.

In 2001, teachers were made accountable through rigorous testing of their students, due to the passing of the No Child Left Behind Act (NCLB) (Linn et al. 2002). School curriculum then changed to meet the content and performance standards. Before the NCLB was passed, teachers had more voice and choice in their curriculum. With the standardized tests, students are measured on memorization skills rather than comprehension. In preparing for the exams, students are taught that only right and wrong answers exist. But the real world is not purely black and white. We must encourage the youth of tomorrow to seek multiple solutions to complex problems. Adding the arts into the STEM fields can combat this issue. Well-rounded art problems never have one answer. Within the art classroom, opportunities arise for students to construct their own learning through the decision-making process. Artistic license forces students to articulate their own interpretation of the material in a unit by exploring possibilities. Integrated art-education prompts can help students better understand or synthesize core content knowledge. Students should not be assessed on how well they can remember the parts of a cell by name, but instead better understand the inner workings or attributes of cells, or how the cells connect to and interact with the individual.

2.2 Student Engagement

We often ask children, "What do you want to be when you grow up?" More often than not, a child will say an athlete, a teacher, or the job of one of his or her parents. For as long as I can remember, I knew I wanted to help others and invest my energy into the art field. I attribute this to the personal connections I developed with a crafty au pair and my elementary and high school art teachers. I consider myself lucky. When I graduated high school, most of my peers did not know what they wanted to be when they "grew up" or even what area they wanted to study. Adults are role models for children, and we need to expose students to more role models from different fields. The jobs we share and demonstrate for students can directly affect their future. Schools should use local community members and parents in respective fields to add "experts" for in-class research and an authentic audience to give feedback on student projects. In addition to face-to-face time with professionals, online experiences exist for schools that empower students to test-drive future STEM careers. LifeJourney, one of these immersive online experiences, provides students with opportunities to learn directly from top industry professionals while practicing the so-called hard and soft skills required for different occupations. "LifeJourney, Inc." (2013–2017) welcomes students, parents, schools, and companies to be part of this tutoring process. By embedding various STEAM-related occupations in lessons, students can take on professionals' roles such as fabrication engineers, marine biologists, architects, and video-game designers. As a nation, we seem too focused on teaching to the standardized tests. It is time to teach to the purpose of engaging and inspiring students.

Table 1 Percentage of students engaged in school, by Grade ($N = 909{,}617$)[a]

[a]2016 Gallup Student Poll results

According to the 2016 Gallup Student Poll of students in grades five through 12, student engagement drops significantly, from 74% at its peak in grade five to 32% at its lowest in grade 11, with a slight increase to 34% in grade 12 (Table 1). What is it about education from fifth grade onward that contributes to the decline of engagement? I attribute this decline to the lack of new experiences, reduced personal connection, and less student choice. As a student gets older, he or she is exposed to more and more information. Over time, this information is taught repeatedly in schools, building on previous knowledge, even if the content area does not interest the student. Imagine where kindergarten would measure on Table 1. What do you think would happen to the percentage of students engaged? Many kindergarten classrooms today involve play, learning centers, and aesthetically pleasing and colourful spaces. Student choice is essential for engagement. When teaching STEAM lessons, posing open-ended questions or creating challenges with opportunities for discovery is of utmost importance.

2.3 Arts Integration and Its Influence on Retention

Have you ever heard anyone refer to himself or herself as a strictly visual learner? According to functional Magnetic Resonance Imaging (fMRI) results of brain research on the different modalities of learning, this is a common misconception. Everyone has visual, auditory, and kinesthetic learning abilities. If an individual is taught the concept of seed germination through a series of images, this person may be able to describe the appearances of different stages of seed germination. When accompanied with an auditory description of how seed germination occurs, a person is then more likely to recall the stages of germination. Finally, if that person illustrates the stages of germination or climate necessary for the germination process, this individual has more access points for that information. In short, when an individual is taught a single concept, the brain creates neural pathways connecting that concept to his or her experience. The more access points or neural pathways established, the greater the chance of retention and recall. This idea underpins the concept of scaffolding. Integrating the arts into core content areas does more than enable students to explore a single concept from different vantage points: it uses all the different modalities of learning previously mentioned, and thus leads to the formation of more neural pathways.

In her TED Talk, *Doodler's Unite*, Sunni Brown challenges the viewer to redefine doodling as an act "of making spontaneous marks to help yourself think." According to her research, people who doodle when exposed to verbal information have a 29% higher retention rate than their nondoodling counterparts (Brown 2011). Brown reiterates the positive effect doodling has on problem-solving through engaging the different modalities of learning. To fully retain or respond to the information, learners need to use at least two of her four defined modalities (visual, auditory, kinaesthetic, reading/writing), or use one coupled with a strong emotion. Brown argues that the biggest contribution of doodling is that it "engages all four areas simultaneously with the possibility of an emotional experience." STEM lessons that include visual reasoning and responses through the incorporation of art result in students that have richer and more engaging learning experiences.

3 Theory to Practice

3.1 Planning for Success

Supporters of the STEAM initiative may theorize how STEAM looks in the classroom, but it is the educators' and administrators' job to develop and implement the content. One of the biggest challenges of teaching STEAM is the amount of planning that goes into curriculum development. With a compartmentalized system that has been in place for over 125 years, it is difficult for teachers to educate students in ways that bridge all STEAM components without having a strong foundation in each content area. At its core, STEAM is cross-curricular collaboration. With every newly formed teaching method, the teachers involved in designing STEAM units must be "team players" willing to co-plan and sometimes even co-teach. To make this outcome a reality, teachers need more opportunities for uninterrupted planning time with a STEAM-focused collaborative learning team.

If a mathematics teacher and an art teacher work together to implement a STEAM unit, the two teachers should introduce new mathematics skills and new art skills through an overarching concept. This overarching concept, or "big idea," is the branching-off point that relates the two content areas. For example, if the big idea was data, the mathematics teacher could introduce skills related to data collection and frequency charts. Simultaneously, the art educator could introduce skills related to visual literacy through the use of color, icons, and style. In the end, the synthesis of learning may be a data visualization representing personally collected data through the use of colors and icons. Other applications of STEAM may take the form of circuit bending, musical compositions, kinetic art, product design, prototype development, and/or performance art.

Five years ago, I would have said that effective STEAM lessons should have an equal amount of learning in each content area. However, this is not always possible or even probable. Some content areas are easier to bridge than others. Many of my educational colleagues hold the belief that a fruitful STEAM lesson should highlight at least three of the STEAM- related capacities. For instance, a STEAM lesson can challenge students to highlight mathematic tessellation in mosaic sculptures through the engineering design process. This objective-based approach could require students to use a digital 3D design application, employing the technology piece of STEAM.

A successful way to plan STEAM lessons is to start with the standards: specify the knowledge you want students to gain by the end of the project, then find connections in the real world. For example, if I want my students to understand simple machines, I should think about what occupations use simple machines. A list could include various professionals: machinist, recreation architect, commercial welder, and fabrication engineers, to name a few. Then, I think about what those professionals create, and I brainstorm ways that a student could mimic a similar process in my classroom while exciting and engaging him or her. With a population of elementary school students, I could have students take on the role of a recreation architect to design a simple machine maze game. As a Project Based Learning (PBL) county-support teacher, I phrase most of my STEAM challenges as driving questions: How can you, as recreation architects, construct a simple machine maze game for other students in your class to enjoy? To the complete the challenge, students must fully understand simple machines and design/build a maze, all while being exposed to a real-world profession. Challenge students to embed artistic design elements and a research-based approach, and make sure students use multiple STEAM areas. Ultimately, students can create a model, or mini-version of his or her maze with a group of peers, collaborating and communicating their ideas in the process. This open-ended, objective-driven project allows for student voice and choice. Teachers can assess along the way using blueprint designs with written components, exit tickets, in-progress critiques, and ultimately, an evaluation of the final product and presentation of the work.

With the availability of personal computing devices and abundance of applications available, one or more technology links almost always exit. Many educators use technology for simple substitutions with no functional change, like slide shows or pen-and-paper responses. The Substitution, Augmentation, Modification, Redefinition (SAMR) Model, developed by Dr. Ruben Puentedura (2006), is another tool to advance STEAM lessons to the next level. Puentedura (op. cit.) highlights the importance of using technology to push the boundaries of student knowledge and capability. A student can simply *substitute* a presentation for a slide show or Prezi and then further *augment* the project by embedding videos and personal voice recordings. The student can then *modify* the original project with the design of a user-directed multimedia document with QR codes or augmented reality. To *redefine* the technological capabilities, students can collaborate to create a commercial in iMovie and then post it to a public forum for a completely new interactive experience.

STEAM challenges can provide a student with the opportunity to completely redefine his or her learning. If teachers are exposed to the SAMR model and have the opportunity to create rough SAMR ladders, they can brainstorm ways to advance the ways students explore and curate content knowledge. For more information, reference Common Sense Education (2016). The most crucial step in STEAM lessons is the planning phase. When teachers are allowed more objective-driven planning time, students have richer learning experiences.

3.2 Major STEAM Initiatives

Nationwide, forward-thinking movers and shakers are developing their own STEAM curricula to help advance our youth. In September, 2010, the Wolf Trap Foundation for the Performing Arts was awarded a 4-year $1.15 million grant (Ludwig et al. 2016). This grant helped launch a "first-of-its-kind" arts initiative, The Early Childhood STEM Learning Through the Arts (Early STEM/Arts). According to the 2016 publication, *Arts Integration: A Promising Approach to Improving Early Learning*, this program was the subject of a 4-year study conducted by the American Institutes for Research (AIR) in partnership with Fairfax County Public Schools in Virginia. The AIR study found that Wolf Trap's Early STEM/Arts program had a significant positive impact on the standardized mathematics test scores of students whose teachers had participated in Wolf Trap's training and related activities in the first 2 years of the study. This effect was attributed to integrating performing arts into the mathematics curriculum.

In 2013, Dr. Maggie Madsen, Potsdam University Provost, worked with a team of professionals on an interdisciplinary STEAM degree program research study in partnership with Lockheed Martin. Faculty for this multidisciplinary research study included instructors in the Art, Music, Theater, Biology, Psychology, Chemistry, Computer Science, Mathematics, Physics, and Business arenas. Unfortunately, due to funding constraints, the STEAM degree program was never officially approved. Nonetheless, professors are now more interested in collaborating outside of their degree program area, leading to an increase in more integrated classes and greater interest for dual degrees among the student population. Kristin Esterberg, Potsdam University President, refers to the University as a "creativity campus."

In 2013, Congresswoman Suzanne Bonamici (D-OR) announced the formation of the bipartisan Congressional STEAM Caucus (Bonamici 2013). Her co-chair, Elise Stefanik (R-NY), joined the push for STEAM in 2015. This group focuses on integrating the arts, and advocates for policy changes that will encourage educators to adopt STEAM into their repertoire. John Maeda, former President of Rhode Island School of Design (RISD) and a major proponent for STEAM, participated in a Capitol Hill briefing to the Congressional STEAM Caucus. At this briefing, the Congressional STEAM Caucus reintroduced House Resolution 51, stating that art and design promote creativity and economic growth. On July 20, 2017, US Representatives Jim Langevin (D-RI), Suzanne Bonamici (D-OR), and Elise Stefanik (R-NY) amended the STEM Education Act of 2015 by requiring the National Science Foundation (NSF) to promote the integration of art and design thinking in STEM education (U.S. Government, House of Congress 2017). The amended bill, STEM to STEAM Act of 2017, entails designing and testing informal STEAM programs by the NSF. This will not only allow for program assessment, but also help improve educational outcomes and promote creativity and innovation. Informal STEAM programs may include afterschool programs, maker spaces, nature labs, science and technology-centered community programs, STEAM specials, and museum workshops.

4 Conclusion

STEM education was created to educate a more realized youth with the "hard" and "soft" high-tech skills necessary for expanding the STEM job market. Professionals across the board appreciate this effort, but even with a focus on STEM, a broad perception is that recent graduates lack the innovative spirit and drive required to advance the USA. In the end, the education system revolves around the students. The STEAM initiative offers students more than high-tech skills. We need to provide school systems with proper funding and teachers with more training. Moreover, teachers need additional focused planning time to develop objective-driven STEAM lessons. Integrating the arts into the STEM curriculum provides pathways for personal meaning making and real-world connections. Educational awareness and funding of the STEAM initiative can enable students to construct their own learning and thus become more resilient, goal-directed, critical and creative thinkers.

References

Bonamici, S. (2013). Reps. *Bonamici and Schock announce bipartisan congressional STEAM caucus*. Retrieved from House.gov website: https://bonamici.house.gov/press-release/reps-bonamici-and-schock-announce-bipartisan-congressional-steam-caucus

Brown, S. (Presenter). (2011, March). *Doodlers, unite!* [Video file]. Retrieved from https://www.ted.com/talks/sunni_brown#t-244080

Common Sense Education (Producer). (2016, July 12). *Ruben Puentedura on applying the SAMR model* [Video file]. Retrieved from https://www.commonsense.org/education/videos/ruben-puentedura-on-applying-the-samr-model

Cornell University, INSEAD and the World Intellectual Property Organization (WIPO). (2016). *Global Innovation Index 2016:* Switzerland, Sweden, UK, U.S., Finland, Singapore Lead; China Joins Top 25. Retrieved from the WIPO website: http://www.wipo.int/pressroom/en/articles/2016/article_0008.html

Immerman, S. D. (2011). *Letting off "STEAM" at Montserrat college of art.* Retrieved from *New England Journal of Higher Education* website: http://www.nebhe.org/thejournal/letting-off-steam-at-montserrat-college-of-art/

LifeJourney, Inc. (2013–2017). Retrieved September 2, 2017, from LifeJourney website: https://www.lifejourney.us/

Linn, R., Baker, E., & Betebenner, D. (2002). Accountability systems: Implications of requirements of the No Child Left behind Act of 2001. *Educational Researcher*, 31(6), 3–16. Retrieved from http://www.jstor.org/stable/3594432

Ludwig, M., Marklein, M. B., & Song, M. (2016). *Arts integration: A promising approach to improving early learning.* Retrieved from Wolf Trap website: http://www.wolftrap.org/~/media/files/pdf/education/arts-integ-brief-2016-final.pdf?la=en

Pentedura, R. R. (2006). Transformation, technology and education. online at http://hippasus.com/resources/tte/

Root-Bernstein, R., Allen, L., Beach, L., Bhadula, R., Fast, J., Hosey, C., Kremkow, B., Lapp, J., Lonc, K., Pawelec, K., Podufaly, A., & Russ, C. (2008). Arts foster scientific success: Avocations of nobel, national academy, royal society, and sigma ii members. *Journal of Psychology of Science and Technology.* https://doi.org/10.1891/1939-7054.1.2.51.

U.S. Government, House of Congress. (2017). *115th congress 1st session H.R. 3344.* Retrieved from Congress.gov website: https://www.congress.gov/115/bills/hr3344/BILLS-115hr3344ih.pdf

Whiteley, G. (Producer, director, writer), Dintersmith, T., Leibowitz, A., Lombroso, D., & Ridley, A. (Producers). (2015). *Most likely to succeed* [Documentary]. United States: One Potato Productions.

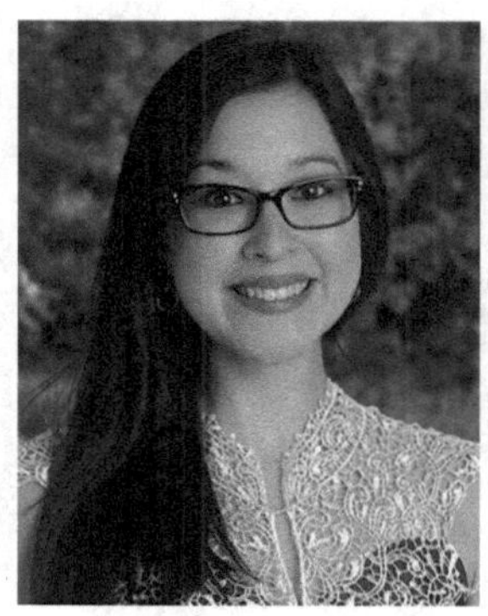

Michelle Land is an Art Specialist, STEAM Teacher, and Project Based Learning (PBL) County Support teacher for Fairfax County Public Schools. She received a Bachelor of Fine Arts degree and a Master of Arts in Teaching degree from Maryland Institute College of Art and has taught STEAM-based summer camps for the National Smithsonian in Washington D.C. She also has presented on STEAM for the National Art Education Association. Her current teaching practice revolves around exposing students to real-world challenges and professions while putting the "A" in STEAM.

Purposeful Pursuits: Leveraging the Epistemic Practices of the Arts and Sciences

Bronwyn Bevan, Kylie Peppler, Mark Rosin, Lynn Scarff, Elisabeth Soep, and Jen Wong

> *The object of art is to give life a shape*
> *Jean Anouilh*

1 Introduction

Decades of research make clear that purpose and meaning are essential to learning (NRC 2012a). Purpose and meaning drive our interests, questions, and persistence; they urge us to branch out and go deeper. They are central to both being and becoming: reflecting and also shaping our identities as individuals, or as members of communities, who value and know a place, a practice, a domain of objects, a body of knowledge.

B. Bevan (✉)
College of Education, University of Washington, Seattle, WA, USA

K. Peppler
The Creativity Labs, University of California Irvine, Irvine, CA, USA
e-mail: kpeppler@uci.edu

M. Rosin
Guerilla Science US, New York, NY, USA
e-mail: mark@guerillascience.org

L. Scarff
National Museum of Ireland, Dublin, Ireland
e-mail: lscarff@museum.ie

E. Soep
YR Media, Oakland, CA, USA
e-mail: lissa@yrmedia.org

J. Wong
Guerilla Science UK, London, UK
e-mail: jen@guerillascience.org

© Springer Nature Switzerland AG 2019   21
A. J. Stewart et al. (eds.), *Converting STEM into STEAM Programs*,
Environmental Discourses in Science Education 5,
https://doi.org/10.1007/978-3-030-25101-7_3

Purpose is highly personal, cultural, and contingent. It can ebb and flow over time. It develops from opportunity and it also creates opportunity. Purpose emerges when individuals recognize, and are recognized for, opportunities to participate in, contribute to, and transform social activities that have meaning and value to the learner (Stetsenko 2017). Implicit in this vision is the idea of entering into and becoming an integral part of communities of practice, with opportunities to observe expertise, to practice one's skills, to develop and float one's ideas, while becoming more and more central participants in the valued activity, including coming to redefine or reinvent the activities themselves (Lave and Wenger 1991).

For decades, research on the arts and education has documented the ways in which the arts create personally meaningful contexts for young people to develop a sense of purpose and therefore of self (e.g., Greene 1977). At the heart of these arts activities are opportunities for the creative production of something—whether temporal, virtual, performed, displayed, or gifted to an authentic audience. The value of creative production with authentic audiences has been explored, for example, in theatre (Heath 2000), gallery exhibitions (Smith 2014), community arts projects (Kafai and Peppler 2012), storytelling and poetry (Soep 2006), and, more recently, in maker communities (Bevan 2017).

Purpose also sits at the heart of current science educational reform movements in the USA. State science standards, adapting the National Research Council's consensus volume *K-12 Framework for Science Education* (NRC 2012b), emphasize the practices of science, such as identifying questions, using computational thinking, creating and using models, and arguing from evidence. These reforms hold that engaging students in science, technology, engineering, and mathematics (STEM) practices creates authentic and purposeful contexts for deeper engagement in and with STEM—driven by questions or needs identified by the learners—leading to greater STEM learning as well as to the development of productive STEM learning identities (Berland et al. 2015).

But many young people face significant economic, cultural, historical, and/or social obstacles that distance them from STEM as a meaningful or viable option—these range from under-resourced schools, race- and gender-based discrimination, to the dominant cultural norms of STEM professions or the historical uses of STEM to oppress or disadvantage socio-economically marginalized communities (Philip and Azevedo 2017). As a result, participation in STEM-organized hobby groups, academic programs, and professions remains low among many racial, ethnic, and gender groups (Dawson 2017). One solution to this imbalance has been to reposition STEM as STEAM—integrating the arts and design in ways that can have wider appeal to a broader cross section of young people. Integrating the arts and sciences is not only a strategy for broadening appeal, it also reflects the ways in which participation in civic, academic, and professional activities is becoming increasingly hybridized, requiring communication, design, and technological skills. A STEAM approach to broadening participation or inclusion can be relevant across the four distinct "discourses of equity" that Philip and Azevedo (2017) posit underpin research on equity in out-of-school STEM. They argue that equity in informal STEM education is seldom articulated but variously conceptualized as (a) supporting student achievement in school STEM, (b) building student STEM learning interest and identity through more authentic engagement with STEM, (c) democratizing STEM by locating its presence and uses in everyday life, or (d) understanding how STEM can be taken up by social justice movements as a tool for achieving transformation and change. How informal STEM programs advance equity can thus vary widely, depending on their conceptualization of equity.

As STEAM becomes more deeply theorized, it may be valuable to consider the evidence base that suggests that engaging in the epistemic practices of a discipline will lead to deeper disciplinary learning and more productive learning identities (e.g., see NGSS or Common Core in the USA). In other words, integrating purpose and meaning into engagement with the disciplines links epistemological and ontological processes of development. Further, an epistemic approach (unlike a concepts-based approach) is potentially relevant across the four distinct discourses of equity articulated by Philip and Azevedo; i.e., it can enrich programs focused on STEM school achievement as well as those concerned with broader social transformation. But what are the epistemic practices of the "discipline" of STEAM?

2 Background and Rationale

In this section we briefly describe what research says about epistemic practices in science and art and explore what these findings might mean for theorizing and enacting STEAM programs that can advance youth purpose and agency.

2.1 *Epistemic Practices in STEM*

In STEM fields, there is a long history of promoting hands-on, problem-based, and inquiry-based learning (Driver et al. 1994). More recently, improvement efforts in the USA have begun to focus on the epistemic practices of these disciplines (NRC 2012b). These reforms emphasize the ways that professional scientists, mathematicians, technologists, and engineers use disciplinary practices, such as evidence-based reasoning, to answer specific questions or solve specific problems. The reform movement posits that students can develop deeper understanding of and about science by using these same practices toward purposeful ends. For instance, students develop conceptual understanding as they engage in epistemic practices to explore local environmental problems, understand community patterns of health and wellness, or design computer games or mobile applications.

Implicit in this approach are two ideas. One is that purposeful learning provides opportunities for deeper learning (NRC 2012a), and another is that by engaging in STEM practices, young people may come to value STEM in the same ways that practicing STEM professionals do—not by reading about it, but by doing it for an authentic purpose (NRC 2012b). This strategy may be especially important for STEM programs that seek to engage learners from communities that have been historically excluded from STEM.

The National Research Council's (2012b) *K-12 Framework for Science Education* identified eight practices related to science and engineering. These practices describe how scientists and engineers build knowledge about the natural and designed world. To help educators, researchers have organized these practices into three conceptually manageable clusters of activities: investigating, sense-making, and critiquing practices (McNeill et al. 2016). A critical distinction between practice-oriented approaches to STEM learning and prior instantiations, such as inquiry-based learning, is an emphasis on the critiquing phase of STEM learning (Engle and Conant 2002). Prior inquiry-based reform efforts focused on the investigating and sense-making phases, but often were organized in ways that, intentionally or not, omitted explicit attention to scientific argumentation—i.e., negotiating between competing explanations through critiquing the nature and validity of evidence marshalled to support explanations—as well as to connecting explanations with a larger scientific discourse related to the scientific phenomena being explored. This key aspect of STEM sits at the heart of peer review processes.

2.2 Epistemic Practices in the Arts

Three broad concepts are important to epistemic practices in the arts: (1) active engagement in the learning process; (2) youth's personal connection to their work, which is posited to inspire a general love of learning and build upon their prior experiences; and (3) the creation of projects that are of value to a larger community. These concepts connect to both sociocultural theories of learning and theories of the arts and aesthetics (Greene 1995; Dewey 1934/1980). Arts learning focuses, on one level, on the design of artifacts rather than on the use of artifacts and tools, and, on another level, it focuses on the bidirectional relationship between an individual and a community of learners. According to Dewey, "[a]rt denotes the process of doing or making," and provides a tool by which we search for meaning (1934/1980, p. 47). Being active in the learning process is important to current conceptions of what it means to be motivated and to engage deeply in the content.

Kafai and Peppler (2011) have investigated the epistemic practices of youth's creative production, illuminating how youth engage in multiple literacies and diverse forms of authentic participation as they engage in interest-driven projects (New London Group 2006; Guzzetti and Yang 2006; Lankshear and Knobel 2011). Their work provides a holistic vision of visual/media culture, new technologies, and traditional art making, and identifies ten practices relating to full participation in communities built around creative production. These ten practices are organized into four clusters of activities: technical, critical, creative, and ethical practices. Technical practices in the arts relate to crafting within the medium selected for a piece, which determine medium-specific skillsets such as coding, debugging, and repurposing of materials. Artists engage in critical epistemic practices—observing and deconstructing media, evaluating or reflecting (i.e., critique), referencing and reworking—as they strive for originality or to thoughtfully break from tradition.

Creative practices, perhaps the most well known in the arts, involve making meaningful and artistic choices (e.g., pertaining to color theory, shape and contour, musical instrumentation, etc.) to transform the intended "meaning" of a piece, or to forge connections between various modalities in contemporary work. Lastly, a range of ethical practices has emerged in the arts around ownership and information. This expanded palette of previously conceptualized practices surrounding participation in the arts includes a broader spectrum of design activities important to contemporary practice, as well as youth culture.

2.3 STEAM: Intersections Between the Arts and Sciences

The promise of STEAM approaches is that by coupling STEM and the arts, new understandings and artifacts emerge that transcend either discipline. Evidence of this potential can be seen through fundamental shifts in both fields. The infusion of the arts into STEM has shown to be transformative, for example, with the emergence of tools and communities that not only engender new content understandings but also invite participation from populations historically underrepresented in STEM fields (Peppler 2013). Similarly, the use of STEM tools and data in the arts have created important bodies of work for artists exploring intersections of the natural and social worlds.

In the UK, Ireland, and Europe, the evolution of the field of art-science (the field in which art and science overlap), particularly in the last 10 years, has been socially and culturally significant. Science now appears in places and spaces ranging from galleries like the Wellcome Collection and Science Gallery Dublin to culturally arts-oriented festivals such as Latitude and Secret Garden Party (Bultitude and Sardo 2012; Dowell 2014). Scientifically embedded arts programs ranging from Collide@CERN and a subset of science festivals (von Roten and Moeschler 2007) represent one end of a spectrum, while hybrid programs like Ars Electronica and Waag Society occupy a more central position. Growing out of the Wellcome Trust's SciArt funding scheme started in the mid 1990s (Glinkowski and Bamford 2009; Born and Barry 2010), art and science programming has moved from a more academic realm into a public space. This work has not instrumentalized the arts or science, but rather sought out a rich, fertile space for producing compelling cultural events and programming through a collaborative-practices approach. Art-science approaches speak to contemporary social, political, and economic concerns demanding transdisciplinary platforms and methods of working that allow for professionals, particularly in science, health and technology, to conceptualize their subject expertise through a broad thematic approach, as opposed to a discipline-specific perspective. Art-science programs have demonstrated the strong effects that such approaches can have on learners—removing them from specific identities of the "artsy" or "mathsy" person and placing them in a context that is purpose driven, offering an opportunity for creative and flexible thinking that maps onto their key concerns.

Increasingly this work is being developed in the USA, supported by organizations such as the Sloan Foundation, the Simons Foundation, and others.

Despite the promising aspects of this work, practice is well ahead of educational research in the context of STEAM. As this volume illustrates, there are many examples of exceptional STEAM or arts-and-sciences programs for young people. But there are also many examples, not detailed in this volume, of programs that style themselves as STEAM but do not do so in any deeply theorized way: in these programs, science activities may have some decorative tasks attached, or arts activities may integrate scientific phenomena, such as color mixing or electrical circuitry. But the epistemological and conceptual aspects of the "added" discipline—for example, the use of evidence-based reasoning in science programs, or of performance and critique in arts programs—typically are left unexplored.

In the next section we describe an approach being taken by an international collaboration of educators and researchers to better understand and theorize the ways in which the thoughtful integration of the epistemic practices of the arts and the sciences enrich learning in both areas, while supporting ontological processes for young people from economically and racially marginalized communities (Bevan et al. 2018; Bevan and Scarff 2015; Peppler and Wohlwend 2017).

3 A Framework for the Intersection of Epistemic Practices in the Arts and Sciences

Funded by joint grants awarded by the National Science Foundation in the USA and the Wellcome Trust in the UK, our study sites include Science Gallery Dublin, a public museum space at the intersection of science and art that targets young adults; Guerilla Science, a program based in London and New York that stages art and science events in unexpected settings, such as music or arts festivals serving primarily young adults; Youth Radio in Oakland, a program serving older youth in which participants explore contemporary issues and produce nationally aired radio segments and various digital platforms; WacArts, a London-based secondary school specializing in the arts; and the Boys & Girls Clubs of Indiana's summer Maker camps, serving young children and tweens. All of these programs are free and serve youth from marginalized communities. As such, they work with youth over time and collect demographic data about participants. Guerilla Science is an exception given its focus on situating programs in live festivals or street-corners.

The underlying sociocultural theory guiding our work conceptualizes learning as a process that develops in supportive and responsive communities in which youth's cultural, intellectual, and emotional resources are recognized and leveraged as they participate in purposeful and consequential activities (Nasir et al. 2006). Under these conditions, youth exercise and expand their agency, key to the development of productive learning identities (Holland et al. 1998). As such, we are using a framework (see Table 1) that builds on the work of the National Research Council's

Table 1 A framework for epistemic practices of the arts and sciences

STEM practices	Epistemic intersections	Arts practices
Investigative practices	*Exploratory practices*	*Technical & critical practices*
• Asking questions/ defining problems • Planning and carrying out investigations • Using mathematical and computational thinking	• Noticing and questioning • Exploring materiality • Defining the problem space/ deconstructing components • Producing tentative representations	• Looking closely • Deconstructing the parts of the text (at a literal level) and the meaning behind the text
Sense-making practices	*Meaning-making practices*	*Creative practices*
• Developing and using models • Analyzing and nterpreting data • Constructing explanations/ designing solutions	• Principled iterations/revisions (responding to feedback) • Considering multiple approaches • Engaging multiple modalities • Finding relevance • Adopting a critical stance	• Applying artistic principles to augment meaning • Designing interrelations within and across multiple sign systems • Referencing or combining existing works and ideas
Critiquing practices	*Critiquing practices*	*Ethical practices*
• Arguing from evidence/peer review • Evaluating and communicating findings	• Sharing results • Hacking the ideas of others • Engaging in critical reviews • Cultivating dissent • Holding commitments to standards of the field	• Negotiating what constitutes a "good" project • Given a particular artistic goal, evaluating how successfully this goal has been met

(2012b) *K12 Framework for Science Education* and of Kafai and Peppler (2011). Through this framework, investigators can explore how engaging in the socio-historically purposeful epistemic disciplinary practices of arts, sciences, and arts-and-sciences (STEAM) advances purpose, agency, and learning for young people from communities historically excluded from STEM and other domains of privilege.

The intersection of these epistemic practices is central to many STEAM-based programs (Bevan and Scarff 2015). For example, Maker programs, involving activities such as e-textiles or kinetic sculptures, entail exploring materiality, producing tentative representations, collecting and responding to feedback, and revising plans and products (Peppler 2013; Bevan 2017).

Digital production programs, such as Scratch (www.scratch.mit.edu), integrate processes of planning and designing, deconstructing components, responding to feedback, and critiquing and explaining within the Scratch community (Resnick and Rosenbaum 2013). Media-related programs, such as Youth Radio, integrate noticing and questioning, collecting data, developing representations of understanding, responding to feedback and critique, and producing and communicating evidence-based explanations (Chávez and Soep 2005).

To illustrate: at one of our study sites, to mark the 60th anniversary of the desegregation of Central High School in Little Rock, Arkansas, a group of present-day Central High students contacted Youth Radio in Oakland, California, to create a social media reenactment of that historic day. With interviews, hand-drawn illustrations, and archival tape and photos from the original Little Rock Nine (LR9), the young people issued this provocation: If Twitter had been available in 1957, how would the Little Rock Nine story have unfolded? Also, how do the themes and conditions of the original Little Rock Nine remain relevant to racial inequality and school resegregation today? With this premise as a starting point, the teams in Oakland and Little Rock designed and coded a website from scratch, programmed a live Twitter reenactment and produced a story that aired on National Public Radio. They went through several iterations for the design of the final site as well as individual assets—for example, using illustration software and data tables to meticulously match hand-drawn portraits of the original "Nine" to archival and contemporary photos. Youth members of the various teams reviewed and critiqued one another's work and devised ways to showcase the strongest materials to bring the past alive while connecting that history to data on racial divisions and disparities in education today.

In another example, every quarter, Science Gallery Dublin transforms its entire exhibition and education programming to focus on a broad theme, such as HUMANS NEED NOT APPLY (artificial intelligence), SECRET (data, privacy and encryption), IN CASE OF EMERGENCY (global challenges through trope of apocalyptic literature). The future-facing topics are chosen so that they can be interrogated by artists, scientists, humanities scholars, and designers. As a part of this work, the museum invites groups of 17–18-year-old "Transition Year (TY) students" to spend a week at the museum developing public programs related to the new theme. These students are predominantly from second level schools that are traditionally under represented in undergraduate student intakes in Irish universities. Students engage with scientists and artists working on the theme, and begin developing their own project. For example, for IN CASE OF EMERGENCY, the museum created a Situation Room in which visitors were presented with a realistic, potentially catastrophic situation such as a viral outbreak or a tsunami. The TY students worked with the researchers and designers to develop, test and adapt the gameplay of the Situation Room. They started by looking at the established game "Cards Against Humanity," critiquing and understanding the gameplay. This was followed by a deep immersion into the themes of the exhibition through Q&A sessions with curatorial advisors and researchers. In this process, participants came to grips with some of the global challenges we face, from viral outbreaks to climate change, and they began to develop scenarios based around these issues. This learning was then mapped onto critiquing and developing the gameplay developed for the IN CASE OF EMERGENCY Situation Room.

As these examples show, such approaches have authentic audiences and consequential purposes. They integrate a range of investigatory, sense-making, and critiquing processes in the context of the creative production of, in one case, a radio segment and website and, in the other case, a museum experience. We follow this overview with two short examples from our early work in the field that may better illustrate the ways in which epistemic practices emerge, interweave, and hybridize in arts-and-sciences programs. While we chose these examples because they highlight a particular cluster of practices, we also note that other arts, STEM, or STEAM practices co-occur in these activities, which are not reflected here. We also argue that while these new types of epistemic intersections frequently occur in this work, it would be unlikely to find all of them well represented in any one STEAM activity. Rather, these examples serve to highlight and illustrate our emerging understandings.

3.1 Example 1 – STEAM-Rich Tinkering in a Weekly Afterschool Youth Program

At the Sunshine Public School (an alias) afterschool Making and Tinkering program (conducted in partnership with the Exploratorium Tinkering Studio), young people explore various physical materials as they engage in everyday versions of "engineering practices" to design and build contraptions of various types. Participants develop design goals inspired by the available materials as well as examples of prior work shared in the introduction of the activities. Negotiating form and function is at the heart of tinkering. Students' initial aesthetic goals inevitably lead to cascading sets of unforeseen constraints forcing improvisational problem-solving and generating new possibilities and ideas. This iterative process may be more reactive and less scripted than most formalized engineering processes, but it similarly involves optimizing performance while maintaining commitments to particular (often evolving) aesthetic goals and design solutions.

An example of this back-and-forth in artistic and scientific/engineering practices of creative production can be found in the story of a young, teenage student, Stephan. We documented Stephan, during a weekly afterschool program, as he designed a marble machine: a 5-foot peg board on which one builds a series of ramps and tracks that can guide a marble's journey from the top to bottom of the board. At the onset of this activity, Stephan developed an aesthetic goal, unique to him among his classmates, to use both sides of the peg board to make the marble's pathway that much longer and therefore that much slower than it would be otherwise (Bevan et al. 2018). His personal solution to the challenge of slowing the descent created a unique set of challenges that other students didn't have to negotiate: at the points where he tried to connect the tracks at the front and back of the board, the marble frequently shot off to the side, or got caught at the joints of connecting ramps. Stephan experimented with many materials and methods for rounding the edges of the board before deciding to use a Slinky as a tunnel that could curve around the board. This elegant solution addressed one problem, but it created a new one: the Slinky's sagging coils created dips or valleys that caught the marble. Stephan needed to create a dense network of supporting dowels to prop up the Slinky at key points. Inspired by the demonstration model shared by the facilitator, he explored how different fabrics could be used to line the track and slow the marble further. Through principled iterations, he discovered that the most textured fabrics slowed the marble the most. Finally, when he placed a metal bowl at the bottom of the ramp to catch the marble, he was surprised by the delightful ring the marble made when it dropped into the bowl. This unplanned discovery led him to go back to the top of the peg board and add metal objects throughout so that the marble's journey became a sound installation as well.

In this STEAM activity we can see epistemic practices of both art and science, as well as a hybrid form inherent in tinkering. Engineering practices included designing solutions, testing and optimizing solutions, and later communicating results (NRC 2012b). Aesthetic practices included looking closely, augmenting meaning through aesthetics (as in the case of the wrap-around track), and even evaluating the success of the project exemplified by his decision to add musical elements to the entire project (Kafai and Peppler 2011). The integrated or hybridized practices of tinkering involved iterative but not necessarily systematic experimentation. Choices were purposeful, and frequently made in response to aesthetic goals. Through these iterations, Stephan engaged with scientific concepts (such as velocity and friction) and gained intuitive feelings for the properties of materials and phenomena that could serve as the foundations for more formalized sense-making experiences. His positive experience may also inspire interest in taking up such formalized experiences in the future. In the example we provide here, sense-making is implicit. The classroom was organized so that young people could observe each other's work, and choose to adopt or adapt particular solutions (Petrich et al. 2013). Facilitators also led student share-outs on a regular basis. These processes allowed students to collaboratively engage in various forms of sense-making about the materials they were investigating or the design practices they were undertaking. The educators introduced forms of criticality by taking students on field trips to visit and speak with local artists who worked with similar materials or mechanisms (Ryoo et al. 2016). In addition to conveying a sense of standards in the field (what professional uses of the same materials or skills looked like and led to), field trips operated to expand horizons by making connections between the students' afterschool experiences and related professional pathways.

Not as evident in our example is the articulation of evidence-based reasoning. However, we posit that its use may be implicit in the evolution of the increasingly optimized as well as aesthetically embellished marble machine. Where evidence-based argumentation is seen as a key aspect of STEM practices, in the context of making, the ways in which productive strategies for getting objects to work are taken up by others could perhaps represent an implicit form of scientific argumentation. Our example of Stephan adopting the method to lining tracks with fabric might suggest that he accepted this "solution" as a better strategy than others to slow down the marble. More specifically, it suggests that Stephan accepted the implicit "argument" that rough materials could create friction that would reduce the marble's velocity. On this view, hacking the ideas of others might serve as an active, embodied form of argumentation. This implicit understanding could be made explicit through the guidance of the afterschool educator.

3.2   *Example 2 – STEAMY Science Engagement at a Music Festival*

Based in London and New York, Guerilla Science offers public engagement with science events that target young adults. They do this primarily through arts events (e.g., concerts, exhibitions) that feature scientific phenomena or through science events (e.g., interactive talks, hands-on activities) held at music, arts, or street festivals. In their science communication activities, the program draws from theatrical practices of cabaret, alternative comedy, and masquerade to stimulate curiosity and immerse their audience in a kind of "figured world" where the nature of scientific questions, evidence, and knowledge is explored and made concrete. At a camping music festival targeting young adults, these immersive and interactive engagements with scientists address topics such as the science of sexual attraction, the role of celebrity in social behavior, knowing your biome, or the effects of psychedelic drugs on the brain. Both in their choices of topics and in the high production quality of their presentations, Guerilla Science signals to its audience that science is a journey, and it invites audiences to learn about and join this journey.

An example of how Guerilla Science interweaves arts and science practices in the festival format comes from the fourth day of a music festival held over a long weekend in the English countryside. A scientist, wearing a mini dress, striped stockings, and a bowler hat—i.e., looking like the other festival goers (and indeed she had spent 3 days camping out at the festival)—introduced herself to the 100 or more people draped about the lounge chairs in the performance tent.

> I am a psychology researcher—a psychology lecturer—at the University of [X]. And I am also really passionate about dreaming. And I do my academic research on dreaming. So I am what they call an "oneirologist," which comes from the Greek word oenirus, which means dream. And I am also an oneironaut, which means I am a dream explorer or a dream investigator or a dream traveller, something like that.

In the interactive presentation, creative production is largely the work of the scientist drawing on elements of dramatic narrative to create a sense of intimacy with the audience that invites them into the social and cultural world of science. In this presentation about the science of lucid dreaming, the dream researcher's scientific narrative integrated personal descriptions of her own lucid dreaming, including flying to explore another planet or watching her face melt in the mirror. She described compelling reasons for why lucid dreaming might be of interest to those in the tent, including providing the opportunity to fly, to have sexual relations with people not normally available (such as celebrities), to improve one's musical skills through practice drills in one's sleep, or to engage in psychic and even physical healing. Thus, through interweaving personally relevant connections with the story of how lucid dreaming was being pursued as a scientific process, she engaged the audience in how questions of relevance guide processes of scientific inquiry.

Scientists who work with Guerilla Science typically stress the tentative, emerging, contested, and socially relevant nature of the scientific enterprise:

> It may surprise those of you that are lucid dreamers that it wasn't until about 50 years ago that the scientific community actually agreed that it was a real phenomenon. ... [Until then it] was viewed very skeptically almost like fringe science, like pre-cognitive dreams or tarot, or these kind of what's considered New Age things. But somebody developed a really ingenious way to show, beyond doubt, that lucid dreams are a real phenomenon. ...They got [a group of lucid dreamers] to go sleep in the Lab. They were hooked up to EEG machines, which means that they're measuring their brain wave activity on the scalp. They were also hooked up to EOG, so you could measure the activity of the eyes. And EMG, which is measuring the activity of the muscles, which is usually on the chin. And the reason that's important is that because using the brain waves and the muscle activity we can tell when somebody is in rapid eye movement sleep [or deep asleep]. [Using these machines, the scientists were able to capture a] signal that [the lucid dreamers had agreed they] would show them to indicate that they were ... [both asleep and also] lucid dreaming, [by moving] their eyes in [a] deliberate way that show[ed] that they were [both asleep and aware that they were asleep because they were in a dream].

The scientist presented and described data representing eye movement of an awake and a sleeping person. She resumed her description of the experiment:

> ... The brain wave activity was showing beyond doubt that they were asleep. And the muscle paralysis was showing that they were asleep. [But the agreed upon eye movement signal also showed that they were conscious in their dreams: They were lucid dreaming.] So 50 years ago they were able to prove that lucid dreaming is a real phenomenon and that's when the research really started to take off.

After describing further benefits of lucid dreaming, she closed her account with the following description of the tentative and contested nature of scientific knowledge:

> ...Finding out the kind of outer limits of what our imaginations can do, which is what we can do with lucid dream research, is a worthy thing to do and needs to be researched... And the last thing I am going to mention is pre-cognition. There are some researchers that think that precognitive dreams are most likely to happen, if they exist, in lucid dreams. So a pre-cognitive dream is when you can foretell the future in a dream. And a lot of people, many many people, think that they have these kind of dreams. It's very controversial in the scientific world because a lot of people have found very good evidence that they do exist and a lot of people have found evidence that they don't. So it's kind of—we're undecided.

In her 15-min narrative, she thus invited the audience to hear about and care about how scientists employ the epistemic practices of science, including developing a question, designing a valid experiment, engaging in evidence-based reasoning. She then engaged them in an activity to induce hypnagogic sleep—a state of dreaming while still awake—essentially turning the audience's own bodies into experimental apparatus to explore some of the ideas she had presented. Afterward, the audience engaged in a sense-making process, where they informally shared the results of their attempts to hypnagogically dream, and asked additional questions relevant to dreaming and lucid dreaming.

This art-and-science event—leveraging the fantastical and transgressive aspects of a cabaret ethos at a music festival, while engaging the audience with current science—integrated hybridized STEAM practices in several ways. Science is both wondrous (e.g., enabling flying through lucid dreaming) and evidence-based, as illustrated through the laboratory experiment described by the scientist. This juxtaposition operates to trigger an emotional openness that can enhance audience engagement with scientific explanations. Her presentation used multiple modalities: a "lecture" provided by a scientific authority, followed by an invitation to each audience member to engage in an embodied experimentation around hypnagogic sleep. Critique was supported through audience members sharing their hypnagogic experiences, or lack thereof. Moreover, the performative presence of the scientist—who seamlessly connected her familiar everyday interests in one's dreams with a larger enterprise of scientific investigations of dreaming, i.e., who clearly communicated the relevance and social nature of science—created the invitation to consider how scientists use evidence to advance understanding, as well as the contested and emerging nature of knowledge in science, i.e., the centrality of a critical stance.

4 Conclusions

As our two examples show, epistemic practices are interwoven, rather than sequential. In Making, designing solutions occur as materials are explored, and the materials themselves prompt aesthetic goals which create new engineering constraints. In the Guerilla Science case, audiences are invited to see the use and value of experimental design, drawn in by the personal passions of the scientist for a subject (dreaming) that is available and meaningful to everybody in the tent.

We are only beginning to explore these connections and expect to elaborate the conjectures shown in Table 1, to add new conjectured practices, and possibly to find that some are either rare or difficult to enact in authentic ways. For example, while critique is a regular part of arts education programs, engaging in scientific argumentation is rare in many science education programs. Yet, bringing critique to science learning promises to open a door not only onto "what constitutes the best evidence" as descried by the NRC (2012b), but also to the possibilities for developing a critical stance toward the culture of science and its history of power, especially with marginalized communities, thus addressing a gap some scholars have noted in current science reform efforts (see Philip and Azevedo 2017). Our Making example provided above suggests that perhaps in a STEAM context there is a kind of embodied critique—the evaluation and adoption, adaptation, or hacking of techniques or ideas from others as representing a consideration of alternatives and the utility or vibrancy of one approach over another. In this sense, critique may be embodied through choices that build on convincing approaches, rather than explicit production and consideration of evidence. In the context of the Guerilla Science presentation, the cabaret-like setting and mode of presentation is designed to upend normative power structures and expectations. The resulting dissonance, or suspension of quotidian patterns of interaction, opens up the contested nature of science in ways rarely available in most mainstream media accounts or school classrooms.

Adopting a critical stance in science is a significant step forward for efforts to address broadening participation in STEM. But to fully realize the potential of STEAM to advance participation in STEM or the arts, it will be important for STEAM educators to explicitly relate program activities and practices to relevant careers, academic pursuits, and other community-learning opportunities where such practices are valued. Brokering opportunities for young people to recognize and take up these opportunities is as important as providing them the initial chance to develop them.

Finally, a focus on epistemic practices may support critical reflection on the part of STEAM educators, which is key to advancing the field of STEAM education. In designing, implementing, and assessing programs, educators can probe the extent to which their programs engage learners in ways of knowing. Drawing on student work examples, video or written vignettes, or other forms of data, educators can reflect on whether or not "art" has operated as a token part of their science program, or vice versa. If deeper learning emerges through engaging in epistemic practices, reflection guided by clarity about what those practices look like can elucidate whether their programs are truly hybridized forms of STEAM, or rather token forms of integration. Educators can be challenged to design for and demonstrate how student experiences in producing knowledge or representations of ideas and insights result through the integrated nature of STEAM practices.

Thus, theorizing the epistemic practices of STEAM may provide educators with a tool for articulating as well as improving practice, leading to a fuller realization of what STEAM can mean for creating purposeful and agentive learning.

Acknowledgements This work was supported by the National Science Foundation 1647150 and the Wellcome Trust. The Sunshine Public School study was supported by the National Science Foundation 162365 and the Overdeck Foundation. The data reported in this work were collected by the field team of Exploratorium researchers, Dr. Jean J. Ryoo and Nicole Bulalacao. Opinions and findings reported in this work do not reflect the views of the funders.

References

Berland, L. K., Schwarz, C. V., Krist, C., Kenyon, L., Lo, A. S., & Reiser, B. J. (2015). Epistemologies in practice: Making scientific practices meaningful for students. *Journal of Research in Science Teaching, 53*, 1–31. https://doi.org/10.1002/tea.21257.

Bevan, B. (2017). The promise and the promises of making in science education. *Studies in Science Education, 53*, 75–103. https://doi.org/10.1080/03057267.2016.1275380.

Bevan, B., & Scarff, L. (2015). *Integration and inclusion: Connecting art and science to broaden participation in STEM*. San Francisco: Exploratorium and Science Gallery Dublin.

Bevan, B., Ryoo, J. J., Vanderwerff, A., Petrich, M., & Wilkinson, K. (2018). Making deeper learners: A tinkering dimensions framework. *Connected Science Learning*, (7). Retrieved from Connected Science Learning website: https://csl.nsta.org/2018/07/making-deeper-learners/.

Born, G., & Barry, A. (2010). Art-science: From public understanding to public experiment. *Journal of Cultural Economy, 3*, 1.

Bultitude, K., & Sardo, A. M. (2012). Leisure and pleasure: Science events in unusual locations. *International Journal of Science Education, 34*, 1–21.

Chávez, V., & Soep, E. (2005). Youth radio and the pedagogy of collegiality. *Harvard Educational Review, 75*(4), 409–434.

Dawson, E. (2017). Social justice and out-of-school science learning: Examining equity in science television and science clubs. *Science Education, 101*(4), 539–547. https://doi.org/10.1002/sce.21288.

Dewey, J. ([1934] 1980). *Art as experience.* Reprint. New York: Wideview/Perigree.

Dowell, E. (2014). Einstein's garden 2009–2014: Unexpected encounters with science. *Journal of Science Communication, 13*, 4.

Driver, R., Asoko, H., Leach, J., Mortimer, E., & Scott, P. (1994). Constructing scientific knowledge in the classroom. *Educational Researcher, 23*(7), 5–1.

Engle, R. A., & Conant, F. R. (2002). Guiding principles for fostering productive disciplinary engagement: Explaining an emergent argument in a community of learners' classroom. *Cognition and Construction, 20,* 399–483. https://doi.org/10.1207/S1532690XCI2004_1. Retrieved on December 21, 2017, from http://www.tandfonline.com/doi/abs/10.1207/S1532690XCI2004_1

Glinkowski, P., & Bamford, A. (2009). *Insight and exchange: An evaluation of the Wellcome Trust's Sciart programme.* London: Wellcome Trust. Retrieved November 1, 2009, from www.wellcome.ac.uk/sciartevaluation

Greene, M. (1977). Toward wide-awakeness: An argument for the arts and humanities in education. *Teachers College Record, 79*(1), 119–125.

Greene, M. (1995). *Releasing the imagination: Essay on education, the arts, and social change.* San Francisco: Jossey-Bass.

Guzzetti, B. J., & Yang, Y. (2006). Adolescents' punk rock fandom: Construction and production of lyrical texts. In B. Maloch, J. V. Hoffman, D. L. Schallert, C. M. Fairbanks, & J. Worthy (Eds.), *54th yearbook of the national reading conference* (pp. 198–210). Oak Creek: National Reading Conference.

Heath, S. B. (2000). Seeing our way into learning. *Cambridge Journal of Education, 30*(1), 121–132. https://doi.org/10.1080/03057640050005816.

Holland, D., Lachicotte, W., Jr., Skinner, D., & Cain, C. (1998). *Identity and agency in cultural worlds.* Cambridge, MA: Harvard University Press.

Kafai, Y., & Peppler, K. (2011). Youth, technology, and DIY: Developing participatory competencies in creative media production. *Review of Research in Education, 35*(1), 89–119.

Kafai, Y. B., & Peppler, K. (2012). Transparency reconsidered: Creative, critical, and connected making with e-textiles. In M. Ratto & M. Boler (Eds.), *DIY citizenship: Critical making and social media* (pp. 179–188). Cambridge, MA: The MIT Press.

Lankshear, C., & Knobel, M. (2011). *New literacies: Everyday practices and social learning.* Maidenhead: Open University Press.

Lave, J., & Wenger, E. (1991). *Situated learning: Legitimate peripheral participation.* Cambridge: Cambridge University Press.

McNeill, K. L., Katsh-Singer, R., Fagan, K., & Lowenhaupt, R. J. (2016, April). Principals' views of "good" science instruction: General pedagogy, hands on science, or science practices. *Paper presented at the annual meeting of the American Educational Research Association (AERA),* Washington, DC. Retrieved on December 21, 2017, from http://www.katherinelmcneill.com/uploads/1/6/8/7/1687518/mcneill_et_al_aera_2016_final.pdf

Nasir, N. S., Rosebery, A. S., Warren, B., & Lee, C. D. (2006). Learning as a cultural process. Achieving equity through diversity. In *The Cambridge handbook of the learning sciences* (pp. 489–504). New York: Cambridge University Press.

National Research Council. (2012a). *Education for life and work: Developing transferable knowledge and skills in the 21st century.* Washington, DC: The National Academies Press. https://doi.org/10.17226/13398.

National Research Council. (2012b). *A framework for K-12 science education: Practices, crosscutting concepts, and core ideas.* Washington, DC: The National Academies Press. https://doi.org/10.17226/13165.

New London Group. (2006). A pedagogy of multiliteracies: Designing social futures. *Harvard Educational Review, 66*(1), 60–92.

Peppler, K. (2013). STEAM-powered computing education: Using e-textiles to integrate the arts and STEM. *IEEE Computer, 46*, 38–43.

Peppler, K., & Wohlwend, K. (2017). Theorizing the nexus of STEAM practice. *Arts Education Policy Review, 119*, 1–12.

Petrich, M., Wilkinson, K., & Bevan, B. (2013). It looks like fun, but are they learning? In M. Honey & D. Kanter (Eds.), *Design, make, play: Growing the next generation of STEM innovators* (pp. 50–70). New York: Routledge.

Philip, T. M., & Azevedo, F. S. (2017). Everyday science learning and equity: Mapping the contested terrain. *Science Education, 101*(4), 526–532. https://doi.org/10.1002/sce.21286.

Resnick, M., & Rosenbaum, E. (2013). Designing for tinkerability. In M. Honey & D. Kanter (Eds.), *Design, make, play: Growing the next generation of STEM innovators* (pp. 163–181). New York: Routledge.

Ryoo, J. J., Kali, L., & Bevan, B. (2016). Equity-oriented pedagogical strategies and student learning in after school making. *Published in the proceedings of FabLearn '16, the 6th annual conference on creativity and fabrication in education* (pp. 49–57). Stanford: ACM. https://doi.org/10.1145/3003397.3003404. Retrieved on December 21, 2017 from https://dl.acm.org/citation.cfm?doid=3003397.3003404

Smith, J. K. (2014). *The museum effect: How museums, libraries, and cultural institutions educate and civilize society*. Lanham: Rowman & Littlefield Publishers.

Soep, E. (2006). An art in itself: Critique: Assessment and the production of learning. *Teachers College Record, 108*(4), 748–777.

Stetsenko, A. (2017). *The transformative mind: Expanding Vygotsky's approach to development and education*. New York: Cambridge University Press.

von Roten, F. C., & Moeschler, O. (2007). Is art a "good" mediator in a science festival. *Journal of Science Communication, 6*(3), A02.

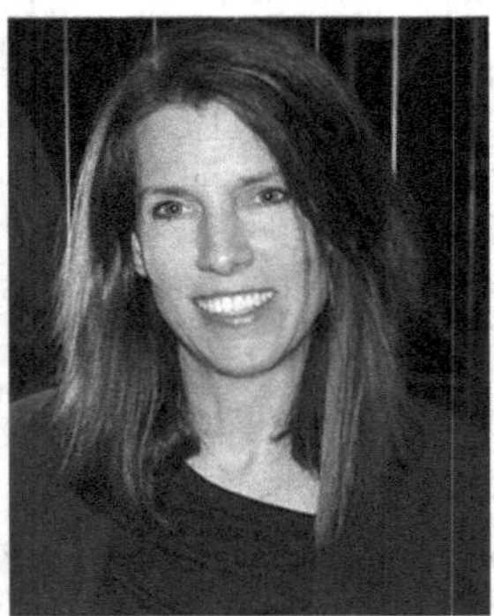

Bronwyn Bevan is Senior Research Scientist at the University of Washington. She is the Principal Investigator of the Research+Practice Collaboratory, and Co-Principal Investigator of the Center for the Advancement of Informal STEM Education. She also leads a pilot professional development program for mid-career public engagement with STEM professionals whose work focuses on broadening participation in STEM. Her research examines how learning opportunities, across formal and informal settings, can be organized to advance equity in education. She served on the National Research Council's *Committee on Out-of-School Time STEM Learning* and is on the editorial board of *Science Education.*

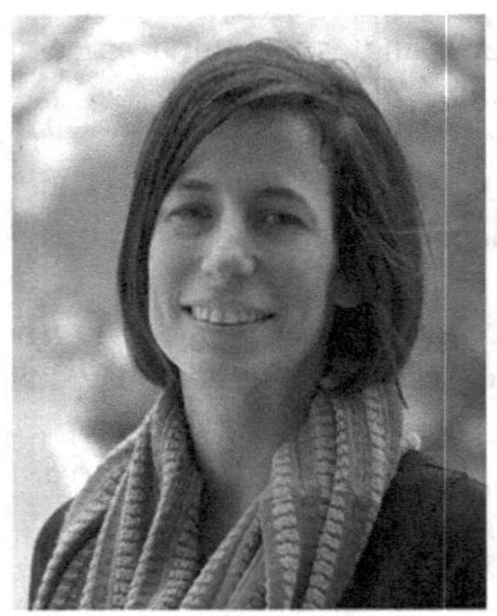

Kylie Peppler an artist by training, engages in research that focuses on the intersection of arts, technology, and interest-driven learning. As Director of the Creativity Labs at University of California Irvine (www.creativitylabs.com), Peppler brings together educators, designers, artists, and learning theorists interested in constructionist and hands-on, design-based learning. A primary focus of this work is on an emerging community of makers creating computational textiles (or e-textiles)—textile artifacts that are computationally generated or that contain embedded computers, like the LilyPad Arduino—which capture youth's pre-existing interests in new media, fashion, and design while supporting learning and creativity in computer science, arts, design, and engineering.

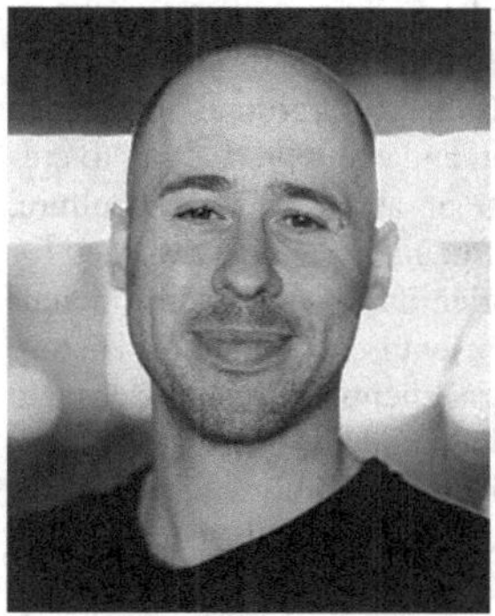

Mark Rosin is an associate professor of mathematics at the Pratt Institute of Art and Design. He is co-founder of Guerilla Science International, an international science and society organization, and director of Guerilla Science US. His research focuses on problems in applied mathematics, and the connections between science, art, and culture. He is the winner of the Early Career Award for Public Engagement with Science from the American Association for the Advancement of Science and is an Ashoka Foundation Emerging Innovator. Mark has worked at the University of California, Los Angeles, and received his Ph.D. from the University of Cambridge. His work with Guerilla Science has been featured in the *New Yorker*, the *New York Times*, and in VICE, and he has been supported by the National Science Foundation, the Smithsonian, Wellcome Trust, Simons Foundation, NASA, and Burning Man.

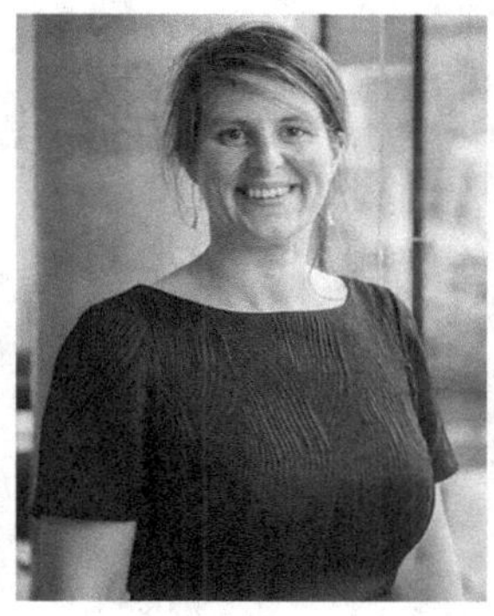

Lynn Scarff was the Director of Science Gallery Dublin. She has extensive experience in developing and leading public engagement projects in science, arts, and education fields. Lynn comes from a background of work in the environmental and not-for-profit sectors and has developed a series of programs, exhibitions, events, books, TV, and radio for these areas. Beginning her role in Science Gallery as the Education and Outreach Manager, Lynn has been involved since its inception. She is passionate about science and arts and the potential of spaces like Science Gallery to be facilitators of transformation in people's lives. Since 2018, she has served as the director of the National Museum of Ireland and is actively expanding her experience of cultural spaces as accelerators of transformative experiences.

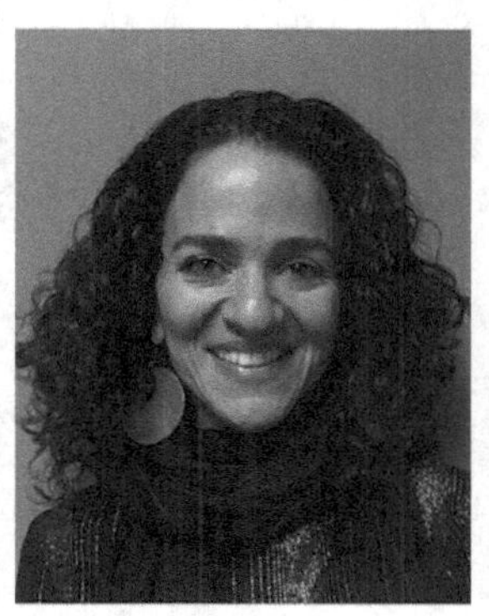

Elisabeth (Lissa) Soep is Senior Editor and Founding Director of the Research and Innovation Lab at YR Media, a youth-driven production company and national network of next-generation storytelling. The YR Media stories Lissa has produced have appeared on NPR, The New York Times, and Teen Vogue and been recognized with honors, including two Peabody Awards, three Murrow Awards, an Investigative Reporters and Editors Award, and the Robert F. Kennedy Journalism Award. With a Ph.D. from Stanford University's School of Education, Lissa has written about digital media and learning for academic journals, popular outlets, and books, including Youthscapes (with Maira, UPenn Press), Drop that Knowledge (with Chávez, UC Press), and Participatory Politics (MIT Press). With Asha Richardson, she co-founded YR Media's Innovation Lab, now an NSF-backed partnership with MIT and Cornell Tech. In 2011, she became one of six members of the MacArthur Foundation's Youth and Participatory Politics Research Network, which explores how young people use digital and social media to express civic voice and agency. Lissa also served more than 10 years on the Board of Directors of the United States' premier youth poetry organization, Youth Speaks (HBO series, 2009 & 2010).

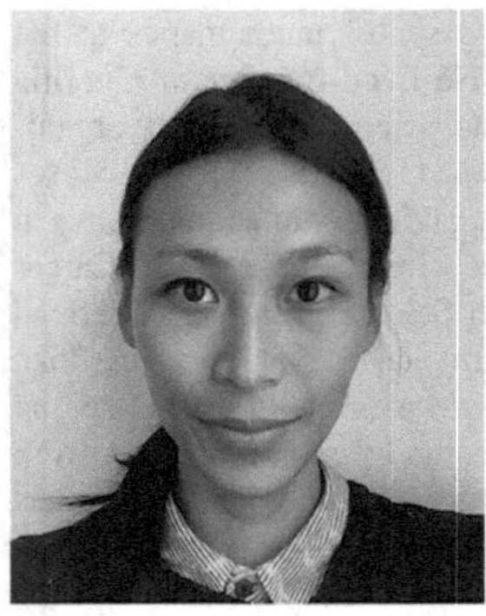

Jen Wong is co-founder of Guerilla Science International and Director of Guerilla Science UK, and Head of Programming at Science Gallery London. Jen uses her background in science communication, exhibition curation, and event production to create participatory experiences that sit across science, culture, engagement, and media platforms. She has previously worked for a range of national museums and scientific organizations, including the Dana Centre, Science Museum, and Natural History Museum, and a wide range of cultural clients from Selfridges and Secret Cinema to Wellcome Collection and Glastonbury Festival. She works with both artists and scientists to deliver participatory art/science collaborations that place audiences at the center of the story. Jen has a degree in Natural Sciences, and an M.Phil. in History & Philosophy of Science.

Investigating the Complexity of Developing STEAM Curricula for K-8 Students

Danielle Herro and Cassie Quigley

1 Conceptualizing STEAM: How Do We Avoid Repeating the Past?

To date, STEAM teaching is not well understood. In fact, the difficulty in moving forward with STEAM instruction, for many educators, parallels issues that continue to plague STEM. Many teachers wonder what *is* and what *is not* considered STEAM, and experience challenges when shifting instructional models to support discipline expertise and integrated curriculum (Portz 2015). Williams (2011) describes how elementary curricula often favors teaching science over technology, and secondary school curricula attend to each discipline from an academic rather than practical focus (i.e., connected to the real world). He suggests that educators need a sound rationale for "why" and "how" integrating various disciplines will lead to quality learning outcomes for students. Advocates suggest that broadening STEM by incorporating the "A" to encompass the arts and humanities in STEM instruction has the potential to reach a wider and more diverse range of young learners (Bequette and Bequette 2012). STEAM is intended to appeal to learners who might be interested in creative, human-centered, and humanitarian ways of thinking about problem-solving through learning-by-doing (Boy 2013). However, it is not clear how STEAM perspectives might translate into teaching practices, or even what declaring a classroom or school as "STEAM" focused means. These conceptualization issues and challenges can translate into poorly designed STEAM curricula or relegating STEAM problem-solving to afterschool or specialized programs that may or may not appeal to broad, diverse populations of students. Further complicating the issue is that predictive reports and classroom examples often conflate STEAM instruction with a makerspace, fablab, or grand challenge. These spaces and challenges can

D. Herro (✉) · C. Quigley
Clemson University, Clemson, SC, USA
e-mail: dherro@clemson.edu

© Springer Nature Switzerland AG 2019

A. J. Stewart et al. (eds.), *Converting STEM into STEAM Programs*,
Environmental Discourses in Science Education 5,
https://doi.org/10.1007/978-3-030-25101-7_4

augment STEAM instruction, but in themselves are really more or less STEAM incubators or activities. To that end, our research-based practices led us to conclude that all of these ideas are valuable in conceptualizing and enacting STEAM instruction in classrooms, but only one part of effective STEAM instruction. Like other instructional ideas, STEAM needs a framework to focus its design that adheres to principles for offering more equitable opportunities, in essence more appealing curricular opportunities that resonate with young learners. This chapter offers a window into how this might be achieved in K-8 classrooms.

2 Positioning Our STEAM Work

Our work on this project began in 2013, with a request from a large, local school district wishing to understand and successfully offer STEAM instruction. Over the next 5 years, we expanded our research project and conducted a longitudinal qualitative study on STEAM professional development (PD), teachers' implementation practices, and students' collaboration when solving STEAM problems. During this time more than 100 K-8 teachers, from various grade levels and subject areas, including science, math, English-language arts, art, music, and physical education, participated in intensive STEAM PD, and nearly half of these teachers allowed us into their classrooms to observe their STEAM teaching in action. The teachers all work in public schools in the Southeast USA serving rural, suburban, and urban students. Throughout each phase of the research, we collected data in the form of teachers' reflective journals, classroom observations, individual and focus group interviews, video data, and surveys. For this chapter, we focus on the teacher PD aspect, and how it informed the creation of a conceptual model, which subsequently guided teachers to create STEAM unit plans. We then focus specifically on two examples of teachers—one elementary and one middle school—who designed STEAM curricula to understand the complexity of creating relevant, problem-based scenarios, how they approached transdisciplinary teaching, ways they included the arts and humanities, and techniques for offering meaningful technology integration. We also discuss challenges these teachers faced when negotiating standards and assessment in rigid school environments.

3 Professional Development to Inform STEAM Unit Creation

When we began this project in 2013, there was very little research on STEAM practices (Henriksen 2014), and only a few predictive reports detailing the potential of STEAM curricula for engagement and learning (Johnson et al. 2015). Consistent with the literature, the teachers we interacted with were either new to STEAM (and in some cases, STEM) or using existing STEM models and attempting to "add on" experiences with the arts or humanities that did not appear relevant or engaging (Kim and Park 2012). For instance, teachers would ask students to propose an invention that might solve an imaginary problem, or draw a poster that depicted a scientific solution, or create a kite out of straws, paper, and tape that could actually fly. In fact, most teachers noted in our pre-surveys that STEAM "meant that the arts were included, but they weren't exactly sure how."

To address this problem, we created an intensive 50-h PD session after thoroughly reviewing the existing research on both STEM and STEAM education. We then delved into what STEAM might look like, based on predictive reports and classroom examples. We noted common themes in the literature and examples that included transdisciplinary teaching, problem-solving, technology integration, attention to the arts and humanities (including social justice issues), and student-centered collaborative work. These ideas and our work as former science and technology teachers with experience in project-based learning (for an understanding of how project-based learning and STEAM connect, see Gary Ubben's chapters in this book) and technology integration in K-12 guided our first PD. The literature and our former work also assisted us in creating a STEAM conceptual model (Fig. 1; Quigley et al. 2017) that was iteratively refined throughout our research. During the PD, teachers created STEAM units after participating as *students* in solving a STEAM problem using collaborative and interactive technologies (e.g., Google Docs/Forms, Presentations, Adobe Spark videos, and Infographics) that was embedded in a larger unit. The STEAM unit was aligned with the conceptual model and served as a template for teachers to then create their own units to be implemented in their classrooms during the fall and spring semesters. The resulting units included a problem scenario aligned with real-world practices, a driving question, standards alignment, and daily activities in which elements of STEAM and technology-enabled learning was to be presented to students.

After operating the 50-h PD session, we reviewed the teacher-created units and provided extensive feedback to guide them as they revised and implemented their units. We observed the teachers at least twice during each semester while they taught their STEAM units (see Quigley and Herro (2016) for additional example units). During the classroom implementations, teachers reflected on their instructional practices weekly, completed a peer observation of other STEAM teachers' instruction, and met with us to discuss their progress. We also hosted two short, follow-up PD sessions to share practices and increase understanding of STEAM teaching among colleagues. Generally, the initial research (see Herro and Quigley 2016) conducted on the teachers' understanding of STEAM instructional practices demonstrated the following: (1) before the PD, STEAM was not well conceptualized by teachers; (2) problem-based[1] learning and collaborative technologies greatly

---

[1] We acknowledge the differences between problem-based and project-based learning. We agree with Ubben's notion that the difference is project-based learning often surrounds a challenge and a product/project at the end. The similarities include a focus on the process in both instances. For our work, we focused on problem-based learning to allow for the absence of a project/product.

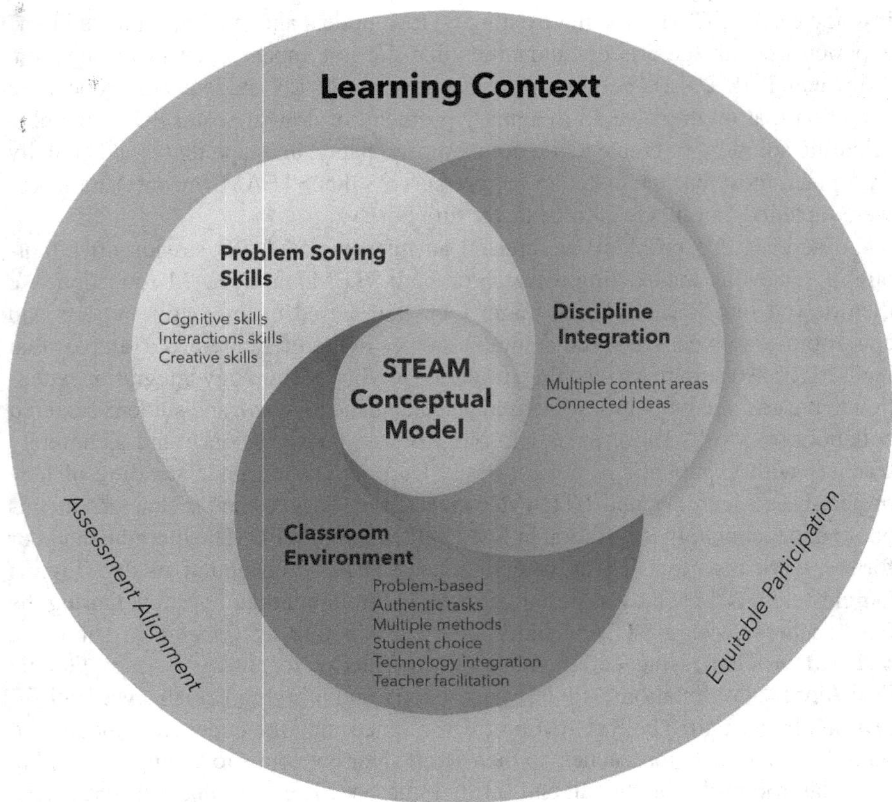

Fig. 1 STEAM education conceptual modele

assisted teachers in understanding STEAM principles and learning content, but understanding how to effectively incorporate the "A" remained challenging; (3) most teachers felt successful at creating and teaching STEAM units posing relevant, local problems using problem-based instruction; and (4) teachers used technology for instruction and assessment far more than facilitating students' collaboration and media creation.

In the next sections, we provide an overview of the STEAM conceptual model and discuss the importance of scenario-based teaching to position the two examples we deconstruct in this chapter.

4 A STEAM Education Conceptual Model

To create this model, we analyzed data from teachers who effectively and systematically enacted STEAM-based curricula (for details on how this model was developed, see Quigley et al. 2017). We also tested this model in vivo to refine it based on the realities of schooling. This conceptual model includes three dimensions that focus on the components of instruction that engage students in STEAM education: discipline integration, classroom environment, and problem-solving skills.

The first dimension, discipline integration, is the way in which teachers connect multiple disciplines or content areas through a problem-based unit. While the goal of this STEAM model is transdisciplinarity, the model also looks at variances of discipline integration (single content, multiple disciplines, and transdisciplinarity). Our research found teachers were more readily able to integrate multiple disciplines when they aligned the disciplines to the problem to be solved. Therefore, the attributes in this dimension can be defined as multiple content areas and connected ideas. In classrooms with a high level of discipline integration, the content selection draws on different disciplines during problem solving by using expert knowledge, multiple sources of information, and various concepts, theoretical approaches, or methods.

The second dimension, classroom environment, examines the ways in which the teachers structure the classroom environment to facilitate problem-solving. This dimension includes a problem-based approach, authentic tasks, multiple methods to solve the problem, student choice, technology integration, and teacher facilitation. When teachers situate the task in a real-world event through a problem scenario, it helps to make the problem and, by extension, the content more relevant to the students. This dimension significantly informs practices teachers focus on during the creation and enactment of STEAM curricula.

The third dimension, problem-solving skills, involves the ways in which teachers support the development of students' cognitive, interactive, and creative skills through various instructional activities. These skills provide students with the means to solve problems. During STEAM lessons, teachers support students in developing higher-order skills, such as abstracting, analyzing, applying, formulating, collaborating, and interpreting; constructing explanations; engaging in argumentation; disseminating evidence; and presenting. The conceptual model looks for teachers to regularly provide opportunities for students to practice these skills in multiple contexts. In this model, teachers encourage students to explore multiple paths to solving a problem, which provides favorable conditions for sparking creativity or for exercising creative skills. These creative skills rely on a teacher's ability to offer concepts, tools, and experiences in open-ended, problem-solving scenarios. By using this conceptual model, the teachers design problem-solving tasks and classroom environments to promote student-guided learning that relies on peer support and collaboration. By developing problems that urge students to reach out to peers for assistance and work collaboratively, teachers also encourage developmentally appropriate levels of social and emotional engagement in learning. In this manner, our STEAM conceptual model is more than just a mashup combination of science, technology, engineering, arts, and math content; it defines an instructional approach by which teachers inculcate a transdisciplinary perspective on real-world problems.

5   The Importance of Scenario-Based Teaching in STEAM Instruction: Examples from the Field

When thinking about the ways that teachers are able to successfully integrate the dimensions of the conceptual model, we needed a way to support teachers in curricular design. Through this work, we found that scenario creation helped to provide a platform for these instructional approaches. Scenario creation also provided the teachers with a starting point to design the unit and daily lesson plans. Below are two examples of these scenarios. In these two examples, however, we are careful to view these as a starting point to the unit. Because one of the goals of STEAM is to encourage multiple ways of solving a problem, students are encouraged to pursue different pathways, and therefore the daily activities and final products can look very different from the ideas proposed in the scenario. By intentionally crafting a problem scenario, there should be attention specifically to the arts. One criticism of STEAM is that art is still viewed as an add-on or an afterthought. However, during scenario creation, we found that this provided teachers with an opportunity to thoughtfully integrate the arts into the problem scenario.

5.1   Example 1: Birds of Prey

The third grade teachers of Parkview Elementary School used a local resource, the Birds of Prey Center, as the backbone for their STEAM scenario. This center provided a real-world problem. Further, by leaning on an existing resource, the teachers could construct a problem that naturally integrated the arts (music and visual as well as English-language arts), science (understanding habitats), technology (3D printing, presentation tools), engineering (designing and building), and math (measuring).

6   Problem Scenario

The Birds of Prey Center has noticed an increase in the number of injured birds. While the Center takes care of injured animals, they are concerned about why this increase is happening. Recently they noticed injuries of mainly migratory birds that have broken wings, have breathing problems, or are malnourished. This is concerning because birds are high up in the food chain, and they also are good indicators of the general state of biodiversity. When birds of prey start disappearing, it means that something is wrong with our environment, and that we probably need to take action. The Birds of Prey Center asked the Parkview Elementary School to help them figure out this problem, and to come up with possible solutions.

Driving question: How can we create an environment that protects and nurtures native birds?

7   State Standards

7.1   Science

3.L.5 The student will demonstrate an understanding of how the characteristics and changes in environments and habitats affect the diversity of organisms.

3.L.5.1 Obtain and communicate information to explain how changes in habitats can be beneficial or harmful to the organisms that live there.

7.2   English-Language Arts

3-RI.5.1 Ask and answer literal and inferential questions to gain meaning of text or topic.

3-C.3.2 Create presentations using video, photos, and other multimedia elements to support communication and clarify ideas, thoughts, and feelings.

7.3   Math

3.MDA.4 Generate data by measuring length to the nearest inch, half-inch, and quarter-inch and organize the data in a line plot using a horizontal scale marked off in appropriate units.

3.MDA.5 Understand the concept of area measurement.

(a) Recognize area as an attribute of plane figures.

(b) Measure area by building arrays and counting standard unit squares.

(c) Determine the area of a rectilinear polygon and relate to multiplication and addition.

7.4   Visual Arts

Standard 6: The student will make connections between the visual arts and other arts disciplines, other content areas, and the world.

7.5 *Music*

Standard 2: The student will improvise, compose, and arrange music within specified guidelines.

8   Daily Activities

1. First, the students will watch a video about saving a bird that has a fishing line stuck around its beak. The students will then be introduced to the problem scenario. Then the students will discuss their observation of birds and record their initial thoughts about birds. The teachers will have an informal bird learning center set up with texts and resources so that students can use this center during their daily center rotations.
2. Bird watching (3 days): Students will participate in bird watching using binoculars and data recording tables noting shape, color, size, and behavior of the birds. Several bird-watching experts will be coming to the classroom to share tips on ways to bird-watch prior to the observation. After the observations, the students will use resources to look up what the types of birds that they saw are.
3. Using the data from the bird-watching, the students will discuss the similarities and differences of the birds as well as their habitat. The teacher will lead discussions about the habitats of birds and what birds need to survive, tied to the science standards. The students will make predictions about the problem of injured birds.
4. The students will visit the Birds of Prey Center and conduct observations.
5. The students will participate in the Lego WeDo activity on extreme habitat. The teacher will lead a discussion on the parallels between these habitats and the areas around the state. The students will design a habitat that would suit a particular bird (of the students' choosing).
6. Students will brainstorm ideas for solutions to the injured birds. In the MakerSpace, students will design and build birdhouses. To do this, the teacher will lead sessions on measurement aligned to the math standards. The students will research designs for different types of birds. Building the birdhouses will take 3–5 days.
7. In music, the students will create bird songs both based on their observations and their research. They will discuss the ways in which bird songs are unique and how they build communities. The students then will discuss the ways music binds their community.
8. The students will read *Charlotte's Web* together and discuss compassion and empathy toward animals. The teacher will lead a discussion on how students are compassionate about birds and what concerns them. Then they will discuss what measures can be taken to ensure the conservation of raptors. Students brainstorm; they create, collaboratively, ways to bring community awareness of Birds

of Prey, through brochures, videos, etc. The students also will create 3-D prints of prosthetics for severed parts of bird and structures that harmfully affect raptors.

9. After the students have decided how they will increase awareness, they will create this educational piece. The Birds of Prey Center will attend the final reveal and will provide feedback to the students. The students will also participate in peer review prior to this reveal.

Ongoing embedded assessment was aligned to the content standards through group discussions as well as peer review of the final products (bird houses, educational materials, songs, and prosthetics). These peer reviews provide opportunities for students to receive feedback; additionally the peer review provides the students with the opportunity to practice giving productive feedback.

8.1 Example 2: DNA STEAM Unit

Audrey is a 7th-grade science teacher. She created a unit in which students were asked to investigate transfer of genetic information through inheritance, and explore how technology might influence this transfer. The unit is intended to address core area standards for 7th graders in her home state, which include "Heredity—Inheritance and Variation of Traits." Her primary discipline expertise is in science, but as noted in the scenario and truncated STEAM unit below, she extends the learning to math, technology, engineering, English-language arts, graphic arts, and the humanities through a transdisciplinary problem. This problem foregrounds the issue and allows the disciplines to emerge naturally through the problem-solving process.

9 Problem Scenario

News.harvard.edu reports that "having a twin sibling diagnosed with cancer poses an excess risk for the other twin to develop any form of cancer" (2016). Miranda and Alecia are 45-year-old identical twins. They were born and raised to wealthy, educated parents in Chicago and lived at home until college. They continue to live in the suburbs of that city and are still close. Recently, Alecia was diagnosed with stage II breast cancer. She is currently undergoing treatment for the disease while her sister Miranda remains healthy. How is this possible? Analysis of the family traits shows that there is no history of breast cancer in the family, yet the general practitioner claims Miranda is still at risk. You and your colleagues are internal medicine doctors who practice in downtown Chicago. Miranda has come to your practice for a second opinion on the clean (yet cautioning) diagnosis of being cancer free. She has many questions about her prognosis and what she might do to lessen her chances of developing the disease. As a group, you must develop a presentation to share with Miranda about your thoughts and suggestions to her as physicians. You must also decide if your team deems it worthy to take a sample of Miranda's DNA to test it for the cancer gene and, if so, determine how you will convince her that the benefits of this information outweigh the risks and cost. Your digital presentation should include a slideshow of the main points your team wishes to discuss with Miranda, including the cost of further studies. Since you are the expert, you have also been charged with creating a digital pamphlet on breast cancer and a 3-D model of DNA that you can show patients like Miranda while you discuss their genetic information.

9.1 Driving Question

How does your DNA determine your risk of developing disease?

9.2 Elements of STEAM

Science: DNA, mutations, epigenetics.
Technology: digital presentation and digital pamphlet, 3-D model.
Engineering: design 3-D model, design solution/test DNA for cancer-causing genes.
Arts: design components of presentation, pamphlet and 3-D model; ethics issues and ELA.
Mathematics: cost analysis of DNA testing.

10 State Standards

Standard 7.L.4: The student will demonstrate an understanding of how genetic information is transferred from parent to offspring and how environmental factors and the use of technologies influence the transfer of genetic information.

7.L.4A.1 Obtain and communicate information about the relationship between genes and chromosomes to construct explanations of their relationship to inherited characteristics.

7.L.4A.2 Construct explanations for how genetic information is transferred from parent to offspring in organisms that reproduce sexually.

7.L.4A.3 Develop and use models (Punnett squares) to describe and predict patterns of the inheritance of single genetic traits from parent to offspring (including dominant and recessive traits, incomplete dominance, and co-dominance).

7.L.4A.4 Use mathematical and computational thinking to predict the probability of phenotypes and genotypes based on patterns of inheritance.

7.L.4A.5 Construct scientific arguments using evidence to support claims for how changes in genes (mutations) may have beneficial, harmful, or neutral effects on organisms.

11 Daily Activities

(90-min class periods)

1. Introduce the unit and have students choose groups; brainstorm questions to ask via Google Classroom; Skype with an Expert (Dr. Harrison); begin research.
2. Continue research, share Google Docs with team and teacher for comments; conduct science labs on DNA and mutation; Skype with another expert (Dr. Decker); begin outline of presentations—determine what is to be included.
3. Finish research and begin developing presentations; begin designing DNA molecule using Google Draw and 3-D printer, post first draft of presentations on Google Classroom for peer feedback.
4. Finish designing DNA molecules; begin cost analysis; begin creating pamphlet.
5. Draft a letter to the editor detailing the ethical issues of epigenetics related to policy—including policies around environmental controls.
6. Finish pamphlets and have students provide peer feedback; complete cost analysis and have groups embed it in their presentation.
7. Share presentations on Google Classroom and have students design peer feedback questions collected via Google Forms.
8. Begin debriefing and discussion of what new questions were raised throughout the unit; have students create short reflections that addresses peer feedback and changes they made.

Formative assessment was used throughout the duration of the project. The teacher used Google Forms, Kahoot, and exit tickets, and she also asked questions about the content students learned and their successes and struggles while completing tasks. As noted in the daily activities, the students critiqued their peers to offer constructive feedback. The summative assessment included rubrics aligned to the pamphlet and final presentation.

12 Discussion

Our discussion focuses on deconstructing STEAM units to understand how they align the dimensions of the conceptual model, including offering relevant problem-based scenarios, transdisciplinary teaching, attending specifically to the "A" in STEAM, and meaningful technology integration. We also want to point out challenges teachers faced, which generally included incorporating requisite standards, developing appropriate assessments, and, at times, addressing disciplines outside of their expertise. Our goal is to increase understanding versus offering generalizable evidence. However it is important to note that the units are abbreviated for this chapter, the full units were more detailed and included numerous links to resources, rubrics, assessments and instructor created media. They are also typical of what teachers created in a majority of classrooms after attending our PD.

13   Relevant, Problem-Based Scenarios

In the two examples above, the scenarios were intended to support students in caring about the learning ahead by offering local, real-world issue appealing to their developmental level. In the Birds of Prey unit, the elementary school was in close proximity to the center, the organization visited the school annually to share protection and conservation efforts, the issue was authentic, and young students are generally interested in animals. Similarly, the DNA unit was involved in an actual news story from a well-known city, using content (e.g., disease, genetics, ethical decision-making) that is fascinating for many teenagers as they become abstract thinkers. In both of these units, the teachers capitalized on the interest of the students and relevance to real-world problem-solving; the problem-solving had a purpose that students could relate to.

14   Transdisciplinary Teaching

As evidenced in the Birds of Prey and DNA examples, science, technology, arts/humanities, engineering, and math were all needed for solving both problems, although not necessarily given equal time or weightage. This approach mirrors what happens in the real world, as complex problems are being addressed. The teachers in both examples were cognizant of requisite content and standards they were expected to cover in their classrooms, and they used the standards to guide the initial creation of the unit. However, they expanded the unit to focus on the problem-solving in a manner that allowed the disciplines to emerge more naturally. While they were careful to develop a scenario that addressed as many disciplines as possible, the teachers approached writing and refining the scenario from two aspects: (1) they were conscious of whether they had the expertise or could partner with other teachers to assist teaching across disciplines and (2) they did not force discipline integration, choosing instead to have the scenario dictate the skills necessary to solve the problem.

15   Including the Arts and Humanities

Through the purposeful design of the daily activities in the Birds of Prey example, students were engaged in design, music, English-language arts, social justice and digital media creation as they built birdhouses, created bird songs, read and compared themes in *Charlotte's Web* to address empathy and compassion and created digital videos and brochures. In the DNA example, students participated in rich discussions about ethics, used the media arts to create aesthetically pleasing presentations and pamphlets and wrote persuasive letters. While not as apparent in these units, a number of teachers we worked with often introduced STEAM units with digital slideshows of images related to the problem at hand (e.g., local injured birds being cared for, DNA helix art) and asked students to react to them. This technique was also used to have students create thought-provoking visual presentations, and this idea adheres closely to suggestions about artwork and images noted in chapter "Transgressing the Disciplines Using Science as a Meeting Place: The Science, Art and Writing Initiative", by Anne Osbourn. Her chapter notes the ability to use theme and images as a starting point for science, art, and writing sessions.

16   Technology Integration

As demonstrated in both STEAM examples, teachers approached planning for technology in a student-centered manner, and students used technology for collaboration, design, data collection, presentation, and other production-focused ways. Teachers' instructional use of technology often included showing videos, sharing resources, and formative assessment, but the primary focus was having students create with technology. In the Birds of Prey example, students used technology for Lego WeDo construction, 3-D printing, and video creation. In the DNA example, students used G Suite (formerly Google Apps for Education) Docs, Forms, Slides, and Draw as part of the data-collection and collaborative problem-solving processes. They also created videos and used design software before printing DNA with the 3-D printer.

17   Challenges

As evidenced in the above examples, STEAM teaching is messy. Lecture-based teaching, multiple-choice and written tests, and providing projects in which all products "look the same" would clearly be easier to implement in classrooms. Thus, STEAM teaching is not without its challenges. In particular, the teachers involved in this research, and the many others we worked with, identify four major challenges when developing and implementing STEAM curricula: pacing/time constraints; student understanding of content or the STEAM process; issues relating to planning and standards alignment; and difficulties adhering to district policies.

Implementing the two units described earlier in this chapter required flexibility of pacing schedules, and both of the teachers reported it took time to align the pacing and time to adhere to a management schedule that did not interfere with other required instructional goals. They also mentioned that students who typically struggled with understanding abstract or more complex processes or concepts often needed additional guidance or assistance, although these teachers—and almost all other teachers (approximately 80%) we worked with—reported a sharp decline in behavior issues because student engagement was greater. Planning with other teachers and aligning standards across disciplines was very challenging for these teachers, and for most other teachers in our longitudinal study, for that matter. A total of 90% of our teachers discussed this issue at some point of the STEAM implementation. To that end, about 25% of the teachers suggested they did not have common planning time or were limited in time to plan, while other teachers (about 15%) mentioned that their colleagues did not want to co-plan or co-teach with them, limiting their ability to use transdisciplinary teaching methods. District policies seldom prevented teachers from enacting STEAM teaching, but policies such as adhering to strict testing schedules, inflexible scheduling/strict pacing guides, and prohibitive access to technology devices or websites often required creative workarounds. Interestingly, integration of the arts/humanities was rarely considered a challenge after writing scenarios in the initial PD (about 10% of teachers struggled with this), but we often felt it was an area that needed better conceptualization and strengthening when we visited classrooms.

Our goal is that with this conceptualization, and a focus on curricular design, we can assist teacher educators supporting teachers enacting STEAM. In this manner, STEAM instruction can be less challenging and more rewarding for teachers and their students. We argue that STEAM education can be a more powerful teaching and learning approach as it encourages students to encounter and ask new questions, sometimes taking them down a completely different line of inquiry. It also emphasizes important social justice or humanitarian elements by focusing on the arts and humanities. Further, when compared to some STEM approaches, the transdisciplinary nature of foregrounding the problem to be solved versus the discipline "to be covered" is more realistic and often more interesting to students. Our hope is to impact teachers with a sound conceptual understanding of STEAM education to guide their instructional practices—and most importantly to affect the students they teach with opportunities for deeper learning.

References

Bequette, J. W., & Bequette, M. B. (2012). A place for art and design education in the STEM conversation. *Art Education, 65*(2), 40–47.

Boy, G. A. (2013, August). From STEM to STEAM: Toward a human-centered education, creativity & learning thinking. In *Proceedings of the 31st European conference on cognitive ergonomics* (p. 3). ACM.

Henriksen, D. (2014). Full STEAM ahead: Creativity in excellent STEM teaching practices. *The STEAM Journal, 1*(2), Article 15. https://doi.org/10.5642/steam.20140102.15.

Herro, D., & Quigley, C. (2016). Exploring teacher perceptions of STEAM: Implications for practice. *Journal of Professional Development in Education, 43*(3), 416–438. https://doi.org/10.10 80/19415257.2016.1205507.

Johnson, L., Adams Becker, S., Estrada, V., & Freeman, A. (2015). *New Media Consortium (NMC) horizon report: 2015 K-12 edition.* Austin: The New Media Consortium.

Kim, Y., & Park, N. (2012). Development and application of STEAM teaching model based on the rube Goldberg's invention. In *Computer science and its applications* (pp. 693–698). Dordrecht: Springer.

Portz, S. (2015). The challenges of STEM education. *Proceedings of the 2015 (43rd) space congress: A showcase of space, aviation, technology, logistics and manufacturing.* Paper 3, 1–9. Daytona Beach, Florida: Embry-RiddleAeronautical University-Digital Commons. Available at commons.erau.edu/space-congress-proceedings/proceedings-2015-43rd/

Quigley, C., & Herro, D. (2016). Finding the joy in the unknown: Implementation of STEAM teaching practices in middle school science and math classrooms. *Journal of Science Education and Technology, 25*(3), 410–426. https://doi.org/10.1007/s10956-016-9602-z.

Quigley, C., Herro, D., & Jamil, F. (2017). Developing a STEAM classroomassessment of learning experiences. *School Science & Mathematics, 117*(1–2), 1–12.

Williams, J. (2011). STEM education: Proceed with caution. *Design and Technology Education: An International Journal, 16*(1), 26–35.

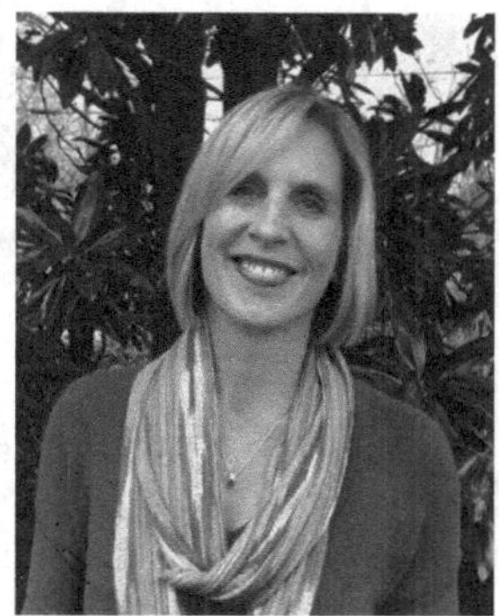

Danielle Herro is Associate Professor of Digital Media and Learning at Clemson University. She teaches courses on social media, games, and emerging technologies.

At Clemson, she co-designed and opened Digital Media and Learning and Gaming Labs in the College of Education. Dani has published numerous journal articles on the impact of bringing game-based, mobile, and other digital media environments to K–12 schools. Her current research interests focus on investigating the intersection of games, identity and social practices, and the efficacy of teacher professional development toward integrating STEAM education.

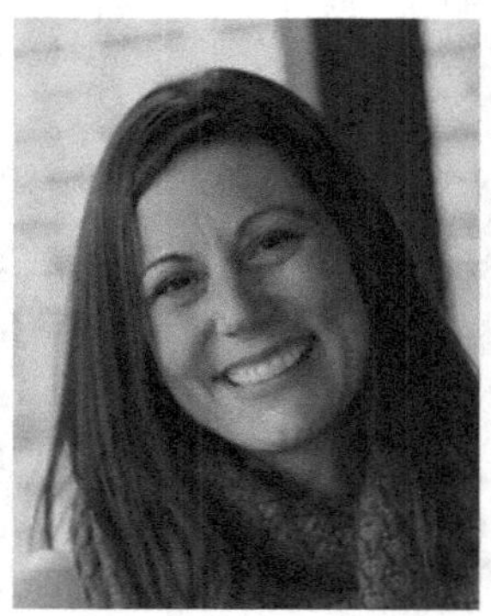

Cassie Quigley is Associate Professor of Science Education in the Department of Teaching and Learning in the College of Education at Clemson University. Her research focuses on broadening the ideas of and participation in science so that all students feel connected to science. Currently, she works with in-service teachers on expanding their current pedagogical practices to include equitable approaches.

We Need More (than) STEAM: Let's Go for Life-Wide and Lifelong Education

Wolff-Michael Roth

1 Introduction

The existence of this book testifies to the fact that there is a new kid on the curriculum-related block: STEAM. It is but the latest innovation, of which there have been many since I entered the field of science education. When I began teaching in 1980, a year after having completed my MSc in physics, those at the forefront of the field used science curricula developed with the aid of funding from the US National Science Foundation later referred to as the alphabet curricula—the many acronyms then current (e.g., SCIS [Science Curriculum Improvement Study], SAPA [Science: A Process Approach], or IPS [Introductory Physical Sciences]) have all but been forgotten. The integration of mathematics and science became of interest in the 1980s and 1990s, followed in the post 2000 years by the integration of the subject areas of science, technology, engineering, and mathematics, now gathered under the acronym of STEM. Before that, science was to be integrated with the social sciences in courses sometimes named "science in society" or something of that ilk. In 2002, I was part of a group of scholars discussing (in the.

Canadian Journal of Mathematics, Science, and *Technology Education*) whether science educators should approach teaching science through societal or citizenship perspectives. At the time, I argued that nothing really changes with the change of a curricular emphasis, as long as schooling itself does not change—*plus ça change, plus c'est la même chose*. This is the reason for the first part of my chapter title: we need more than STEAM, more than yet another presumably better mousetrap for the perennial question of the curricular content and medium. We need more than STEAM so that we do not run out of steam while trying to improve the education of current and future generations. In my view, education should not be organized

W.-M. Roth (✉)
University of Victoria, Victoria, BC, Canada
e-mail: mroth@uvic.ca

© Springer Nature Switzerland AG 2019

A. J. Stewart et al. (eds.), *Converting STEM into STEAM Programs*,
Environmental Discourses in Science Education 5,
https://doi.org/10.1007/978-3-030-25101-7_5

around subject matter in whichever combination. Instead, education needs to be a life-wide and lifelong endeavor. Here, life-wide means that the subject matter areas take their appropriate place in the life of the person as a whole—which may mean that STEM or STEAM are at the bottom of the hierarchy of object/motives of a person; and the person is continuously learning and developing, so the object or motives important today may be of minor importance tomorrow.

2 A Brief History of Curriculum Evolution

In my own science classrooms in the 1980s, forms of mathematics were integrated to help students analyze, interpret, and present data. Student groups prepared contracts determining what they wanted to achieve over a one- or two-week period, and then pursued accomplishing the goals they had set for themselves. A decade later, upon my initiative, the school I worked at made available computers in the physics laboratory so that students could use statistical software to find best-fit curves, or statistically analyze or mathematically model the data; and they used the first calculators capable of plotting and analyzing data. In my first few years, I had all students do the same experiments. But over the course of my first decade of teaching science, I realized that students might not be interested in those experiments. Over time, I increasingly asked students to design their own experiments within the confines of the provincial curriculum. Thus, when the curriculum detailed 12 weeks to be spent on the physics of motion, the students designed experiments on the topic of motion. But it was completely up to them to determine the specific phenomenon they wanted to research. Thus, for example, some students investigated the role of friction on the speed and acceleration of moving vehicles by constructing vehicles of different sizes or chose a series of objects of different shapes and then designed experiments for measuring the variables they planned to investigate. By 1991, 10 years after starting to teach, I had arrived at a point where the students used widely varying means to learn curricular concepts. Thus, for example, in a unit on electricity, one group of students in my junior year physics course did a series of experiments on conductors, semiconductors, and superconductors: they organized everything, even purchasing liquid nitrogen, renting the Dewar flask, and adjusting their school day when they were doing the superconductor experiments (they had to do this because the liquid nitrogen could not be stored long). Another group, interested in theater arts, wrote the script for short plays to teach electrical concepts; for their evaluation, they performed these pieces. A third group designed, wrote, and drew comic scripts designed to teach those concepts that they themselves were learning. A fourth group produced a series of curriculum materials to teach hands-on lessons to fifth graders; they taught two lessons using their materials, and this lesson was part of their evaluation. Across the three parallel classes that I taught during that year, there were still other activities in which students engaged and in the course of which they learned the pertinent concepts.

This approach was highly successful, and students appreciated being involved in the curriculum design and being able to bring their own interests into the physics lessons. There was more integration across the areas covered by the acronyms STEM and STEAM. Students who investigated physical systems designed and built the objects to be researched, always involving technological and engineering questions. In later years, when they became more widely available, computers always were part of students' projects, mathematical or statistical modeling, presentation and design of their reports. Eventually, the arts became an integral part of the classroom, too, as seen in the preceding example. Because I had students read philosophical essays and books, and discussed these with them, this subject, too, was an integral part. Importantly, no student was forced to draw on this or that discipline included in the acronym STEAM.

3 Hitting on Some Wicked Problems of Schooling

Despite all of these advances over what the other teachers did in the schools where I taught, I realized I had not yet gone far enough. Even though I did not have access to the acronym, I realized that more than STEAM was needed. One of the wicked problems of schooling lies in the gap between in-school life and out-of-school life, where there are different required competencies, interests, and so on. Thus, for example, in a research project that I conducted together with a high school student on student epistemologies, we found that some of the high school students' mundane discourses "actively interfere with the curriculum offered by the formal educational setting" (Roth and Alexander 1997, p. 125). In our published article (completed in 1992, while both of us were still teacher and student), we argued for even more interdisciplinarity than is captured in STEAM: "There is a definite place for a high school course in which science, religion, philosophy, arts and other subjects are treated as parallel forms of inquiry towards the development of a 'good world'" (p. 143). We further suggested that "such a course needs to be truly interdisciplinary, in which no single discourse is privileged, such as in some 'science in society' courses, where scientific discourses dominate at the expense of the social and humanistic discourses" (p. 144).

Over more than a decade of teaching in middle and high schools, I have come to realize that science needed to be part of a broader engagement of students in the everyday world, and, I argued, "following Marx … that students (as philosophers and science educators) have only interpreted the world, in various ways; the point is to change it" (Roth 2002, p. 46). When I was writing those words, I was teaching, for the third time, an experimental curriculum in a local middle school that I had designed following my own call for deinstitutionalizing science education. My hunch was that educators needed to do more than just make changes in curriculum content, because education itself was the problem—built as an industrial model, where students were prepared to contribute to a compliant workforce on assembly lines. The second wicked problem I had come to understand lies in the fact that schooling is designed for producing grades and stratifying access to higher learning; schools also produce dropouts, whose life possibilities are generally constrained (Roth and McGinn 1997). Because engineering, agriculture, environmentalism, medicine, and all the other activity systems have object/motives very different from those of schooling, they demand very different "skills" and "competencies" to succeed. Studies show that (a) there is little correlation between number of or success in high school courses (e.g., mathematics) and the competencies that people exhibit in everyday life pursuits (e.g., Saxe 1991) and (b) in many professions, school knowledge is not mobilized at work and work knowledge is not called for at school (e.g., Roth 2014). I was suggesting that the Roman's slogan *Non scholae sed vitae discimus* was not enough, for the studying still occurred for school, and therefore was geared toward obtaining grades rather than toward changing the world and learning by expanding ones agency.

What may be an even more wicked problem is the rapidly changing world. The wicked problems of today and even more the wicked problems of tomorrow do not require some fixed scientific literacy, whether taught within the scope of STEM or STEAM. Rather, what we need for the future are "competencies in creating ever new forms of literacies that are suited to deal with the ever-new forms of problems scientific and technological 'advances' generate" (Roth 2007, p. 396). I also argued that in the face of controversies, we need citizens who can take a critical and ethical-moral stance against STEM personnel over controversial issues when scientists themselves do not do it or cannot do it because of the interests of the company they work for. Some issues that are controversial in some but less controversial in other countries include the production and (unlabeled) sale of genetically modified organisms (e.g., "Frankenfood"), the unfettered use of herbicides and fungicides that are damaging the environment (e.g., neonicotinoids and bees), the production of harmful consumer goods (e.g., cigarettes), or the creations of the pharmaceutical industry that often do more harm than good (e.g., thalidomide). The list can be extended. Even if a substance is legal, education should at a minimum enable students to ask, for example, (a) Why some countries ban a substance whereas another permits its use? or (b) Why do some companies legally export substances that are banned in the countries where they are headquartered?

4 An Example of How to Address the Wicked Problem of Schooling

Wicked problems require thinking outside of the box. My particular position on schooling likely is due to the fact that I am not a science or mathematics educator and that I conducted much research on (situated) cognition outside of schools, in leisure, workplace, and political settings. Two competencies for being successful in the face of wicked problems turn out to be especially important (Roth and van Eijck 2010): (a) the ability to work with people of very different competencies and levels of expertise (e.g., lawyers, scientists, politicians, public stakeholders) and (b) *debrouillardise* (defining and getting out of the problems) and competent *bricolage* (making do with available resources). Over the course of a 3-year span, I taught three iterations of a course where middle school students (seventh grade, 12 years) participated, in one or another way, in the environmental issues of their municipality.

The course I taught got students to work with the environmental group rather than focusing on and discussing scenarios and case studies already prepared. Because this approach involves students in real-world issues I have previously argued that it differs from curricula attending to *socio-scientific issues* or *consensus projects* models that tend to deal with prepared scenarios (Roth 2002). We began with something like the visit of an environmentalist to the class, or with the reading of an article in the local newspaper focusing on the health of the watershed and the activities of the environmental group. The political boundaries of my municipality by and large fall together with two watersheds, the more important one of which goes by the name of Hagan Creek/ḰENES (the indigenous name meaning "whale," because of the whales that used to feed near the mouth of the creek). The municipality itself is semirural, with more densely populated settlements and areas zoned agricultural. This is especially the case for the lower reaches of the creek, where farms and a largely wooded indigenous reserve dominate the landscape (Fig. 1). In other areas, the impacts on the land that began with the white settlers is conspicuous. For example, there used to be wetlands (Maber Flats) where the indigenous peoples hunted for birds and small game. But the settlers created a system of ditches, culverts, berms, and a weir, and the creek was made an integral part to move the water away to create arable land (Fig. 2b)—though nature periodically returns and floods Maber Flats (Fig. 2a). Nearby industries used to release their wastewaters into one of these ditches and thus contributed various heavy metals to the watershed—leading to the name Stinky Ditch for one of them (Fig. 2b). Despite nearly 20 years of continuing environmental efforts in the watershed, Maber Flats remains in the news because, after 70 years of agriculture, flooding has increased. There is an effort now to create a lake to better manage the water; other efforts focus on creating a nature reserve because, according to the region's community green map, it already supports various charismatic species, including peregrine falcons, bald eagles, northern harriers, and short-eared owls.

The health of the Hagan Creek watershed, the quality and quantity of water, and the ecological and human well-being clearly are important issues in this community, and they are addressed in a variety of public forums. Integrating student learning and development in environmental issues was the centerpiece of funded research program aptly entitled "Learning in Heterogeneous and Overlapping Knowledge-Building Communities."

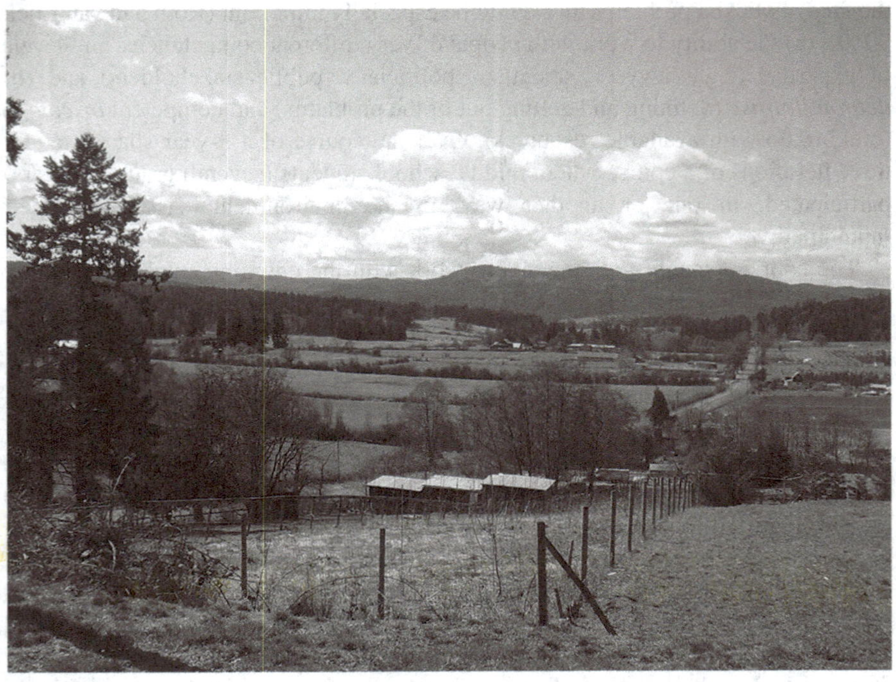

Fig. 1 Almost the entire lower part of the watershed, Mount Newton Valley, is zoned "Agricultural Land Reserve." The forest center left is part of a First Nation community

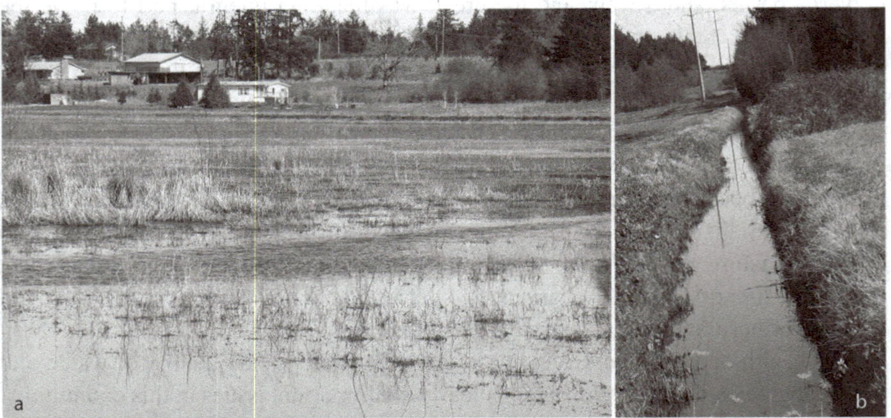

Fig. 2 (**a**) Part of the contested Maber Flats, which some farmers want to drain whereas environmental groups seek to preserve as wetlands. (**b**) During colonization, ditches, culverts, berms, and a weir were used to drain the wetlands that were a main source of food for the indigenous peoples in the area. "Stinky Ditch" (shown on the right) also previously received effluents from an industrial area near the horizon

Through this program, we encouraged students to engage with watershed issues on their own terms and according to their own interests, choosing data-collection methods and representational tools that best fit their interests and needs. There were student groups, often boys, conducting studies on the physical characteristics of water flow; some groups focused on sampling organisms that lived in the creek; other groups focused their efforts on plant and animal life in riparian areas. A group of students, under the watchful eyes of a water technician, contributed to building water-oxygenating riffles and to monitoring changes in the number of macroinvertebrate organisms (Fig. 3). Particularly girls and indigenous students, who felt disenfranchised by the traditional approach to science that they had experienced in the past, preferred to generate different forms of knowledge such as through films, narratives, photographs, and interviews. For example, an indigenous student had been completely disengaged in his regular school classes, as were many of his peers from the reserve. He did not want to do science in the way others did. Instead, he declared an interest in using a film camera to document what his classmates were doing. We gave him one of our research cameras and suggested that he might want to conduct interviews as well (Fig. 4). Our overall endeavour was greatly enhanced because parents, aboriginal elders, scientists, graduate students, and other residents of the municipality joined these efforts, supervising and coaching the students. De facto, this led to an interpenetration of school and village life.

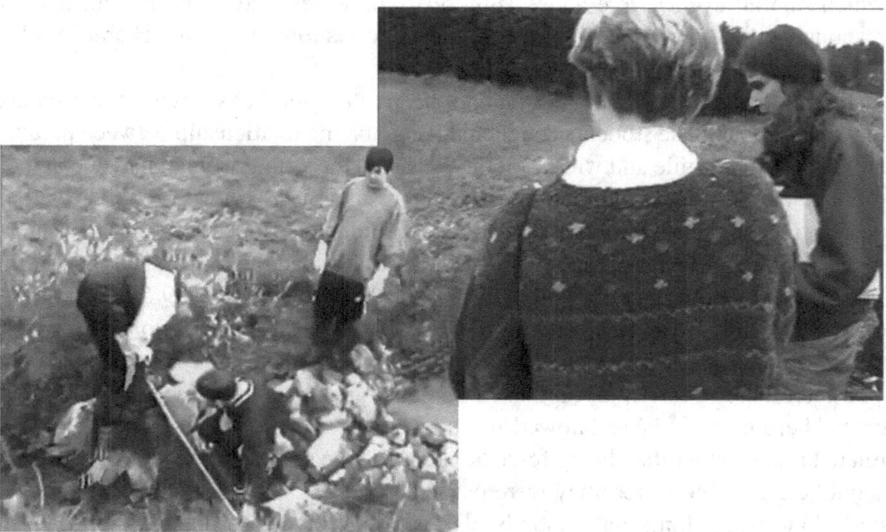

Fig. 3 This photo, which stitches together two shots of the panning camera, features students who contributed to the watershed-related efforts by working with a water technician of a local farm (far right) to construct riffles that oxygenate the water in the creek and to monitor the abundances of aquatic macroinvertebrates

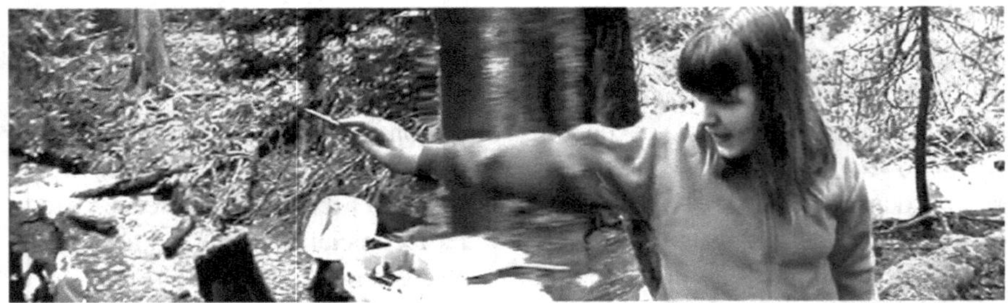

Fig. 4 One student of indigenous origins did not want to contribute to the environmental efforts, but chose instead to make a documentary. This image was stitched together from several parts of a video recording that was panning from an interviewed student to the study area that her group (lower left) investigated

In the end, many individuals and groups participated in the open-house event organized by the environmental group near the end of the respective school year. Jamie was particularly taken by the spectrophotometer used to determine the turbidity of water, an instrument that also included the capability to measure other variables. At the open-house event, he showed visitors how to use it, using in his demonstration some samples he had prepared previously (Fig. 5a). The students' stands were spread out among those of other contributors, which included a graph produced by a strip-chart recorder that mapped the water levels in the creek and to which rainfall events were added (Fig. 5b). The water technician of a local farm responsible for the graph explained to interested visitors the annual characteristics of the water flow through the watershed.

Michelle and her group mates had not been interested in "sciency" experiments that many of the male students conducted—e.g., on the relationship between parameters of stream profile and water velocity. Instead, the young women had decided to produce a report, including photographs, the results of interviews, and their verbal descriptions of the environment, which they recorded on a tape recorder. Their interviews included dignitaries, such as indigenous elders and the mayor, and local folk. In the end, they produced a poster to present the results of their work during the open-house event (Fig. 5c). We noticed that there were several pages of written text, in addition to the photographs. We know from independent interviews conducted a year later that the curricular unit had a tremendous impact on the women—to great extent because they were allowed to choose their ways of engaging with creek-related issues rather than being forced to do the same tasks that are characteristic of school classes. Moreover, they were particularly interested in the creative aspects, both doing something that "nobody else is doing" and in producing particularly interesting displays for the open house. They did so even though they were not enrolled in a course or program named STEAM, and even though nobody told them that they had to integrate the arts into science or the reverse.

One of the remarkable students was Davie. I could see how he behaved in other subjects in ways that were consistent with the learning disability label that the school system had stuck to him. When assigned some graphing task, he spent less

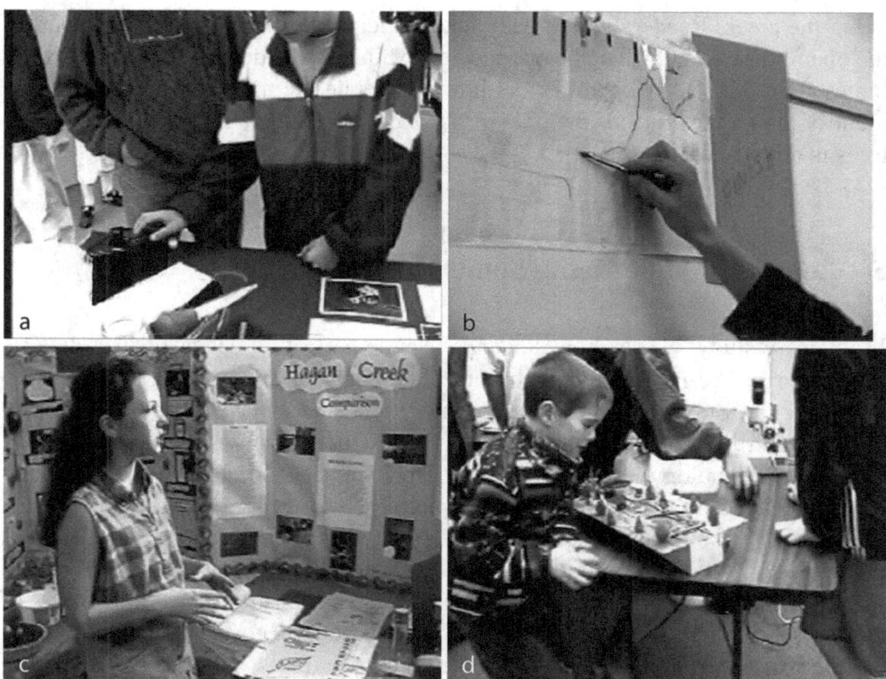

Fig. 5 Video images from an open-house event where the students presented the result of their contributions at open-house events organized by an environmental group. (**a**) Jamie shows an adult visitor how to use the spectrophotometer to test the water for turbidity (cloudiness). (**b**) A water technician talks about the graphs generated from data collected on her employer's farm, featuring amounts of rainfall and water levels in the creek. (**c**) Michelle had interviewed local politicians and elders, taken photographs, and written reports on the state of different parts of the creeks and contributors. (**d**) Stevie (right) teaches a child about the watershed using a model that he built as part of his project, whereas Davie (background) reorganizes the microscope and other materials that he used as part of his demonstration

than 18% of the designated time on the task (which he did not solve) and the remainder in talking to others, walking around the classroom, and in other "off-task" manners. That same Davie, on the other hand, was highly engaged in the experimental curriculum. He studied relationships, such as that between water speed and the abundance of different invertebrates; and he represented these relationships in types of graphs that he did not produce for his mathematics teacher—for lack of want, interest, capability, or any other reason we might discern. More so, when I was looking for students interested in joining me to get another class enrolled in the same program, Davie was the first to raise his hand. Not only did he present what he and his classmates had done, but also he assisted the teacher of this class familiarize herself with pertinent issues in teaching such a unit. When we were looking for students to assist us with teaching this second class by becoming a chaperon of a student group, Davie was again the first student to sign up. He supervised not just one but indeed two groups of students while they were doing their research in and along the creek. Readers will already have anticipated that Davie was also part of the open-house event, where he operated an exhibition on identifying macroinvertebrates (Fig. 5d). He showed visitors how to operate the microscope, helped them identify the key features of the organisms, and showed them how to use these features in classifying the specimen.

5 Learning While Doing Something Important

In this case study, we see how students became part of an endeavor that engages people across the life-span, who become familiar with and expand their room to maneuver (become competent contributors) with respect to community-scale environmental issues. There is no distinction and classification of what they do currently as science, technology, engineering, arts, or mathematics—or, for these matters, any other relevant field of endeavor, such as politics, ethics (e.g., ecojustice, environmental justice), law, animal husbandry, leisure specialist, etc. Students can learn that engagement is not a one-time thing, and that contentious issues do not necessarily get settled. In one instance, a water main was constructed after 15 years of political wrangling, ultimately involving three levels of government financing it. But the issues were complicated, involving lawyers, citizens, geologists, the regional medical board, town engineers, financial planers, town council, the water advisory board, and other stakeholders. They can also learn that some issues do not get settled. Thus, at the time of this writing (2017), the wrangling concerning Maber Flats still is ongoing, with a lawsuit having been filed to turn the area into arable land again; and efforts are mobilized against the municipality, which has been more tentative in the face of the ecological importance of the area, as advanced not only by environmental groups but also by a regional initiative focusing on sustainable living, nature, and culture and society. It is a joint community-university initiative that is governed by community groups, nongovernmental agencies, aboriginal peoples (First Nations), local government, regional funders, and researchers from a university in the region.

In this unit, the children do something that is real and important. Though also in school, they participate in a societal activity—environmental activism—where environmental and human health and their improvement are the defining objectives or motives. The activity does not inherently distinguish among participating subjects by age. If it did, this would only be another form of discrimination, for "the convenient belief that their age in years in and of itself disqualifies some people from being taken seriously or insures their 'natural irresponsibility' is surely as fundamentally immoral as the similar beliefs about race, gender, and nationality" (Lemke 1990, p. 85). Indeed, the activity itself is not defined by learning as such. Participants' learning is co-extensive with their power to act, and increasing this power (i.e., agency) inherently is in the interest of the individual or collective subject. Thus, if someone encounters a snag, they do whatever it takes to resolve it, sometimes turning to others, and in the course of so doing, their competencies of dealing with this type of snag specifically and with snags more generally. They become competent in *debrouillardise*, the capacity to get oneself creatively out of (wicked) life problems and the dispositions (ability) necessary to learn as one goes (i.e., knowledgeability).

Much money has been spent over the last 50 years since Sputnik shocked the USA into developing STEM education. The Organization for Economic Cooperation and Development reports that the relative share of students engaging in science and technology has declined over the last several decades, and female student representation still lags that of male representation (OECD 1997). Despite all the expenditures, however, the wicked problems of education and society only seem to grow in number. Some of today's wicked problems include: (a) failure to appropriately appreciate the scope and limitations of scientific facts (leading to climate change denial and acceptance of "alternative facts"); (b) inability to see that there cannot be continued (economic) growth on a planet with finite resources; (c) STEM discipline-specific hubris that fails to recognize the need for collectively achieved, transdisciplinarity solutions; and (d) recognition that wicked problems are by nature ill-defined and continuously emerging (thus transforming). We need to gather steam to make STEAM, and beyond, relevant to life. In this chapter, I suggest that some form of deinstitutionalizing schooling by making students' actions relevant to the community is one way of achieving this goal. However, the point of this chapter is not that schools are unnecessary. There is a role that schools can play—e.g., as designated physical spaces. But this role will be different from the role they have had in the past. A cultural-historical activity theoretic analysis shows that cognition—thus development, learning, and personality—is a function of the activity system, and different activity systems lead to different forms of cognition. Thus, a school curriculum focusing on the environment with a STEM or STEAM perspective leads to different forms and contents of learning than participating in some environment-related, community-based initiatives. If the aim of schooling is to learn for life, then schooling, if it is to be effective, must take students into that very life for which it is intended to prepare them.

References

Lemke, J. (1990). *Talking science: Language, learning, and values*. Westport: Ablex.

Organization for Economic Co-operation and Development (OECD). (1997). *Science and technology in the public eye*. Accessed at http://www.oecd.org/science/sci-tech/2754356.pdf

Roth, W.-M. (2002). Taking science education beyond schooling. *Canadian Journal of Science, Mathematics and Technology Education, 2*, 37–48.

Roth, W.-M. (2007). Toward a dialectical notion and praxis of scientific literacy. *Journal of Curriculum Studies, 39*, 377–398.

Roth, W.-M. (2014). Rules of bending, bending the rules: The geometry of conduit bending in college and workplace. *Educational Studies in Mathematics, 86*, 177–192.

Roth, W.-M., & Alexander, T. (1997). The interaction of students' scientific and religious discourses: Two case studies. *International Journal of Science Education, 19*, 125–146.

Roth, W.-M., & McGinn, M. K. (1997). Deinstitutionalizing school science: Implications of a strong view of situated cognition. *Research in Science Education, 27*, 497–513.

Roth, W.-M., & van Eijck, M. (2010). Fullness of life as minimal unit: Science, technology, engineering, and mathematics (STEM) learning across the lifespan. *Science Education, 94*, 1027–1048.

Saxe, G. B. (1991). *Culture and cognitive development: Studies in mathematical understanding.* Hillsdale: Lawrence Erlbaum Associates.

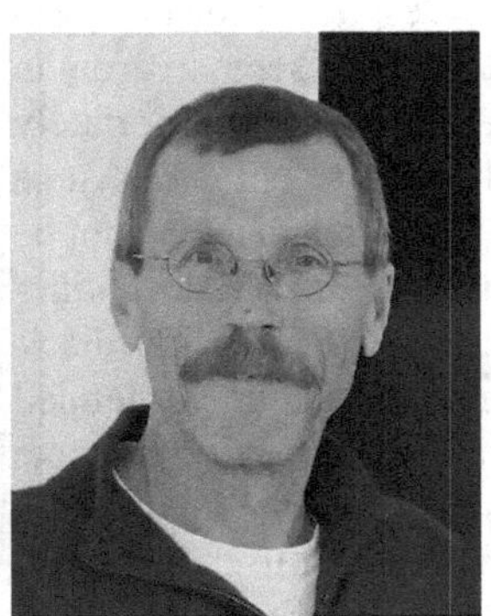

Wolff-Michael Roth is Lansdowne Professor of Applied Cognitive Science in the Faculty of Education at the University of Victoria. His research concerns knowing, learning, and development across the life-span in workplace, formal educational, and leisure settings with a special focus on science, technology, engineering, mathematics, and language.

Using Project-Based Learning to Teach STEAM

Gary Ubben

"Students who are in real contact with problems which are relevant to them, wish to learn, want to grow, seek to discover, endeavor to master, desire to create, move toward self-discipline" (Rogers 1969, p. 158)

1 Introduction

The class had been discussing the hurricane in Puerto Rico and the resulting clean water shortage problem. The people there had water, but it was polluted. How do you make polluted water clean? If it looks clean, is it OK to drink?

The students had small plastic food containers on their table, each containing a different material, including pebbles, sand (course and fine), dirt, charcoal, rice, marbles, and several kinds of fabric materials, such as cotton balls, gauze, scouring pads, window screen, disposable face masks, etc., and a half-liter bottle of very cloudy polluted water. Each team also had a 1-foot length of plastic pipe. The class challenge was to build a gravity-driven water filtration system using some or all of the supplied materials, to get the water as clear as possible. The final water product would be judged on the clarity of the water and the pH level. Working in groups of four or five, the students were to decide which materials to use and the order to place them in the foot-long plastic pipe, marking their location on the outside of the pipe.

In their project report, each group was asked to defend their decisions of which materials to use and the order in which they placed them in their pipe. Some groups started with the courser material first, gradually moving to finer filtering materials, while other groups alternated between course and fine materials, separating them with fabric materials. Yet others went from fine to course materials, thinking of the

G. Ubben (✉)
Department of Educational Leadership, University of Tennessee, Knoxville, TN, USA
e-mail: gubben@utk.edu

© Springer Nature Switzerland AG 2019
A. J. Stewart et al. (eds.), *Converting STEM into STEAM Programs*,
Environmental Discourses in Science Education 5,
https://doi.org/10.1007/978-3-030-25101-7_6

rocks they had seen in mountain streams. The groups were allowed to do internet searches for guidance. At the end of the class period, a half-liter of dirty water was poured slowly through each pipe filter into a clear glass container, where it would be analyzed the following day for clarity and pH. Each group would then give a presentation on their project the following day.

Each group was given a set of guiding questions to help them in their problem-solving and final report preparation. These questions were open-ended, such as "why did you choose the filtering materials you used and why did you place them in this order? Did you seek any expert advice? Do you think your cleansed water is now drinkable? What about water-borne diseases? Do you think your filtering system eliminates them? What else might you need to further reduce this risk?

Excitement permeated the classroom while the teacher circulated through the room answering students' questions. Some of the groups had to rush at the end of the hour to get their filtration system finished and pour the water through their pipe.

It was obvious the next day that some of the teams did better than others in designing an effective system. Then the discussion began: why?

1.1 The Interrelationship of STEAM Subjects in Solving Complex Problems

One evening my youngest daughter was doing her middle school science homework on electricity. I asked her if she was familiar with Ohm's Law, the mathematical formula for determining the flow of current. Her reply startled me: "Dad," she said, "I am studying science. This is not my math course." It made me realize how, at least in the eyes of students, we have compartmentalized learning with each subject existing in its own silo of knowledge. We have done this, of course, to employ specialists in each subject—but as a result, we have often denied ourselves the opportunity for integrating knowledge across disciplines, particularly with complex problems.

An underlying rationale for STEM and STEAM curricula is more than the importance of teaching each of these subjects: it is the integration of these disciplines and how, together, each might contribute to solving complex problems. The solutions to many real-world complex problems must draw on several of the STEAM disciplines, including the Arts. Project-based learning (PBL), with carefully chosen projects, is an excellent vehicle to utilize multiple disciplines in solving complex problems. A well-designed PBL brings together the application of the skills and the knowledge needed in the workplace to solve complex problems. It requires students to draw on multiple disciplines for good solutions. And, by adding the Arts to STEM, we greatly broaden the range of analytical tools available to our problem-solving.

2 Project-Based Learning

Project-based learning has been given many different names over the years, such as experiential learning, inquiry-based learning, maker projects, problem-based learning, challenge-based learning, etc. All are found in the educational literature for lesson plans and activities, involving individual or student-group activities seeking solution to a communal problem or task.

2.1 *Experiential Learning*

During the progressive movement in education of the 1930s, a leading American educational philosopher, John Dewey, concluded that all learning comes through experience and promoted the idea of experiential education (Experience and Education, Dewey 1938). Educational research projects of that period, such as the eight-year study by Wilford (1942) demonstrated the effectiveness of what we now call project-based learning. Other writers, such as Kurt Lewin (1943), proposed ideas of experiential learning as an active process, rather than a passive process. Lewin, and later his student David Kolb (1984), envisioned learning as a four-step process: (1) concrete experience, (2) observation and reflection, (3) abstract conceptualization, and (4) testing of concepts in new situations. Kurt Lewin also promoted ideas of the importance of task interdependence to learning. He suggested that interdependence in the goals of group members creates a powerful learning dynamic (Brown 1988). These learning processes, including the soft skills of group cooperation are built into PBL designs of today.

2.2 *The Foxfire Experience (2002)*

Experiential learning fell from favor in most American schools after WWII as behavioral objectives, learning standards, and testing grew in popularity. But in 1966, a young teacher, Eliot Wigginton (1986), began his teaching career in a rural mountain community of north Georgia. He soon realized his effort to teach grammar and composition to these mountain children was a disaster. His students did not share his passion for the subject. With his new career almost in ruins, he turned to his students for their input as to what to do. Together they came up with a project, the idea of publishing a student literary magazine about things that they knew from their local culture. Wigginton insisted that his students choose their own projects to write about, but that they must learn basic English skills through the process. They named the magazine *Foxfire* after the luminous fungus found on the trunks of rotting trees in the Appalachian Mountains. Student-written articles were about the culture and often the handmade items these mountain children knew. The arts were incorporated holistically into projects reported in *Foxfire*, in the form of stories and projects drawn from the local culture of the community such as "How to dress out a hog" or "Building a log cabin." The magazine grew in popularity among educators nationwide for its practical application of knowledge. The *Foxfire* magazines later were bound by year of publication and widely distributed as books of projects that could be replicated by other students.

Later, as Wigginton (1986) was looking for a philosophical base for the success he was experiencing in the classroom, he made a startling discovery as he was reading John Dewey's *Experience and Education*. All the discoveries he thought he was making about education, Dewey had elucidated into complete clarity 50 years and more before! By the mid 1970s the *Foxfire* magazine and books brought national attention to Wigginton and his students, and many teachers were attempting to create similar cultural journalism programs (Starnes 1999). Wigginton and his colleagues had unwittingly helped revive the concept of *experiential learning*, which had been near death from the progressive era of the 1930s (Ratvitch 2000). Today's version of *Foxfire* might use a YouTube format with the PBL directing students to demonstrate how to build something such as a simple computer or a home security system.

2.3 What Has Project-Based Learning Become Today?

PBL is an instructional strategy in which project work is central—not peripheral!—to the learning process. The PBL lessons described later in this chapter nudge students into roles in which they assume much responsibility and control over their own learning.

A PBL lesson plan directs the activities of the students, rather than the activities of the teacher. The role of the student is as a participant engaged in actively trying to resolve the complex situation. The lesson usually begins by presenting a project to design or a problem that needs to be resolved, ultimately with the solution being reported to a larger audience. The exercise may begin as a task to investigate, plan, organize, or create a product to be used by someone. Students may work individually on a problem, but it is more common for them to work on tasks in small teacher-assigned groups. PBLs often are highly motivating to students—especially when students have an authentic problem to solve and the opportunity to work cooperatively with other children. It is learning by doing. PBLs have been successful with all ages of students, from early elementary school through graduate school. It is common practice in medical schools to use PBLs to help future doctors learn how to diagnose illnesses.

There is no common standardized format for PBLs in the pedagogical literature, but most PBL lesson plans have a common set of features. For example, PBLs typically engage the students in authentic tasks that are modeled after the work of professionals in scientific, technical, business, or community organizations. These tasks are often messy and ambiguous—just like many problems in the real world! Part of the student's task is to cut through the ambiguity and organize the problem. Projects require students to locate resources, gather and analyze data, establish time lines, seek solutions, prepare and edit reports, and present information through various media.

Some examples of PBL topics are:

- The teacher would like to form a school symphony, but lacks the funds necessary to purchase the instruments. Can a group of 5th- and 6th-grade students research, design, and build their own musical instruments, including percussion, string, wind, and synthesized instruments? The final products will be incorporated in a school symphony that will perform for the school in the annual talent show.
- As a group of filmmakers, how can we encourage our community to recycle its waste?
- How can we design an attractive and functional ape (or reptile, or butterfly) house for our local zoo?
- We hear of frequent charges of fake news or the denial of scientific findings. How do we prove what we think to be true? Give examples.

3 Project Versus Problem-Based Learning

Note that some of the topics listed above suggest a specific "project" in solving a problem, such as constructing musical instruments. Other topics, such as the recycling question, identify an open-ended "problem" of "How" where the students must also determine the direction to the solution. This is often the distinction made between the two terms—project or problem. To create a better distinction between the two terms sometimes the term *problem* is changed to *challenge*-based learning (CBL).

The Buck Institute for Education (BIE 2019), a leading nonprofit organization dedicated to promoting project-based learning, defines PBL as "A teaching method in which students gain knowledge and skills by working for an extended period of time to investigate and respond to an authentic, engaging and complex question, problem, or challenge." This definition leans more toward problem-based learning. In its gold standard, BIE identifies a set of essential project design elements that should be included in a well-written PBL lesson plan. These include:

- *Key Knowledge, Understanding, and Success Skills*: The project is focused on student learning goals, including standards-based content and skills such as critical thinking/problem-solving, communication, collaboration, and self-management.
- *Challenging Problem or Question*: The project is framed by a meaningful problem to solve or a question to answer, at the appropriate level of challenge.

- *Sustained Inquiry*: Students engage in a rigorous, extended process of asking questions, finding resources, and applying information.
- *Authenticity*: The project features real-world context, tasks and tools, quality standards, or impact—or speaks to students' personal concerns, interests, and issues in their lives.
- *Student Voice and Choice*: Students make some decisions about the project, including how they work and what they create.
- *Reflection*: Students and teachers reflect on learning, the effectiveness of their inquiry and project activities, the quality of student work, and obstacles and how to overcome them.
- *Critique and Revision*: Students give, receive, and use feedback to improve their process and products.
- *Public Product*: Students make their project work public by explaining, displaying, and/or presenting it to people beyond the classroom.

3.1 My Place, Your Place, Our Place

In 2002, a US Department of State–University partnership project between the University of Tennessee and Bourgas University in Bulgaria used PBLs to work with local public schools in both countries. In this project, teachers were trained to write PBLs about local culture; their students completed the projects and then placed the project reports on the My Place, Your Place, Our Place (MYOP) web page, where they could be shared with classrooms worldwide. The PBL lesson plan format presented in this chapter was first developed in the MYOP project (Ubben, MYOP 2005).

4 How PBLs Work in the Classroom

4.1 The Teachers' Role

The teachers have at least seven distinct roles when working with PBLs.

Role one. The first is that of selecting one or more problematic situations for the students to resolve. It may be done by an individual teacher or by a team of teachers bringing different subject matter ideas and skills to the table. The exercises are often obtained from a library of PBLs to match a prescribed curriculum or set of Standards. There are many libraries of free-domain PBL situations in books and on the internet. Several library sources of PBLs also are listed at the end of Chapter "How to Structure Project-Based Learning to Meet STEAM Objectives". Alternatively, teachers may wish to design and construct and use a new PBL. Here a standard format should be used by all teachers such as the one suggested in this chapter. This is important so that nothing is left out and that other teachers may easily find their way through it.

Role two. The next role of the teacher is to set the stage for the project by introducing to the students the issue, dilemma, or problem to be solved. This may be a real situation such as the example of the water crisis in Puerto Rico or may be a totally contrived but realistic scenario.

Role three. This happens after the teacher has introduced the PBL. The role involves providing the initial organizational structure for student teams or individuals, in which parameters for the project are laid out, and the teacher sets dates and times for progress reports, and informs students about completion dates.

Role four. Now the teacher's role shifts dramatically to that of guide. While students work on the PBL, the teacher becomes a *coach*, an *advisor*, a *resource person*, or *a subject matter expert*, depending upon what is needed by the students—but only as needed to help the students resolve the problem or complete certain tasks where they must learn new skills to succeed. The teacher remains critical to the success of the student-directed activity, but becomes "a guide by the side rather than the sage on the stage." In short, the role of the teacher as presenter of knowledge is greatly reduced. However, they can function as a resource librarian to help students locate information they need. This role is sometimes difficult for teachers who are more accustomed to a lecture/recitation delivery. Often the teacher can best provide guidance by simply asking the group a directing question.

Role five. If the students have not had much experience with self-directed learning, the teacher must help the students develop these soft skills. This is particularly necessary with initial PBL efforts. Depending on the expected outcomes for the project, students will need to be able to exercise a series of research, process, and presentation skills. The research skills may be developed though data collection, recording, and/or analysis. Organizational and planning skills are likely required, as well. If children are to work in teams, they will need group-process skills and strategies such as brainstorming or consensus decision-making. And, if they are to prepare reports of their efforts for a public presentation of their results, they will need to know about digital photography, spreadsheets, multimedia presentation, and web publishing.

Role six. Soft-skills assessment. With the project under way, the teacher should observe student progress and reassure them of their direction and, if needed, redirect their efforts. When students are working in teams, it is particularly important for the teacher to observe individual student effort. This can often best be done with rubrics. Students should also be taught self-assessment and peer assessment—and rubric examples for doing this are given in Chapter "How to Structure Project-Based Learning to Meet STEAM Objectives", on PBL structure.

Role seven. The teacher always has the ultimate responsibility of assessing the content learning that occurs, but this should be done using predetermined criteria and a rubric based on the *learning objectives* and *product specifications* set out for the project. Rubric examples are located in Chapter "How to Structure Project-Based Learning to Meet STEAM Objectives", on PBL structure.

4.2 The Student's Role

In well-conducted PBL projects, students are accountable to themselves as well as their peers and should be given considerable autonomy. This skill set on the part of the student may not be well developed, if most of their previous learning has been highly teacher-directed. Life skills and process skills such as self-management, group-process, and problem-solving are to be practiced by students as they engage in the PBL activities. Opportunities for student creativity and ingenuity also abound in well-designed PBLs. Students are responsible for synthesizing and constructing knowledge that is needed for project success, either in a way the project suggests or in a way the students set for themselves. They develop their own strategies to enable and direct their own work. Students also become stakeholders in the PBL as they immerse themselves in the process: they find pride and ownership in the success of their work. This becomes a great motivator for continued learning. The rubrics suggested in the previous paragraphs can help direct this learning.

4.3 The PBL Classroom

PBL classrooms rarely look like other classrooms. PBL classrooms need flexible and variable arrangements for their activities. Students are both working independently and interacting with each other in pursuit of their mission. While direct instruction occurs in PBL, it usually takes place within other activities. It may be done with one of the small teams of students working on a particular aspect of the project. Classrooms that are equipped with tables that seat four to six students and can be moved to different configurations seem to work best.

Some schools are creating shared *maker spaces* providing students with the infrastructure required for the creation of professional-level solutions, software applications, models, and products. Some of the more elaborate maker spaces include design areas for animation, simulations, and programing and fabrication areas for 3D printing, laser cutting, metal working, vacuum forming, textile fabrication, resin casting, wood working, and robotics. Of course, a middle or elementary school would have a less elaborate arrangement. Because of the cost of some of this equipment, partnerships with local industry frequently occur.

4.4 Benefits of and Outcomes from PBLs in a STEAM Curriculum

PBL fosters in-depth understanding of subject matter content and develops students' skills, strategies, and dispositions associated with professional activities in different subject areas. It is of value in integrating learning across what we have artificially created as separate disciplines or knowledge categories. Adding the arts to establish STEAM broadens opportunities for integration. How can we look at a bridge and not appreciate its beauty or elegance of design, or the lack thereof? Or, view the new One World Trade Center in New York without appreciating the historical significance of its design and location?

Students learn not only the knowledge and skills that are essential to solve problems, but also the life skills and habits that are essential for success in the world beyond the classroom. Such skills include problem identification, problem-solving, group interaction and process, communication with others (using multiple media types), and the ability to prepare reports, edit, revise their writing, manage time in a group setting, and meet deadlines.

Finally, students develop self-assessment skills. In most PBLs, assessment is a responsibility of the students themselves—it is not just the teacher who assesses! Criteria that allow students to determine how well they have met the expectations for the project are laid out in the PBL section on *product specifications* as well as the *assessment rubric*. In addition, the presentation of projects to the entire class or larger adult audience provides students with peer-assessment opportunities that, in turn, improve their own skills at self-assessment.

5 Research Supporting the PBL Approach

Enthusiasm for PBL has been growing for many years with some of the best research coming from the field of medical education (Albanese and Mitchell 1993; Berkson 1993; Colliver 2000; Davies 2000; Vernon and Blake 1993). The field of psychology has also produced some valuable insight into the use of PBLs. (Hmelo-Silver 2004; Hmelo-Silver et al. 2007; Kirschner et al. 2006).

However, the research supporting PBL-type instruction has been very limited, especially in K–12 cases. Foundational, of course, is the work of John Dewey (1916, 1938, 1981). More recent work includes a controlled experimental study of PBL in a middle school population (Wirkala and Kuhn 2011) at Teachers College Columbia University used between- and within-subject comparisons of students learning the same material under three instructional conditions: (1) lecture/discussion, (2) characteristic small-group PBL, and (3) solitary PBL. Through assessments of comprehension and application of concepts in a new context 9 weeks after instruction, this study showed superior mastery in both PBL conditions, relative to the lecture condition. The equivalent performance in the two PBL conditions suggested that the social component of PBL is not a critical feature of its effectiveness. However, this research only measured subject content. No indicators were included to measure soft skills, the main expected outcome of the social component.

Many other studies have been conducted on PBL. Some of these research studies with significance for STEAM are listed in the reference section of this chapter (Dochy et al. 2003; Foxfire Fund Inc 2002; Gijbels et al. 2005; Pease and Kuhn 2011; Starnes 2000; Starnes et al. 1999; Schmidt et al. 2007; Walker and Leary 2009).

6 How to Structure a PBL to Meet STEAM Objectives

PBL lesson plans can be organized in various ways, but a standard format ensures all necessary elements are included for a good PBL. A standard lesson plan format also helps teacher and student participants quickly become familiar with subsequent PBLs. However, teachers using these PBLs should adapt them to their classroom and the maturity and experience level of their students.

Each PBL lesson plan, after the title, should include the following eight sections:

- Topic/Problem/Grade/Length
- Topic Paragraph—(introduction and setting for the study)
- Curriculum Standards
- PBL Objectives
- Guiding Questions
- Product Specifications
- Resources
- Assessment Rubric

6.1 Topic/Problem/Grade/Length

Projects can vary greatly as to difficulty, complexity, and length. It is important to provide identification information to aid the teacher in searching for appropriate PBLs for their classroom. This information can help teachers quickly identify possible activities. For example, a PBL for elementary children might carry the heading *Transportation/Building a Strong Stick Bridge/Grade 3 or 4, 2 hours.*

6.2 Topic Paragraph

Describe, in one paragraph, the project to be completed or the problem to be solved and direct it to the students. An example is, *Working in a group of four and given a stack of popsicles sticks and a tube of wood glue, build a wooden bridge 5 inches wide and spanning at least 12 inches and capable of holding a two-pound weight.* The activity might not give any other instruction, relying on the creativity of the students, or it might include pictures of bridges illustrating successful stick bridges to provide the team with basic ideas. Obviously, additional instructions could be given for this activity using the students' successes and failures as illustrations. Often, setting the stage for an activity with a reason for needing the solution can help provide focus. For example, the discussion of the aftermath of the hurricane and need for drinking water in Puerto Rico and the difficulty in supplying it provided the backdrop for the water-filtration project.

6.3 Curriculum Standards

Aligning projects to curriculum standards can be challenging. If the bridge project is for 3rd and 4th-grade students, the standard may be drawn, for example, from those addressing transportation, communication, or safety. For higher grades, the standard might be drawn from geometry or physics.

The content for each PBL should be selected such that the projects address important curricular goals and standards appropriate for the age and grade of the students involved. Most states in the USA have adopted curriculum standards such as the Common Core Curriculum, and some states use the newer Next Generation Science Standards in science. The Next Generation Science Standards (2017), adopted by 18 states and the District of Columbia, demand a three-dimensional approach to instruction. Each lesson is to combine "practices," or the behaviors of real scientists and engineers; "cross-cutting concepts," which clarify connections across science disciplines and help students create a coherent view of the world based on science, and "disciplinary core ideas," or the fundamental ideas students must know to understand a given science discipline.

In searching for appropriate PBL topics, it might seem logical to begin with the list of standards to which teachers are expected to teach. In practice, however, it is difficult to design a good PBL around these lists. Alternatively, one can reverse the process: first select a good PBL, and then see which standards can be taught through the PBL. For example, assume the topic is gardening. From raised bed squares to aquaponics to aeroponics, horticulture offers creative and innovative projects that incorporate many of the standards we are asked to teach. Beginning with the activity, PBL projects can be designed by finding standards within national, state, and local curriculum and student performance lists. Since our focus is on the integration of STEAM subjects, it would be appropriate to review standards lists from each of these subjects to see how they may contribute to the project's framework. The best projects to spur student interest usually include those based on a realistic challenge, investigation, or problem for the students to resolve. And often, competition among teams adds motivation.

6.4 PBL (Learning) Objectives in the Context of STEAM

Learning objectives describe what the student is going to be able to do at the end of the activity. Objectives are often best written if they begin with a verb such as understand, know, practice, write, or develop skill at, per Bloom's Taxonomy (see https://www.teachthought.com/critical-thinking/249-blooms-taxonomy-verbs-for-critical-thinking/). It is appropriate to indicate basic and advanced levels when a class holds a range of student ages or abilities.

Learning objectives can be divided into a several categories. *General objectives* are those that apply to the entire project and are often skills related; they may deal with research, organization, group process, and reporting. Here is an example.

By the end of this project the student will be able to:

- Formulate questions for effective interviews, conduct interviews, and organize their responses
- Demonstrate their use of new media technologies by preparing photographs and videos for a web presentation
- Practice their writing skills by preparing written illustrated reports for posting on the web
- Develop teamwork skills through participation in student work committees
- Practice their presentation skills by giving illustrated multimedia oral reports to their classmates and others

But in addition to general learning objectives, *content or subject-specific learning objectives* must be included. If the PBL is to serve a multidisciplinary purpose, each discipline should be reviewed to determine what its contributions might be to desired learning outcomes. For example:

- Demonstrate your mapping skills by showing latitude and longitude locations for the areas you are studying.
- Demonstrate your understanding of certain geometric shapes in designing strong bridge structures.
- Identify the sound patterns of at least four different musical instruments using an oscilloscope.

6.5 Guiding Questions

Guiding Questions are used to direct the students' thinking by asking questions. They can encourage students to explore directions or intricacies of which they might not have thought or in preventing the project from wandering off in an unintended direction. They are not intended to be answered directly: rather, they only guide the students' thinking. They are important both in PBL and for meeting STEAM objectives.

PBLs should include a series of questions and subquestions that are provocative, significant, and realistic, and these questions should drive the students' investigation in directions that result in the essential core lesson designed to be taught by the project. For example, the simple bridge-construction PBL discussed earlier might ask students about types of bridges or about pictures of bridges. Which of these bridge types might they be able to copy with a popsicle sticks construction? Or, looking at a picture of a bridge, what are some of the geometric shapes they see and what are their functions? Are some bridges prettier than others? If so, why? Are new bridges designed the same way bridges were built 50 or 100 years ago? If they are different, why?

6.6 Product Specification

Product specification tells the student exactly what their final products are to look like and the form in which they are to be submitted. In the example of the model wooden bridge, the "Product Specification" for elementary school children was to "build a wooden bridge with popsicle sticks 5 inches wide and spanning at least 12 inches and capable of holding a two-pound weight." The class discussion after the presentation and testing of the completed projects can be expanded to questions about which bridges were the strongest, and why, or which bridges used the least amount of material, or which bridges were the prettiest—but the model bridge was what was required of the students.

The same project could be made more challenging as a PBL for middle-school students by asking more complex guiding questions such as, "How we can, as mathematicians, use geometry and pre-algebra to help us design and build a model bridge?" In this case, we need to describe how we wish the students to explain their use of pre-algebra and geography. This can be done by adding a product specification: "Explain, in a written and final oral report, your use of pre-algebra and geometry in designing and building your bridge."

And even higher-order thinking can be achieved by presenting the project as a problem to be solved, and adjusting the product specifications accordingly. Rather than asking the class to build or design a bridge, the teacher could frame the problem as a need to move a minimum of 500 cars a day across a 500-foot-wide fast-flowing river, with banks that are 50 feet above the river. Now the questions become

"How? A ferry? A tunnel? A bridge?" Other questions emerge immediately thereafter. Bridge style? The most cost effective method? The most aesthetically appealing? etc. The ideas to explore can be built into the guiding questions. The product specifications in this case will now include a written report and public presentation that incorporates the team's recommendations, the rationale for their conclusions and explanation about how they arrived at them, pictures or drawings of their suggested solution(s), and estimated costs. The way the questions were asked also classifies this lesson as *problem*-based learning rather than simply project based.

Public feedback to students is an important and integral part of every PBL and should be written into most product specifications. Through PBLs, students often develop new knowledge, and the task of presenting it in a meaningful way to others helps solidify that learning. The presentation may be to classmates, but often presentations to parents and other community members such as subject matter experts or government officials have great benefit. Yet a broader audience can sometimes create some anxiety as well. PBLs may involve a partner school located some distance away, so PBL reports prepared as web-based reports can provide easy access to partner schools.

6.7 Resources

List any material or equipment needed by either the teacher or the students to complete the PBL. For the several PBL examples used in this chapter, necessary materials range from popsicle sticks, glue, and a two-pound weight to pictures of bridges (internet), to an oscilloscope, camera, a computer-based GPS application, presentation software, etc.

Of particular importance are specific information reference sources that students need to best understand the project concepts. These sources can be in the form of teacher-made worksheets, YouTube videos, copies of textbook pages to read, journal articles, or the instruction to students to do a general internet search for specific suggested topics.

6.8 Student Assessment Rubrics

The way students are evaluated for their work should be predetermined in every PBL. To help determine the basis for that assessment, the teacher should review the two major components of the PBL: its *learning objectives* and the *product* (outcomes) *specifications*. These components are given to the students at the beginning of the project. Teachers can develop assessment rubrics or checklists from these two lists. These rubrics can include both quantity and quality measures.

Assessment areas should include higher-order thinking skills, organization of the project, and collaboration among team members. Higher-order thinking skills can be assessed in not only the students' written works, but also in their conversations with team members and the teacher, and in the graphic organizers of their presentations.

All rubrics have three basic components and are generally presented as tables. The vertical dimension lists the skill or standard, the horizontal lists three or four criteria along a continuum (e.g., from low to high). Content within the table describes what the behavior or skill "looks like" at a given performance level.

One set of rubrics useful for most PBLs can be developed around the *general objectives* stated earlier in this chapter: interviewing, using new media technologies, writing skills, teamwork skills, and presentation skills. Similar rubrics can be developed for each of the other general objectives of PBL. Additional rubrics can be created for student self-assessment and for peer assessments from other team members. Examples of these rubrics can be found in the next chapter on PBL structure.

Rubrics can be constructed for each of the content specific objectives stated in PBL. As noted previously, these should be based on the learning objectives and on the product specification. Some can be very specific. For example, the continuum for the stick bridge: spans less than 12 inches, 12 inches, or more than 12 inches. It supports less than 2 pounds, up to 2 pounds, or more than 2 pounds. The top headings might reflect performance levels such as novice, mastery, and distinguished, and the side headings reflect the product specification content. Rubric categories also can have scoring numbers assigned so that a numeric total may be assigned.

6.9 *In Closing*

Even a cursory online search conducted using google.com with the search phrase "STEAM project-based learning" will return dozens of results. The astute reader that is interested in converting her or his classroom from a STEM-focused approach to a PBL-centered STEAM approach will find no shortage of online resources encouraging the conversion. Rather, a STEM-to-STEAM conversion at the classroom level may focus most productively on ensuring the selected projects properly align with appropriate standards for the discipline areas that are considered (see Ubben, Chapter "The Role of STEAM in a Sustainable World" on PBL structure), and that the selected projects provide good opportunities for the students to engage meaningfully with real problems.

References

Albanese, M., & Mitchell, S. (1993). Problem-based learning: A review of literature on its outcomes and implementation issues. *Academic Medicine, 68,* 52–81.

Berkson, L. (1993). Problem-based learning: Have the expectations been met? *Academic Medicine, 68*(Suppl 10), S79–S88.

Brown, R. (1988). *Group Processes. Dynamics within and between groups.* Oxford: Blackwell.

Buck Institute for Education. (2019). http://www.pblworks.org

Colliver, J. (2000). Effectiveness of problem-based learning curricula: Research and theory. *Academic Medicine, 75,* 259–266.

Davies, P. (2000). Approaches to evidence-based teaching. *Medical Teacher, 22*(1), 14–21.

Dewey, J. (1916). *Democracy and education.* New York: The Free Press.

Dewey, J. (1938). *Experience and education.* New York: Macmillian.

Dewey, J. (1981). The school and social progress. In J. McDermott (Ed.), *The philosophy of John Dewey.* Chicago: University of Chicago Press.

Dochy, F., Segers, M., Van den Bossche, P., & Gijbels, D. (2003). Effects of problem-based learning: A meta-analysis. *Learning and Instruction, 13,* 533–568.

Foxfire Fund Inc. (2002). *Foxfire mission.* Mountain City, GA: Foxfire.

Gijbels, D., Dochy, F., Van den Bossche, P., & Segers, M. (2005). Effects of problem-based learning: A meta-analysis from the angle of assessment. *Review of Educational Research, 71*(1), 27–61.

Hmelo-Silver, C. (2004). Problem-based learning: What and how do students learn? *Educational Psychology Review, 16*(3), 235–266.

Hmelo-Silver, C., Duncan, R. G., & Chinn, C. A. (2007). Scaffolding and achievement in problem-based and inquiry learning: A response to Kirschner, Sweller, and Clark 2006. *Educational Psychologist, 42*(2), 99–107.

Kirschner, P., Sweller, J., & Clark, R. (2006). Why minimal guidance during instruction does not work: An analysis of the failure of constructivist, discovery, problem-based, experiential, and inquiry-based teaching. *Educational Psychologist, 41*(2), 75–86.

Kolb, D. A. (1984). *Experiential learning: experience as the source of learning and development.* Englewood Cliffs, NJ: Prentice Hall.

Lewin, K. (1943). Defining the "Field at a Given Time." *Psychological Review, 50,* 292–310.

Next Generation Science Standards. (2017). http://www.nextgenscience.org/

Pease, M., & Kuhn, D. (2011). Experimental analysis of the effective components of problem-based learning. *Science Education, 95,* 57–86.

Ravitch, D. (2000). *A century of failed school reforms.* New York: Simon & Schuster.

Rogers, C. R. (1969). *Freedom to learn* (2nd ed., 358 p). C. E. Merritt Publishing. ISBN-13: 978-0675095792.

Schmidt, H. G., Loyens, S. M. M., van Gog, T., & Paas, F. (2007). Problem-based learning is compatible with human cognitive architecture: Commentary on Kirschner, Sweller and Clark 2006. *Educational Psychologist, 42*(2), 91–97.

Starnes, B. A. (1999). *The foxfire approach to teaching and learning: John Dewey, experiential learning, and the core practices.* Charleston, WV: ERIC Clearinghouse on Rural Education and Small Schools.

Starnes, B. A. (2000, February). On dark times, parallel universes, and de'ja'vu. *Phi Delta Kappan.*

Starnes, B. A., Paris, C., & Stevens, C. (1999). *The Foxfire core practices: Discussions and implications.* Mountain City, GA: Foxfire.

Ubben, G. (2005). *MYOP, education for the neighborhood and the world.* University of Tennessee.

Vernon, D. T. A., & Blake, R. L. (1993). Does problem-based learning work? A meta-analysis of evaluative research. *Academic Medicine, 68,* 550–563.

Walker, A., & Leary, H. (2009). A problem-based-learning meta-analysis: Differences across problem types, implementation types, disciplines, and assessment levels. *The Interdisciplinary Journal of Problem-Based Learning, 3*(1), 12–43.

Wigginton, E. (1986). *Sometimes a shining moment: The foxfire experience. (Twenty years teaching in a high school classroom).* New York: Doubleday.

Wilford, A. (1942). *The story of the eight year study.* New York: Harper.

Wirkala, C., & Kuhn, D. (2011). Problem-based learning in K-12 education: is it effective and how does it achieve its effects? *American Educational Research Journal, 48*(5), 1157–1186.

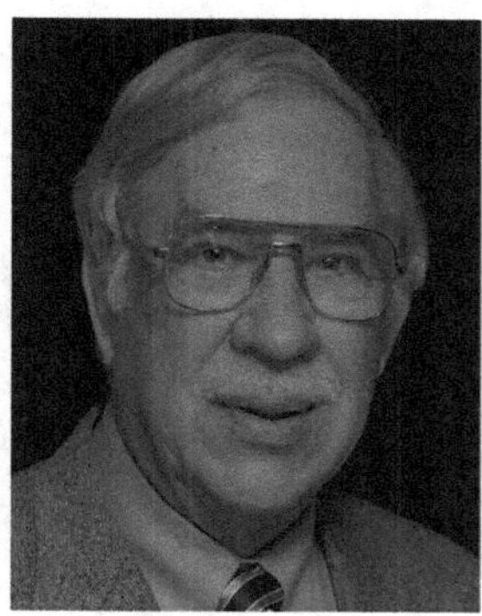

Gary Ubben became Professor Emeritus in the College of Education at the University of Tennessee, Knoxville, after a 40-year career. He retired from the University in 2010. He is the author of over a dozen books in Education and Educational Leadership. Dr. Ubben directed several international projects for the college, including a USDS-sponsored partnership with Bourgas Free University in Bulgaria, where Project-Based Learning (PBL) became the vehicle for children of the two countries to engage in a cultural exchange over the internet. His two chapters in this book are based on the PBL protocols developed for that project.

How to Structure Project-Based Learning to Meet STEAM Objectives

Gary Ubben

1 Introduction

While project-based learning (PBL) lesson plans can be organized in numerous ways, a standard format ensures all necessary components for a good PBL are included. A standard lesson plan format also helps teachers and student participants quickly become familiar with each PBL they encounter. However, every teacher using the PBLs in this chapter will need to adapt them to their classroom and to the maturity and experience level of their students. This chapter provides a structure and examples for a good PBL design.

Each PBL lesson plan, after the title, should include the following eight components:

1. Topic/Problem/Grade/Time
2. Topic Paragraph – (introduction and setting for the study)
3. Curriculum Standards
4. PBL (Learning) Objectives
5. Guiding Questions
6. Product Specifications
7. Resources
8. Assessment Rubrics

G. Ubben (✉)
University of Tennessee, Knoxville, TN, USA
e-mail: gubben@utk.edu

© Springer Nature Switzerland AG 2019

A. J. Stewart et al. (eds.), *Converting STEM into STEAM Programs*,
Environmental Discourses in Science Education 5,
https://doi.org/10.1007/978-3-030-25101-7_7

2   Topic/Problem/Grade/Time

Projects can vary greatly as to difficulty, complexity, and length (usually measured as average time to complete). It is important to provide identification information to aid the teachers in their search for an appropriate PBL for their classroom. This information can help teachers quickly identify a possible activity.

Examples: *Transportation/Building a strong stick bridge/Grade 3 or 4/2 hours.*

Engineering/Design and fabricate a simple, externally powered prosthetic hand/ Grade 11/36 hours/9 weeks.

3   Topic Paragraph

Describe, in one paragraph, the project to be completed or the problem to be solved and direct it to the students. Examples:

Working in a group of four and given a stack of popsicle sticks and a tube of wood glue, build a wooden bridge 5 inches wide and spanning at least 12 inches and capable of holding a two-pound weight.

Using 3D printers make a prosthetic hand for kids using freely available STL files. This project applies 3D printing technology to solve other patient needs. The teams will generate and analyze candidate solutions using criteria formed through their research.

The activity might not give any other instruction, relying on the creativity of the students, or it might include pictures of finished products such as bridges illustrating successful stick bridges or successful prosthetic hands to provide the team with basic ideas. Obviously, additional instructions could be given for this activity using the students' successes and failures as illustrations. Often, setting the stage for an activity with a reason for needing the solution can help provide focus. For example, in Chap. 6, the discussion of the aftermath of the hurricane and need for drinking water in Puerto Rico and the difficulty in providing it provided the backdrop for the water-filtration project.

4   Curriculum Standards

Aligning projects to existing curriculum standards sometimes can be a real challenge. If the bridge project is for third and fourth grade students, the standard may be drawn, for example, from those addressing transportation, communication, or safety. For the prosthetic 3D printed hand the standard might come from geometry, physics, or engineering.

The content for each PBL should be selected such that the projects address important curricular goals and standards appropriate for the age and grade of the students involved. Most states in the USA have adopted curriculum standards such as the Common Core Curriculum, and some states use the newer Next Generation Science Standards in science. The Next Generation Science Standards (2017), adopted by 18 states and the District of Columbia, demand a three-dimensional approach to instruction. Each lesson is to combine "practices," or the behaviors of real scientists and engineers; "cross-cutting concepts," which clarify connections across science disciplines and help students create a coherent view of the world based on science; and *disciplinary core ideas*, or the fundamental ideas students must know to understand a given science discipline.

In searching for appropriate PBL topics it might seem logical to begin with the list of standards to which teachers are expected to teach. In practice, however, it is difficult to design a good PBL around these lists. An alternative method is to reverse the process (this is sometimes referred to as backward planning) by selecting a good PBL topic first, and then see which standards can be taught through the PBL. For example, assume the topic is gardening. From raised bed squares to aquaponics to aeroponics, horticulture offers creative and innovative projects that incorporate many of the standards we are asked to teach. Beginning with the activity, PBL projects can be designed by finding applicable standards within national, state, and local curriculum and student performance lists. Since our focus is on the integration of STEAM subjects, it would be appropriate to review standards lists from each of these subjects to see how they may contribute to the project's framework. The best projects to spur student interest usually include those based on a realistic challenge, investigation, or problem for the students to resolve.

5 PBL (Learning) Objectives in the Context of STEAM

Learning objectives describe what the student is going to be able to do at the end of the activity. Objectives are often best written if they begin with a verb such as understand, know, practice, write, or develop skill at. It is appropriate to indicate basic and advanced levels when a class holds a range of student ages or abilities.

Learning objectives can be divided into several categories. *General objectives* are those that may apply to almost all projects. These objectives are ones that apply to the entire project and are often soft skills related; they may deal with research, organization, group process, and reporting.

For example: By the end of this project the student will be able to:

- *Formulate questions for effective interviews, conduct interviews, and organize their responses.*
- *Demonstrate their use of new media technologies by preparing photographs and videos for a web presentation.*

- *Practice their writing skills by preparing written illustrated reports for posting on the web.*
- *Develop teamwork skills through participation in student work committees.*
- *Practice their presentation skills by giving illustrated multimedia oral reports to their classmates and others.*

In addition to the general learning objectives, *content or subject-specific learning objectives* must be included. If the PBL is to serve a multidisciplinary purpose, as it should in STEAM PBLs, each discipline should be reviewed to determine what its contributions might be to desired learning outcomes. For example:

- *Demonstrate your mapping skills by showing latitude and longitude locations for the areas you are studying.*
- *Demonstrate your understanding of certain geometric shapes in designing strong bridge structures.*
- *Identify the sound patterns of at least four different musical instruments using an oscilloscope.*
- *Present your project to an adult panel of judges using well-designed graphic images to highlight your story.*

6 Guiding Questions

Guiding Questions are used to direct the students' thinking along the proper path by asking questions. They are helpful in encouraging the student to explore directions or intricacies of which they might not have thought or in preventing the project from wandering off in an unintended direction. They are not intended to be answered directly but only to guide the students' thinking. They are important both in PBL and for meeting STEAM objectives.

PBLs should include a series of questions and subquestions that are provocative, significant, and realistic, and these questions should drive the students' investigation in directions that result in the essential core lesson designed to be taught by the project. For example, the simple bridge PBL discussed earlier might ask students:

What are some different style bridges you have seen?
Where might you find pictures of bridges?
Looking at a picture of a bridge, what are some of the geometric shapes you see and what are their functions?
Which of these bridge types do you think you might be able to copy with popsicle stick construction?
Are some bridges prettier than others? If so, why?
Are new bridges today designed the same way bridges were built 50 or 100 years ago? If different, why?
How can you use engineering design process in your project?

7   Product Specification

Product Specification tells the student exactly what their final products are to look like and the form in which they are to be submitted. In the example of the model wooden bridge, the "Product Specification" for elementary school children was to *build a wooden bridge with pop cycle sticks 5 inches wide and spanning at least 12 inches and capable of holding a two-pound weight.* The class discussion after the presentation and testing of the completed projects can be expanded to questions about which bridges were the strongest, and why, or which bridges used the least amount of material, or which bridges were the prettiest – but the model bridge was what was required of the students.

The same project could be ratcheted up a notch as a PBL for middle-school students by asking a more complex guiding question such as, *How can we, as mathematicians, use geometry and pre-algebra to help us design and build a model bridge?* In this case, we need to describe how we wish the students to explain their use of pre-algebra and geography. This can be done by adding a product specification: *Explain, in a written and final oral report, your use of pre-algebra and geometry in designing and building your bridge.*

And even higher-order thinking can be achieved by presenting the project as a problem to be solved, and adjusting the product specifications accordingly. Rather than asking the class to build or design a bridge, the teacher could frame the problem as *a need to move a minimum of 500 cars a day across a 500 foot wide river that is fast-flowing, with banks that are 50 feet above the river.* Now the questions become *How? A ferry? A tunnel? A bridge?* Other questions emerge immediately thereafter. *Bridge style? The most cost effective method? The most aesthetically appealing, etc.?* The ideas to explore can be built into the guiding questions. The product specifications in this case will now include a written report and public presentation that incorporates the team's recommendations, the rationale for their conclusions and explanation about how they arrived at them, pictures or drawings of their suggested solution(s), and estimated costs. The way the questions were asked also classifies this lesson as *problem*-based learning rather than simply *project* based.

Public feedback to students is an important and integral part of every PBL and should be written into most product specifications. Students through PBLs often develop new knowledge, and the task of presenting it in a meaningful way to others helps solidify that learning. The presentation may be to classmates, but often presentations to parents and other community members, such as subject matter experts or government officials, have great benefit but sometimes generate some anxiety as well. PBLs may involve a partner school located some distance away, so PBL reports prepared as web-based reports can provide easy access to partner schools.

8 Resources

List any material or equipment needed by either the teacher or the students to complete the PBL. For the several PBL examples used in this chapter, necessary materials range from popsicle sticks, glue, and a two-pound weight, to pictures of bridges (internet), to an oscilloscope, a camera, a computer-based GPS application, 3D printers and software, presentation software, etc.

Of importance are specific information reference sources that students need to best understand the project concepts. These sources can be in the form of teacher-made worksheets, YouTube videos, copies of textbook pages to read, journal articles, or the instruction to students to do a general internet search for specific suggested topics.

9 Student Assessment Rubrics

The way students are evaluated for their work should be predetermined in every PBL. Because much of the work of students in a PBL is self-guided, the rubrics become the benchmark for students in guiding their actions and assessing their performance. The two major assessment-related components of PBL are its *Learning Objectives* and *Product Specifications* (outcomes). Teachers should develop assessment rubrics or checklists from these two lists. These rubrics should include both quantity and quality measures. Assessment rubrics should be shared with students at the time the PBL is being introduced to them along with the *learning objectives* and *product specifications*. Since much of the learning in a PBL environment is self-directed, the assessment rubrics are invaluable in helping the student stay properly focused on what is expected of them.

All rubrics have three basic components and are generally constructed as two-dimensional tables. The vertical dimension lists down in the left margin the skill or standard to be assessed. The horizontal dimension across the top of the table lists three or four performance levels along a continuum, for example, from low to high or performance levels such as novice, mastery, and distinguished. Content within the table describes what the behavior or skill looks like at a given performance level. For example, the left margin categories on the stick bridge project might be *Length of Bridge* and *Weight supported* with the performance categories headings across the top of the rubric being: *spans less than 12 inches, 12 inches, and more than 12 inches*, and supports *less than 2 pounds, just 2 pounds, or more than 2 pounds*. Rubric categories also can have scoring numbers assigned so that a numeric total can be calculated for the student's performance.

9.1 Stick Bridge Product: Performance Rubric

Product spec/performance	Beginner (1)	Mastery (2)	Distinguished (3)
Bridge length	*Spans less than 12 inches*	*12 inches*	*More than 12 inches*
Bridge weight	*Less than 2 pounds*	*Just 2 pounds*	*More than 2 pounds*

Rubrics need to be constructed for each of the *content-specific objectives* stated in the PBL. These rubrics should be based on the PBL learning objectives and the product specifications. If the PBL has been developed by an interdisciplinary team of teachers or it is being taught by such a team, each member should contribute to the rubric from their discipline. Be sure to include other learning objectives such as higher-order thinking skills and problem-solving skills in the rubrics to encourage students to develop those skills. A rubric developed around the engineering design process can have this focus.

One set of rubrics useful for most PBLs can be developed around the *general objectives* stated earlier in this chapter. These general objectives were interviewing, using new media technologies, writing skills, teamwork skills, problem-solving, and presentation skills. Additional rubrics can be created for student self-assessment and for peer assessments from other team members. Examples of these rubrics can be found at the end of this chapter.

9.2 What to Do Next

This chapter has presented a useful structure for PBL design and implementation incorporating the components of a PBL lesson plan most commonly used. While PBLs can be taught by individual teachers at any grade level, STEAM PBLs, because of their multidisciplinary nature, have some unique challenges particularly as disciplines become more complex at the upper grade levels. Team-designed and team-implemented PBLs focusing on the expertise of teachers from multiple disciplines provides opportunity for schools to address more complex projects or problems drawing the diverse expertise of multiple teacher experts.

While this chapter has focused on the original design and development of a PBL there are many PBL resources already available from which teachers can borrow, adopt, or adapt. Many local schools that focus on STEM or STEAM curriculum have posted on the school web page their PBL curriculum units such as the Chattanooga, TN STEM school. In a similar fashion, State Departments of Education have posted on their state website PBL materials for their teachers. These sites often include extensive rubric ideas for PBL assessment (see West Virginia Department of Education PBL Library (2017)). Finally, a number of private profit and nonprofit agencies that specialize in PBL instruction offer professional development opportunities for schools in PBL teaching and also have extensive libraries of previously developed and tested PBLs (BIE 2019). The reference list at the end of this chapter

Larmer et al. (2015) and Pete and Fogarty (2018) give examples for each of these type sources.

Here I provide an example of a middle school STEAM PBL, a content-assessment rubric, and a soft-skills presentation rubric.

10 PBL Example for Middle School STEAM Curriculum

Topic/Problem/Grade/Time
Robotics: Construct and code a Parallax Boe-Bot (Fig. 1) to navigate an unknown maze/Grade 9/5 weeks (20 hours).

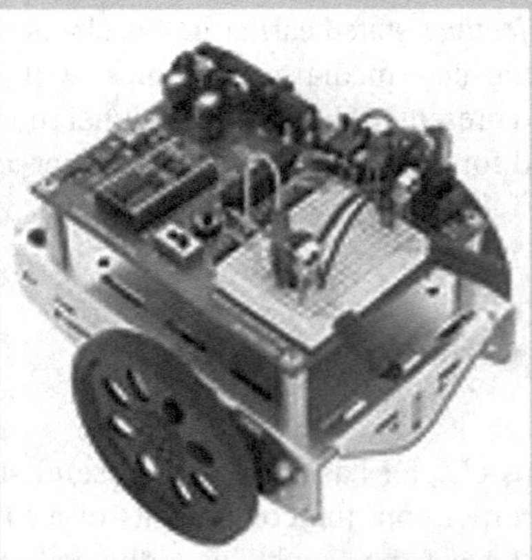

Fig. 1 Boe-Bot

Topic Paragraph (Introduction and Setting for the Study)
This PBL on Robotics will introduce students to the essential concepts underlying the principles of electrical circuitry and coding with robotics. Along with the study of circuitry, students will apply critical thinking to collaboratively assemble and code a Parallax Boe-Bot. Through the use of various types of sensors and coding, students will successfully maneuver the robot through a maze. Classroom design teams will compete with each other in a final event. Students will also create a digital Troubleshooting Guide that includes tips for constructing, wiring, coding, and testing their robot. The Guide is to use at least two types of procedural text, including a rationale of the effectiveness of the images used in the guide. Students will also design a labeled schematic of the circuitry, including a digital Pop-Up History Blurb of a chosen component.

Curriculum Standards
- **Graphic Arts:** Choose and apply images to communicate an idea. Include formatting, graphics, and multimedia for comprehension.
- **Mathematics:** Create an algebraic equation and use it to solve coding problems.
- Develop geometric definitions of transformations and represent them in the plane.
- **Technical writing:** Introduce topics and organize information. Use domain-specific vocabulary. Demonstrate command of Standard English grammar. Integrate information into text and avoid plagiarism.
- **Physical World Concepts**: Identify, describe, and calculate magnetic and electrical forces, charges, and fields. Use Ohm's law to design and build series and parallel circuits.
- **History:** Gather relevant information on robotics and particularly on Boe-Bot from multiple sources.

PBL Objectives
- **General(soft) skills**

  - Demonstrate use of new media technologies by preparing photographs and videos for a web presentation.
  - Practice writing skills by preparing written and graphic illustrated oral reports and for graphic posting on the web.
  - Develop teamwork skills through participation in student work committees using engineering design strategies.
  - Practice presentation skills by giving illustrated multimedia oral reports to their classmates and others.

- **STEAM-specific skills**

  - Graphic Arts:

    - I can choose and apply subject matter and symbols to communicate an idea.

  - Mathematics:

    - I can create algebraic equations and use them to solve problems.
    - I can define geometric transformations and represent them in a plane.

  - Technical writing:

    - I can introduce a topic; organize complex ideas, concepts, and information to make important connections and distinctions; and include

(continued)

- formatting, graphics, and multimedia when useful to aid comprehension.
  - I can use precise language and domain-specific vocabulary to manage the complexity of the topic.
  - I can demonstrate command of the conventions of Standard English grammar and usage when writing or speaking.

- Physical World Concepts:

  - I can use mechanics to measure, calculate, describe, and represent the motion and energy of an object.
  - I can identify, describe, and calculate work, force, and power.
  - I can identify, describe, and calculate magnetic and electric forces, charges, and fields.
  - I can use Ohm's law to design and build series and parallel circuits.

- History.

  - I can gather relevant information from authoritative print and digital sources using advanced searches effectively.
  - I can assess the usefulness of each source in answering the research question; integrate information into the text selectively to maintain the flow of ideas.
  - I can avoid plagiarism and follow a standard format for citation.
  - I can draw evidence from informational texts to support analysis, reflection, and research.

Guiding Questions (Not Complete-Examples Only)
- Are you using the engineering design process to plan your work?
- What software will you be using to program your robot, such as PBASIC, Scratch, or Unity?
- Have you learned to program in this language? If not, what are your learning options?
- What are the steps to program your robot to go to more than one destination?
- How do you create an equation to determine the distance the robot has traveled?
- Are your suggestions in the Guide presented sequentially and are they helpful for troubleshooting the construction, wiring, coding, and testing of the Boe-Bot?
- Does your written history of the Boe-Bot component give an organized summary of the history and development of the invention?
- Does your written Blurb answer the who, what, where, and when questions of the history?
- Have you used proper industry vocabulary and jargon in your Guide?

Product Specifications
- Student in teams of three will construct and code a Parallax Boe-Bot with the goal of successfully navigating the robot through an unknown maze at a robotics competition with other class teams. The robot should also be capable of producing sound.
- Students will design and exhibit a digital Troubleshooting Guide for their robot, outlining helpful suggestions for constructing, wiring, coding, and testing the robot.
- Students will design and exhibit a digital labeled schematic, including a digital Pop Up History Blurb on a chosen component.

Resource (Partial List)
- Lessons and worksheets on PBASIC coding from your math teacher.
- Lessons and worksheets on geometric terms from your math teacher.
- Vocabulary worksheet provided by your writing teacher.
- Internet search for YouTube videos of Boe-Bot options to help solve specific problems.
- Robotics with the Boe-Bot, Student Guide, download at https://www.parallax.com/sites/default/files/downloads/28125-Robotics-With-The-Boe-Bot-v3.0.pdf
- List of available Boe-Bot accessories.

This PBL was adapted from the STEM School of Chattanooga, TN; see www.stemschoolchattanooga.net.

10.1 Assessment Rubrics

Robotic PBL Product Specification: Student Project Assessment Rubric
This PBL rubric for student presentation was adapted from the STEM Academy of Chattanooga, TN, website.

| Boe-Bot robotics PBL rubric | | Student name:_______________________ | |
| Date:________ | | Grade ___9___ | |
Subject category	Advanced (3)	Proficient (2)	Needs improvement (1)
Art: graphic arts	The visual images in the guide improve the effectiveness, clarity, and understanding of the assembly and operation of the Parallax Boe-Bot, which is described in the rationale	The visual images (still photos or video) used in the troubleshooting guide and discussed in the rationale are appropriate for the task and purpose	Graphics detract. Do not help clarify content
Math: algebra	Using calculations from the proficient section, students will calculate the number of pulses they would need to run the robot in order for it to travel 100 cm	Students will program their Boe-Bot to go to one or more destinations and return to starting point Students will create an equation to calculate the distance the robot travels	Equation is wrong or only partially correct
Math: geometry	Students will generate a program to control a robot to perform the basic maneuvers: forward, backward, rotate left, rotate right, and pivoting turns	Students will explain how speed and direction are controlled for continuous rotation servos	Explanation or program is incorrect
Science: physical world concepts	Students will explain how speed and direction are controlled for continuous rotation servos	Students will draw their Boe-Bot circuit to scale as an appendix to the manual. this drawing will include the total voltage, source, all switches, and all resistances	Drawing is incorrect. Labels not correct
		Appropriate labels of the schematic, including the associated voltage, amperes, or ohms, are included in the drawing	Ohm calculations not correct
		A calculation using Ohm's law will be shown at the bottom of the schematic	Current discussion in error
		A discussion of how the differing resistances affect the current in both parallel and series wiring schemes is included in a written piece below the drawing	B/B discussion weak
		A description of how the Boe-Bot takes advantage of angular momentum is included	

(continued)

| Boe-Bot robotics PBL rubric | | Student name:________________ | |
| Date:________ | | Grade ___9___ | |
Subject category	Advanced (3)	Proficient (2)	Needs improvement (1)
Language arts: technical writing	The formatting and graphics make the Guide's instructions effective and easy to follow	The suggestions in the guide are presented sequentially and are helpful for troubleshooting the construction, wiring, coding, and testing of the Boe-Bot	Not sequential. suggestions not clear. Vocabulary weak. Manual not focused on consumer
	The different types of multimedia and procedural text aid in the understanding and effectiveness of the guide's suggestions	All of the suggestions in the guide are written clearly and are easy to understand, using a problem-solution format	
	The guide uses proper parallel structure and comma usage with complex sentences	Examples of domain-specific vocabulary related to robotics and procedural text are used	
	The use of domain-specific vocabulary and proper grammar, spelling, and punctuation create a professional product that could be used by the Parallax Boe-Bot Company	The point of view of the manual is consistent and focused on the consumer, not the author	
Social studies history	The written history of the component gives an insightful look at the invention and also an understanding of its development over the years	The written history of the component gives an organized summary of the history and development of the invention	History research on sequence weak. Did not cover the 5 Ws.
	The written blurb answers who, what, when, where, and why it is important	The written blurb answers the who, what, where, and when questions of the history	Vocabulary weak. Grammar errors
	Grammar and spelling are free from errors	Proper industry vocabulary and jargon is used correctly	
		The history blurb contains no more than 1–2 errors in grammar or spelling	

(continued)

Student Presentation Rubric

Oral presentation rubric	Name_____________________		Date	
_____________ Class period		_________		
Category/ points	4	3	2	1
Preparedness	Student is completely prepared and has obviously rehearsed	Student seems reasonably well prepared but might have needed several more rehearsals	The student is somewhat prepared, but it is clear that rehearsal was lacking	Student does not seem at all prepared to present
Presentation of information	Stays on topic all (100%) of the time. Ideas were presented sequentially with a definite beginning, middle, and an interesting conclusion	Stays on topic most (99–90%) of the time. Topics are organized so ideas can be followed easily	Stays on topic some (89–75%) of the time. The sequence of the topics made them difficult to follow	It was hard to tell what the topic was. It was difficult to follow the line of thought
Eye contact	Makes eye contact with the audience throughout the presentation. Works independently of notes	Makes eye contact with the audience during the presentation. Is not dependent on notes or prompts from others	Makes some eye contact with the audience during the presentation. Is somewhat dependent on notes or prompts from others	Is unable to make eye contact with the audience during much of the presentation. Is dependent on notes or prompts from others
Speaking skills	Speaks clearly and distinctly all (100–95%) the time, has good inflection and volume; mispronounces no words. Uses appropriate body language to convey meaning	Speaks clearly and distinctly all (100–95%) the time, uses some inflection and has good volume; mispronounces few words. Uses body language to convey meaning	Speaks clearly and distinctly most (94–85%) of the time, volume may be inappropriate. Mispronounces some words. Uses some gestures to convey meaning	Often mumbles or cannot be understood OR mispronounces several words. Uses no body language or gestures to convey meaning
Time-limit	Presentation is 5–6 minutes long. _________ minutes	Presentation is 4 minutes long. _________ minutes	Presentation is 3 minutes long. _________ minutes	Presentation is less than 3 minutes OR more than 6 minutes. _________ minutes

Comments:

Student Peer Team Member Evaluation Rubric

Collaborative Work Skills Peer Evaluation Rubric		Name:	
Date: _______		Class	
Period____			
Category	4	3	2	1
Contributions	Partner routinely provided useful ideas when participating in the group and in classroom discussion	Partner usually provided useful ideas when participating in the group and in classroom discussion	Partner sometimes provided useful ideas when participating in the group and in classroom discussion	Partner rarely provided useful ideas when participating in the group and in classroom discussion
Quality of work	Partner provided work of the highest quality	Partner provided high-quality work	Partner provided work that occasionally needed to be checked/redone by other group members to ensure quality	Partner provided work that usually needed to be checked/redone by others to ensure quality
Time-management	Partner routinely used time well throughout the project to ensure things got done on time. Group did not have to adjust deadlines or work responsibilities because of me	Partner usually used time well throughout the project, but may have procrastinated on one thing. Group did not have to adjust deadlines or work responsibilities because of me	Partner tends to procrastinate, but always got things done by the deadlines. Group did not have to adjust deadlines or work responsibilities because of me	Partner rarely got things done by the deadlines AND the group had to adjust deadlines or work responsibilities because of my inadequate time management
Preparedness	Partner brought needed materials to class and was always ready to work	Partner almost always brought needed materials to class and was ready to work	Partner almost always brought needed materials but sometimes needed to settle down and get to work	Partner often forgot needed materials or was rarely ready to get to work
Working with others	Partner almost always listened to, shared with, and supported the efforts of others. I tried to keep people working well together	Partner usually listened to, shared with, and supported the efforts of others. I did not cause "waves" in the group	Partner often listened to, shared with, and supported the efforts of others, but sometimes I was not a good team member	Partner rarely listened to, shared with, and supported the efforts of others. I often was not a good team player

From Bomine Adventures UP3267WS11, West Virginia PBL Library (2017)

References

Bomine Adventures. UP3267WS11 West Virginia PBL Library.
Larmer, J., Mergendoller, J., & Boss, S. (2015). *Setting the standard for project based learning.* Alexandria: ASCD.
Next Generation Science Standards. (2017). https://www.nextgenscience.org/search-standards
Pete, B., & Fogarty, R. (2018). *Everyday problem-based learning.* Alexandria: ASCD.

STEM School Websites Using PBL

Buck Institute for Education. (2019). http://www.bie.org/about
Google search for STEAM PBL Schools. www.stemschoolchattanooga.net
West Virginia Department of Education PBL Library. (2017). http://wvde.state.wv.us/teach21/PBLTools.html

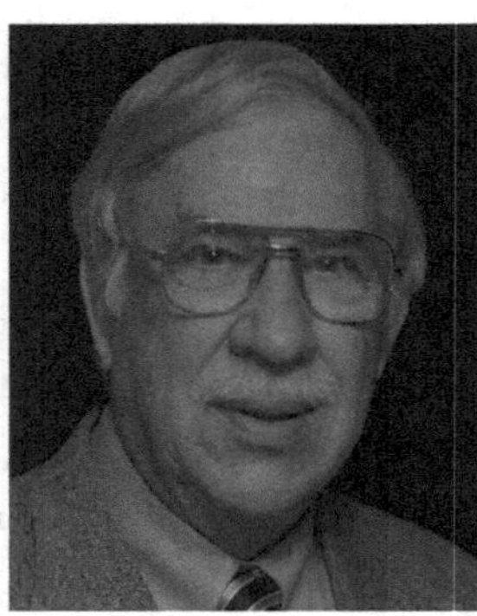

Gary Ubben became Professor Emeritus in the College of Education at the University of Tennessee, Knoxville, after a 40-year career. He retired from the University in 2010. He is the author of over a dozen books in Education and Educational Leadership. Dr. Ubben directed several international projects for the college, including a USDS-sponsored partnership with Bourgas Free University in Bulgaria, where Project-Based Learning (PBL) became the vehicle for children of the two countries to engage in a cultural exchange over the internet. His two chapters in this book are based on the PBL protocols developed for that project.

Enhancing STEAM Education Through Cultivating Students' Savoring Capacity

Shu-Hsuan Chang, Li-Chih Yu, Jing-Chuan Lee, and Chih-Lien Wang

Interdisciplinary integration and the cultivation of innovation skills are important core competencies in STEAM education. Given major sociotechnical challenges of our twenty-first century, a state of wide-awakeness is a central component in processes of problem framing and creative thinking. Savoring, an aspect of wide-awakeness, is a self-regulatory response that involves a person's awareness/perception of his/her inner states: savoring has been shown to affect creativity. In this chapter, we discuss, in two parts, how school support systems can help cultivate their students' savoring and creative abilities within the context of interdisciplinary STEAM education. First, we present results of an investigation of factors driving the demand-pull innovation in cultivating creativity. These factors involve students' ability to savor (through anticipation, enjoying the moment, and reminiscence). We then show how a student's savoring capacity affects individual creativity, and how student perceived support for creativity affects their individual creativity through savoring. Second, we explore relationships between students' creative self-efficacy, savoring capacity, and individual creativity, and explain how school support for creativity enhances these relationships. We conclude with theoretical and practical implications for maximizing student creativity through STEAM education.

S.-H. Chang
Department of Industrial Education and Technology, National Changhua University of Education, Changhua, Taiwan
e-mail: shc@gm.ncue.edu

L.-C. Yu
E-Learning Center, National Changhua University of Education, Changhua, Taiwan

J.-C. Lee · C.-L. Wang (✉)
Graduate Institute of Educational Leadership and Evaluation, Southern Taiwan University of Science and Technology, Tainan, Taiwan
e-mail: jclee@stust.edu.tw; gellien1@stust.edu.tw

© Springer Nature Switzerland AG 2019

A. J. Stewart et al. (eds.), *Converting STEM into STEAM Programs*, Environmental Discourses in Science Education 5, https://doi.org/10.1007/978-3-030-25101-7_8

1   Savoring and Creativity

Positive psychology is the key to promoting competition among higher education institutions; it is a cornerstone for inspiring students to cultivate knowledge, realize potential, and contribute to society during their college career (Lopez and Snyder 2009). Savoring is a concept developed by positive psychology; it is a type of ability to appreciate and enjoy an experience (Seligman 2002), including recognizing and treasuring life's positive aspects and consciously directing attention to a pleasant experience and maintaining mindfulness on said experience (Bryant and Veroff 2007). Bryant (1989), who first introduced the concept of savoring, indicated that there are four types of perceived control when facing an event: avoiding, coping, obtaining, and savoring, in which avoiding and coping are negative perceived abilities to external events while obtaining and savoring are positive perceived abilities related to internal self-awareness. Bryant and Veroff's (2007) study showed that savoring correlates to simple pleasure, but not to obtaining. In other words, a positive event (e.g., obtaining) does not necessarily bring forth feelings of pleasure, but the duration of pleasure can be prolonged through immersion and enjoyment (i.e., savoring), which strengthens the benefits of the emotion. As savoring can extend positive emotions, it is believed to be a form of emotional regulation (Quoidbach et al. 2015). Fredrickson's (2001) "broaden and build" theory indicated that positive emotions (1) broaden the likelihood of thinking, (2) undo negative emotions, (3) enhance upward spirals, and (4) build lasting resources to promote happiness and well-being. The important significance of idea this lies in that positive emotions such as joy and love can induce humankind's behavioral and cognitive capacities because it can expand the scope of one's attention and cognition. Expanding the scope of these mental activities is tantamount to increasing elements of innovation, which in turn enhances creativity (Amabile et al. 2005). In the context of the complex sociotechnical challenges of the twenty-first century, the process of investigating problem framing and creative thinking requires one to open up to acknowledging the wide-awakeness of the inner world. Savoring is the mental capacity of detecting and expanding the enjoyment of positive experiences or perception of positive outcomes. Its capabilities include (1) using cognitive or behavioral strategies to expand or extend the enjoyment of positive events, (2) looking forward to future positive outcomes based on feeling happy in the present moment, (3) increasing happiness by recalling past positive events, and (4) enjoying a positive event through the help of a friend or relative. In light of this, the ability to savor is conducive to improving creativity (Bryant and Veroff 2007). Next, we will further expound on the implication of the relationship between savoring, creative self-efficacy, and individual creativity.

2   Savoring and Creative Self-Efficacy

Cognitive learning is important in the intellectual process of learners. According to the self-regulation perspective of social cognitive theory, self-efficacy is central in the self-regulatory system (Bandura 1989, 1991). Creative students are successful self-regulated individuals who can control and monitor their learning environment (Sternberg 2004). In addition, Tierney and Farmer (2002) indicated that creativity requires the support of personal inner strength; this force motivates one to persevere to the will for action, which turns into confidence in engaging in creative work. Ford (1996) views creative self-efficacy as the belief that can influence creative behavior. There are four main ways by which one can perceive self-efficacy: cognition, motivation, emotion, and optimism. In response to these perceptions, one may produce behaviors such as choosing, performing, and persisting to regulate human functioning. In other words, witnessing the success of others, gaining encouragement and positive feedback from important people in their lives, or being in a good mood helps learners enhance self-efficacy (Bandura 1977).

Learners with high self-efficacy are more likely to face challenges proactively, accurately evaluate their efforts, and show a higher perseverance to their choice of action. Hannula (2002) argued that learners' emotions exert a larger influence over their learning attitudes than cognition. From his original theory of authentic happiness to the flourish theory of well-being, Seligman, the founder of contemporary psychology, has referred to positive education in his works and advocated for educating students on the concepts and ways of well-being at school (Seligman 2011). Seligman believed that well-being can aid in learning and increase creative thinking. He also believed that authentic happiness is first and foremost a pleasant life, which holds in store positive emotions involving the past (such as fulfillment and satisfaction), present (such as pleasure, calmness, flow), and future (such as hope, optimism, and faith) (Seligman 2002). Savoring involves the self-regulation of positive emotions. One classic example is the self-enjoyment and positive experience from savoring through reminiscing, savoring the moment, or anticipation of such savoring activities, which produce, maintain, or strengthen positive emotions (Bryant 1989, 2003). Studies have shown that the intervention of savoring the pleasant life has a lasting emotional effect (Seligman et al. 2005). Besides improving investments in young adults and students' learning, documenting the good things in life and reflecting on them (positive ruminations) also has long-term benefits (Seligman et al. 2009). Bryant and Veroff (2007) believe that during the process of savoring, unexpected observations of new images, ideas, and insights are gained, which can enhance creativity. Further, meta-analyses of efficacy beliefs have confirmed that motivation and learning, sociocognitive functioning, emotional well-being, and performance accomplishments all contribute significantly (Katz and Stupel 2015). These points are important to the education of students and young adults (Bandura 2005; Seligman 2011; Zimmerman 2000).

3 STEAM Educational Environment and Support for Creativity

In the future, the most competitive learners will be those who are compassionate, creative and armed with interdisciplinary skills. The STEAM integrative education model is tailored to fit modern educational trends. Arts is integrated into STEM education. The "A" in STEAM can be interpreted as any of three different educational outcomes: arts learning, aesthetic education, and/or creativity (Clapp and Jimenez 2016). Csikszentmihalyi (1988) proposed the systems view of creativity, which asserts that creativity is the process by which the three elements of individual, domain, and field interact with each other. This model suggests that creativity is neither fixed nor constant; it is not an individual trait, but a skill that can be shaped by environment; it can be taught, and learned.

Developing a creative idea from its inception to its execution is an eternal challenge for creators. Environmental factors may shape the creative process. For college students majoring in design, creative performance is a commonly anticipated and highly valued goal (Cartier 2011). In a study on creativity training, Amabile and Pillemer (2012) suggested that creativity should be studied from the perspective of social psychology and that efforts be put forth into exploring the comprehensive development of individual creative behaviors in the social context. Studies have found that constructive environments with encouraging and supportive members prompt stronger positive emotions, stimulate creative thinking, and promote their creativity (Amabile et al. 1996, 2005; Chang et al. 2015; Lee et al. 2016). Bryant and Veroff (2007) proposed the cognitive construct that social support increases savoring, thus emphasizing that social support is the most appropriate method for managing stressful and unfortunate events. The effects of positive emotions can be especially enhanced through sharing personal experiences on social media (Bryant and Veroff 2007; Gable et al. 2004). Past studies have tended to overlook environmental factors' influence on individual result variables through the intermediary effects of individual factors (Amabile 1996). While emotions are nonroutine, they are important in self-regulation, focusing attention, and cognitive bias. Under social contexts such as interpersonal relations and social coordination of collaborative action, the additional functions of emotions are more pronounced (Hannula 2015). Thus, based on the foundations of positive psychology, we explored the mediated effect of savoring capacity in perceived support for creativity and individual creativity (Fig. 1: Model 1).

Creative self-efficacy is an inevitable topic in the field of creativity training research. Sternberg (2004) believed that all students have the ability to become creators and experience joy associated with creation and production. Thus, if students lack faith, they will not invest time and effort (Sternberg 2004). Creative thinking is

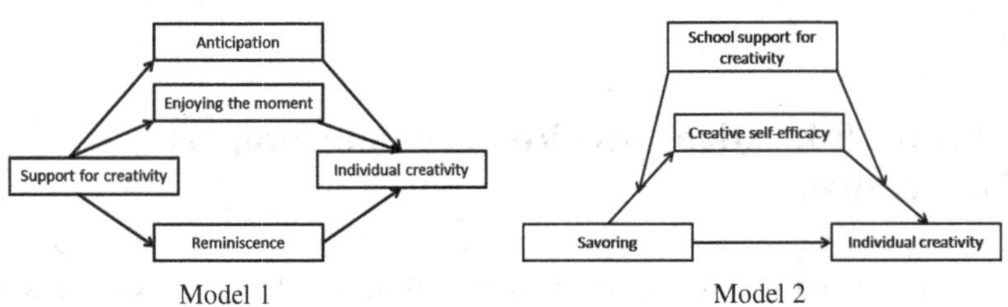

Fig. 1 Conceptual model

viewed as a type of meta-cognitive process that requires the ability to self-regulate sustained awareness for creativity (Katz and Stupel 2015). A highly creative student who also has the ability to self-regulate can control the process and display high confidence in self-efficacy (Katz and Stupel 2015). Therefore, in the context of ever-changing sociotechnical challenges which students face, the mental processes of their affects and creative thinking has long been regarded as indispensable elements in creativity (Amabile et al. 2005).

The framework of the STEAM educational environment sets the stage for interdisciplinary learning and collaborative learning, where students can prompt and promote dialogue between professional fields and create and shape a universal language. This interdisciplinarity in turn will be displayed in his/her work, which becomes the source and motivation for progress for humanity. It also offers a space and setting for stimulating creativity and innovation (Guyotte et al. 2015; Sochacka et al. 2016). Studies have shown that school support for creativity has a moderated mediation effect on creative self-efficacy and individual creativity in award-winning students, but the moderated mediation effect was not observed in non-award-winning students (Chang et al. 2016). School is a work environment for creativity, where students can gauge the levels of their school's encouragement and recognition of their creativity, their teacher's assistance and support, and the degree of mutual collaboration and support from their team members. The school's creative support directly or indirectly affects an individual's creative process (Amabile 1996). These direct or indirect effects need to be further clarified. Muller et al. (2005) indicated that if both moderated and intermediary effects can be taken into account simultaneously, theoretical and empirical studies will have more significance. Accordingly, we also explored how the intermediary mechanism/indirect effects of creative self-efficacy influence savoring capacities and individual creativity in the context of social cognitive theory. We also studied the moderated mediation effect (also known as conditional indirect effect) of school support for creativity in this intermediary effect (Fig. 1: Model 2).

4 Multiple-Mediation Effects of Savoring

4.1 Purpose

In our first study, we sought to better understand how students' perceived support for creativity affects individual creativity through savoring capacities (savoring through anticipation, being in the moment, and reminiscing), and whether there is a difference between the effects of different forms of savoring capacities on individual creativity.

4.2 Study Design and Methods

4.2.1 Scope

Our study subjects were fourth-year college students majoring in design (industrial and commercial design, industrial design, commercial design, architectural and interior design, digital media and animation design, and fashion design) at science and technology universities located in Taiwan. We recruited this group as our research subjects because they have attended capstone courses and have experience in writing a graduate thesis. Using stratified purposive sampling, we received consent from 13 universities that agreed to participate in the study. A total of 851 effective surveys were received, 34% of which were from males and 66% from females.

4.2.2 Measures

To ensure content validity, the scales we used in the study were from other relevant research (Amabile et al. 1996; Bryant 2003; Tierney et al. 1999). During compilation, we invited experts to give suggestions on the items of the questionnaire and revised the questionnaire accordingly.

In the Support for Creativity Scale, we used the "Encouragement of Creativity" section from "Assessing the Climate for Creativity, KEYS" proposed by Amabile et al. (1996) as a measuring tool to assess students' perceived support for creativity. This scale includes three dimensions: organizational encouragement, supervisory encouragement, and work group supports. Employing a four-point Likert scale, a higher score indicates greater support for creativity. An example of school encouragement includes "My school encourages using creativity to solve problems." An example of teacher encouragement is "The teacher communicates well with our work group." Classmate support items include "There is trust among my team members." The Creativity Scale survey component includes 24 questions.

In the Savoring Capacity Scale, we used Bryant's (2003) "Savoring Belief Inventory (SBI)" to measure students' savoring capacity. The three dimensions of the scale include anticipating (ANT), savoring the moment (MOM), and reminiscing (REM). Employing a seven-point Likert scale, a higher score indicates a stronger savoring capacity. This scale includes 12 positive questions, such as "Anticipating that good things will happen brings me joy," "I know how to make good memories," and "I like to reminisce about the good times I've had."

In the Individual Creativity Scale, we used a six-point, "Individual Creativity" Likert scale proposed by Tierney et al. (1999) to measure students' creativity. A higher score indicates more individual creativity. The scale includes nine questions, including "I try to use new methods and new ways of thinking to face a problem," "I think of novel and practical ways to go about tasks," and "I can find new uses for existing equipment and find new applications for existing methods." Overall, after the aforementioned research tools underwent confirmatory factor analysis (CFA) in AMOS 22.0 statistical software, the measurement models demonstrated excellent convergent validity, discriminate validity, and goodness of fit.

4.2.3   Statistical analysis

Using the multiple mediator model proposed by Preacher and Hayes (2008) as a theoretical framework for our study, we explored the mediated/indirect effects of the three types of savoring capacity on perceived support for creativity and individual creativity in college students of science and technology in a STEAM educational environment. We further explored the significance of the moderation of creative self-efficacy in this mediated effect, i.e., estimation of the moderated mediation effect. We used PROCESS macro for SPSS (Model 4) provided by Hayes (2013) to perform path analysis for multiple mediation effects. Through this effort, we conducted a path analysis for all paths in one attempt, because we wanted to avoid increasing the 1 error. A 95% confidence interval that does not include 0 (as determined by the Sobel test and the bootstrapping procedure) is considered to have a significant mediated effect (also known as indirect effect).

4.3   Results

As shown in Fig. 2, Lee et al. (2016) found that (1) support for creativity has a positive effect on savoring capacity (the capacity to savor through anticipating, being in the moment, and reminiscing) and individual creativity; (2) among the savoring capacities, only the capacity to savor through being in the moment has a positive effect on individual creativity; and (3) support for creativity can exert a positive effect on individual creativity through the mediated effects of the capacity to savor the moment. It is notable that among the three savoring capacities, only the capacity to savor through the moment has a positive indirect influence (partial mediated effect) on support for creativity and individual creativity, which establishes its indirect effects. This is consistent with Amabile et al.'s (1996) emphasis that environmental factors can affect individual outcome variables through the mediated effects of individual factors. Our results also support Amabile et al.'s (2005) research: a constructive environment with mutual encouragement leads to greater positive emotions and stimulates creative thinking and ideas. On the other hand, our do support Fredrickson's (2001) assassertions in the broaden-and-build theory, that positive emotions can broaden the likelihood of creative thinking. Savoring can maintain and enhance a person's positive emotional experience (Bryant 1989, 2003; Seligman et al. 2006); individuals can use cognitive or behavioral strategies to expand or extend their ability to enjoy positive events (Bryant and Veroff 2007).

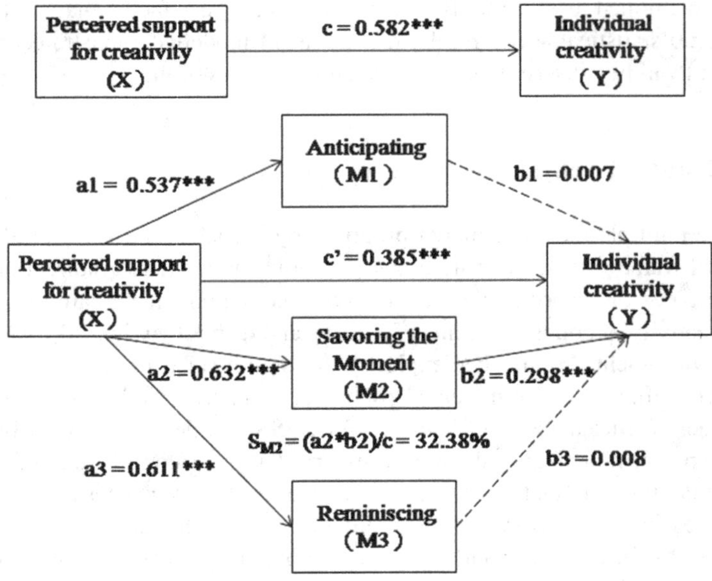

Fig. 2 Multiple-mediation model of savoring. (From Lee et al. 2016, p. 241)

Why does only savoring the moment have a positive impact on individual creativity? Just as Snyder and Lopez (2007) stated, savoring can be applied when perceiving pleasure from past or future positive events, but enjoying the moment is the most stable form of savoring. This highlights the uniqueness and importance of the capacity to savor the moment in creativity training. However, further research is necessary to elucidate whether this result is affected by other factors (e.g., spirit of the times and values).

5 Moderated Mediatory Effects of Perceived School Support for Creativity

5.1 Purpose

In the second study, we aimed to explore the mediated effect of creative self-efficacy on savoring capacity and individual creativity and to determine the moderated mediation effect of school support for creativity in this mediation mechanism (also known as conditional indirect effect).

5.2 Study Design and Methods

5.2.1 Scope

As with the first study, we recruited design major students in their fourth year at science and technology universities in Taiwan as our study subjects. These students differed from the research subjects in the first study in that in addition to experience in attending capstone courses and writing a thesis, they also had experience in participating in both domestic and international creative design competitions. We conducted our survey in the form of questionnaires, and we received an effective sample size of 720 surveys. Females accounted for 67% of the sample; and males accounted for 33%. Award winners accounted for 47% of the sample.

5.2.2 Measures

We added to the four items by Tierney and Farmer (2002), and used a seven-point Likert scale to measure students' creative self-efficacy in the Creative Self-Efficacy Scale. A high score indicates more confidence in creative self-efficacy. Items in the questionnaire include "I think I am good at coming up with novel ideas" and "I believe I have the ability to solve problems creatively." To measure school support for creativity, the Support for Creativity Scale uses "Organizational Encouragement" proposed by Amabile (1996) in "Assessing the Climate for Creativity." This scale contained eight items, including "Our school has a free and open atmosphere," "Our school rewards creative work," and "Our school has a good system put in place that encourages creative ideas." Using a four-point Likert scale, a higher score indicates that students perceive a greater support, encouragement, and recognition from their school. Overall, the measurement models had excellent convergent validity, discriminate validity, and goodness of fit after the aforementioned measurement tools underwent confirmatory factor analysis.

5.2.3 Statistical analysis

We used the Path Diagram Model 5 of the moderated mediated model proposed by Preacher et al. (2007) as the theoretical model of our research framework. We explored the mediated effects of creative self-efficacy on savoring capacity and individual creativity under a STEAM educational environment in students at universities of science and technology and to further explore school support for creativity in this mediation mechanism, namely, the estimation of moderated mediated effect.

We conducted path analysis analyzed empirical data of the mediated effect and moderated mediation effect using PROCESS macro version model 4 and model 58, respectively. To verify the research results, we used Sobel test and bootstrap method to test the 95% confidence interval.

5.3   Results

As shown in Fig. 3, Chang et al. (2015) found that (1) savoring capacity, creative self-efficacy, and school support for creativity all have significant positive effects on creativity; (2) both savoring capacity and school support for creativity have significant positive effects on creative self-efficacy; (3) school support for creativity has a significant positive moderated effect between creative self-efficacy and individual creativity; (4) the mediated effect of creative self-efficacy can cause savoring to have a significant positive effect on individual creativity – this is the partial mediation effect; and (5) school support for creativity had effects of positively moderated mediation relationships between savoring capacity and individual creativity via creative self-efficacy. In particular, the moderated mediation effect of school support for creativity occurred only between creative self-efficacy and individual creativity (Stage 2), but not between savoring capacity and creative self-efficacy (Stage 1). This result illustrates Ford's (1996) theory of creative action, which proposes that motive determines one's performance in creativity or adherence to the habit of action, in which capability beliefs and emotions are the most important motivational

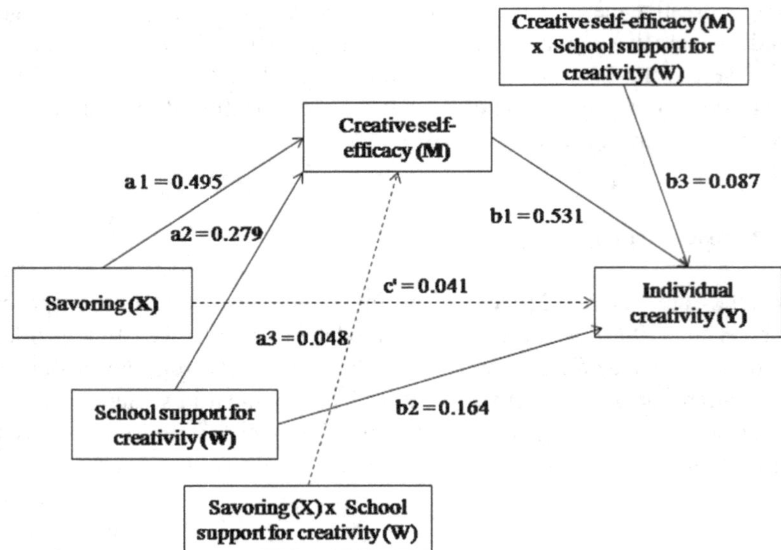

Fig. 3 Moderated mediatory model of perceived school support for creativity (Chang et al. 2015, p. 408)

factors. Just as savoring is one of the regulatory strategies of positive emotions, it also correlates strongly with state and trait positive emotions (Livingstone and Srivastava 2012), and enhancing savoring can promote the individual utility of those with few positive experiences (Hurley and Kwon 2013). One possible reason is college students need more positive psychological source material to build cognitive structure. At the same time, there needs to be more practice so that there are more opportunities to reconstruct cognitive structure – in other words, so that the reconstructed cognitive structure has a chance to grow. Thus, college students could benefit from more opportunities to learn positive mental capacities and to develop state and trait positive emotions and rich positive experiences. This idea highlights the urgency of implementing positive education during learning in college.

6 Theory and Practical Implications

STEM and STEAM show that future education continues to improve learning through various routes. As educators, we can understand today's social progress by viewing from the life course perspective. From promoting educational practices, it is clear that in addition to building a supportive external environment, arousing students' subjectivity is even more important, because it activates thinking and anticipation for a better future, becoming a force in social progress and a source for goodwill and intellect. Root-Bernstein et al. (2011) proposed the ArtScience concept, which integrates all human knowledge through the process of exploration and invention and combines synthetic experience with analytical exploration to provide new ways to explore cultural, social, and human experiences. Savoring is a type of capacity that controls inner self-perception. Just as Greene (1995) calls for teachers to train students to actively learn, critically inquire, and be wide awake, its value is its outwardly moving inner mental activity that reflects in the self but is also aware of the interactions between the self and others, envisaging new possibilities for our society through social practice.

Based on these results, we concluded from our first study that savoring capacity has its roots in positive psychology. From the perspective of Amabile's componential theory of creativity and social psychology, we conclude that external environmental factors (e.g., perceived support for creativity) exert influence over one's creativity performance through an individual's emotional factors (i.e., savoring capacity). Because savoring ability varies from person to person, we aimed to explore the intermediary role of its multiple facets in creative support and individual creativity. The results confirmed that the mediated effect of savoring the moment on support for creativity and individual creativity. This conclusion more importantly highlights the significance and importance of savoring capacity in cultivating creative talents in addition to providing support for empirical studies on savoring capacity in creativity. We based our second study on the importance of building self-efficacy in students, cross-fertilizing ideas, and encouraging creative collaboration in creative education (Sternberg and Williams 2003). Self-efficacy is an important factor in predicting behavioral performance and has garnered the attention of other researchers. We continue the discussion of our first study to elucidate the mechanism of the creative process, by which savoring capacity increases creativity through creative self-efficacy. First, we discovered that savoring capacity has a positive influence over creative self-efficacy, and this finding leaves a concrete contribution to the accumulated empirical studies on savoring and creative self-efficacy. Second, and more importantly, our study confirmed the significance of the mediated effects of creative self-efficacy on savoring capacity and individual creativity. We believe that the present study further highlights the significance of savoring in theories of creativity and contributes to training creative talent. Third, our study by the dual stage moderated mediation effect of school support for creativity further explored savoring capacity research.

The results of our two studies have at least three educational implications for cultivating interdisciplinary/creative talents. Our first study showed that support for creativity has a positive impact on all three aspects of savoring capacity. However, only the capacity to savor the moment had an important mediated effect in supporting creativity and individual creativity. Creative self-efficacy plays an important mediated role between savoring capacity and individual creativity. Under this mediated mechanism, school support for creativity is conducive to moderating the influence between creative self-efficacy and individual creativity. From this, we recommend that schools mold an environment that supports creativity – an environment where teachers model this attribute and can provide students encouragement and support, and in which team members collaborate and share and celebrate the outcomes of their creativity. A mechanism and method for encouraging creativity should be established, and enhancing the school's creative atmosphere should become the norm for inspiring students to achieve creativity. Second, we recommend that schools establish a positive psychology course so that students can strengthen their positive psychological experience and pursuit during their studies in order to enhance students' self-efficacy.

We especially recommend giving more support and encouragement or a positive perspective to view competition results to students who have not won awards (e.g., a more extensive cognitive significance and value to view participation/nonparticipation and award-winning/non-award winning). Additionally, methods that enhance savoring capacity can be applied to the curriculum design of creative education to sustain improvement in cultivating students' savoring capacity, including (1) savoring the details of everyday life (daily vacation practice), (2) making connections between relevant positive thinking from positive events that happened at different times (i.e., retrospect on life), and (3) focusing attention on savoring (camera exercise). Finally, we recommend that teachers and students participate in courses that offer new approaches for empowering teachers/students to cultivate STEAM qualities and innovative, calculated ways of thinking. We encourage students to participate in intra or interschool and international creativity competitions to enhance their confidence. We suggest that schools organize intercollegiate creativity exchange to learn from each other's work and lay out plans for improvement, thus continuing to enhance students' confidence.

"A swallow cannot beget a season of spring." We hope that this chapter arouses the attention of industries, education, and research to cultivate savoring and innovation according to the different stages of a talent's career and to provide effective training strategies.

References

Amabile, T. M. (1996). *Creativity in context*. Boulder: Westview Press.

Amabile, T. M., & Pillemer, J. (2012). Perspectives on the social psychology of creativity. *The Journal of Creative Behavior, 46*(1), 3–15.

Amabile, T. M., Conti, R., Coon, H., Lazenby, J., & Herron, M. (1996). Assessing the work environment for creativity. *Academy of Management Journal, 39*(5), 1154–1184.

Amabile, T. M., Barsade, S. G., Mueller, J. S., & Staw, B. M. (2005). Affect and creativity at work. *Administrative Science Quarterly, 50*(3), 367–403.

Bandura, A. (1977). Self-efficacy: Toward a unifying theory of behavioral change. *Psychological Review, 84*(2), 191–215.

Bandura, A. (1989). Human agency in social cognitive theory. *American Psychologist, 44*(9), 1175–1184.

Bandura, A. (1991). Social cognitive theory of self-regulation. *Organizational Behavior and Human Decision Processes, 50*(2), 248–287.

Bandura, A. (2005). Evolution of social cognitive theory. In K. G. Smith & M. A. Hitt (Eds.), *Great minds in management* (pp. 9–35). Oxford: Oxford University Press.

Bryant, F. B. (1989). A four-factor model of perceived control: Avoiding, coping, obtaining, and savoring. *Journal of Personality, 57*(4), 773–797.

Bryant, F. B. (2003). Savoring beliefs inventory (SBI): A scale for measuring beliefs about savoring. *Journal of Mental Health, 12*(2), 175–196.

Bryant, F. B., & Veroff, J. (2007). *Savoring: A new model of positive experience*. Mahwah: Lawrence Erlbaum Associates.

Cartier, P. (2011). Most valuable aspects of educational expectation of the students in design education. *Social and Behavioral Sciences, 15*, 2187–2191.

Chang, S. H., Wang, C. L., & Lee, J. C. (2015). The effects of savoring on individual creativity: The mediatory effects of creative self-efficacy and the moderated mediatory effects of perceived school support for creativity. *Chinese Journal of Science Education, 23*(4), 397–419. https://doi.org/10.6173/CJSE.2015.2304.04.

Chang, S. H., Wang, C. L., & Lee, J. C. (2016). Do award-winning experiences benefit students' creative self-efficacy and creativity? The moderated mediation effects of perceived school support for creativity. *Learning and Individual Differences, 51*, 291–298.

Clapp, E. P., & Jimenez, R. L. (2016). Implementing STEAM in maker-centered learning. *Psychology of Aesthetics, Creativity, and the Arts, 10*(4), 481.

Csikszentmihalyi, M. (1988). Society, culture, and person: A systems view of creativity. In R. J. Sternberg (Ed.), *The nature of creativity* (pp. 325–339). New York: Cambridge University Press.

Ford, C. M. (1996). A theory of individual creative action in multiple social domains. *The Academy of Management Review, 21*(4), 1112–1142.

Fredrickson, B. L. (2001). The role of positive emotions in positive psychology: The broaden-and-build theory of positive emotions. *American Psychologist, 56*(3), 218–226.

Gable, S. L., Reis, H. T., Impett, E. A., & Asher, E. R. (2004). What do you do when things go right? The intrapersonal and interpersonal benefits of sharing positive events. *Journal of Personality and Social Psychology, 87*(2), 228–245.

Greene, M. (1995). *Releasing the imagination: Essays on education, the arts, and social change.* San Francisco: Jossey-Bass.

Guyotte, K. W., Sochacka, N. W., Costantino, T. E., Kellam, N. N., & Walther, J. (2015). Collaborative creativity in STEAM: Narratives of art education students' experiences in transdisciplinary spaces. *International Journal of Education & the Arts, 16*(15), 1–39.

Hannula, M. S. (2002). Attitude towards mathematics: Emotions, expectations and values. *Educational Studies in Mathematics, 49*(1), 25–46.

Hannula, M. S. (2015). Emotions in problem solving. In S. J. Cho (Ed.), *Selected regular lectures from the 12th international congress on mathematical education* (pp. 269–288). Springer. https://doi.org/10.1007/978-3-319-17187-6_16.

Hayes, A. F. (2013). *An introduction to mediation, moderation, and conditional process analysis: A regression-based approach.* New York: Guilford Press.

Hurley, D. B., & Kwon, P. (2013). Savoring helps most when you have little: Interaction between savoring the moment and uplifts on positive affect and satisfaction with life. *Journal of Happiness Studies, 14*(4), 1261–1271.

Katz, S., & Stupel, M. (2015). Promoting creativity and self-efficacy of elementary students through a collaborative research task in mathematics: A case study. *Journal of Curriculum and Teaching, 4*(1), 68.

Lee, J. C., Wang, C. L., Yu, L. C., & Chang, S. H. (2016). The effects of perceived support for creativity on individual creativity of design-majored students: A multiple-mediation model of savoring. *Journal of Baltic Science Education, 15*(2), 232–245.

Livingstone, K. M., & Srivastava, S. (2012). Up-regulating positive emotions in everyday life: Strategies, individual differences, and associations with positive emotion and well-being. *Journal of Research in Personality, 46*(5), 504–516.

Lopez, S. J., & Snyder, C. R. (2009). *Oxford handbook of positive psychology.* New York: Oxford University Press.

Muller, D., Judd, C. M., & Yzerbyt, V. Y. (2005). When moderation is mediated and mediation is moderated. *Journal of Personality and Social Psychology, 89*(6), 852–863.

Preacher, K. J., & Hayes, A. F. (2008). Asymptotic and resampling strategies for assessing and comparing indirect effects in multiple mediator models. *Behavior Research Methods, 40*(3), 879–891.

Preacher, K. J., Rucker, D. D., & Hayes, A. F. (2007). Addressing moderated mediation hypotheses: Theory, methods, and prescriptions. *Multivariate Behavioral Research, 42*(1), 185–227.

Quoidbach, J., Mikolajczak, M., & Gross, J. J. (2015). Positive interventions: An emotion regulation perspective. *Psychological Bulletin, 141*, 655–693. https://doi.org/10.1037/a0038648.

Root-Bernstein, B., Siler, T., Brown, A., & Snelson, K. (2011). ArtScience: Integrative collaboration to create a sustainable future. *Leonardo, 44*(3), 192.

Seligman, M. E. P. (2002). Positive psychology, positive prevention, and positive therapy. In C. R. Snyder & S. J. Lopez (Eds.), *Handbook of positive psychology* (pp. 3–9). New York: Oxford University Press.

Seligman, M. E. P. (2011). *Flourish: A visionary new understanding of happiness and well-being.* New York: Simon and Schuster.

Seligman, M. E. P., Steen, T. A., Park, N., & Peterson, C. (2005). Positive psychology progress: Empirical validation of interventions. *American Psychologist, 60*(5), 410–421.

Seligman, M. E., Rashid, T., & Parks, A. C. (2006). Positive psychotherapy. *American Psychologist, 61*(8), 774–788.

Seligman, M. E. P., Ernst, R. M., Gillham, K. R., Reivich, K., & Linkins, M. (2009). Positive education: Positive psychology and classroom interventions. *Oxford Review of Education, 35*(3), 293–311.

Snyder, C. R., & Lopez, S. J. (2007). *Positive psychology: The scientific and practical exploration of human strengths.* Thousand Oaks: Sage.

Sochacka, N. W., Guyotte, K., & Walther, J. (2016). Learning together: A collaborative auto-ethnographic exploration of STEAM (STEM+ the arts) education. *Journal of Engineering Education, 105*(1), 15–42.
Sternberg, R. J. (Ed.). (2004). Handbook of creativity. New York: Cambridge University Press, pp. 3, 450.
Sternberg, R. J., & Williams, W. M. (2003). *Teaching for creativity: Two dozen Tips.* The center for development and learning. Retrieved from www.cdl.org/articles/Teaching-for-creativity-two-dozen-tips/
Tierney, P., & Farmer, S. M. (2002). Creative self-efficacy: Its potential antecedents and relationship to creative performance. *Academy of Management Journal, 45*(6), 1137–1148.
Tierney, P., Farmer, S. M., & Graen, G. B. (1999). An examination of leadership and employee creativity: The relevance of traits and relationships. *Personnel Psychology, 52*(3), 591–620.
Zimmerman, B. J. (2000). Attaining self-regulation: A social cognitive perspective. In M. Boekaerts, P. R. Pintrich, & M. Zeidner (Eds.), *Handbook of self-regulation* (pp. 13–41). San Diego: Academic.

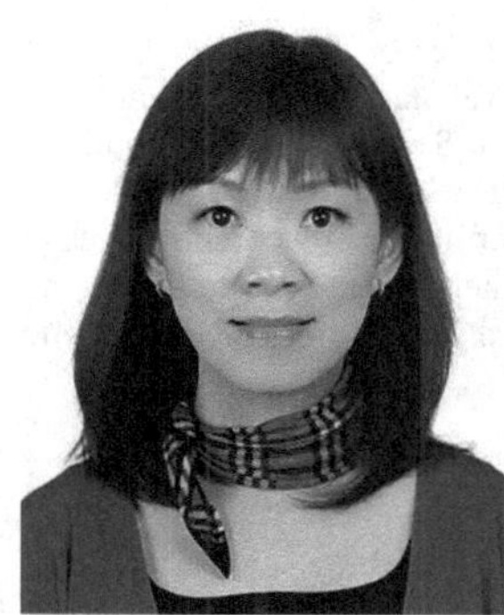

Dr. Shu-Hsuan Chang is currently a distinguished professor of the Department of Industrial Education and Technology as well as the dean of College of Technology, National Changhua University of Education (NCUE), Taiwan. Her research interests include the cultivation of creativity/imagination, innovation and entrepreneurships, and engineering & technological education.

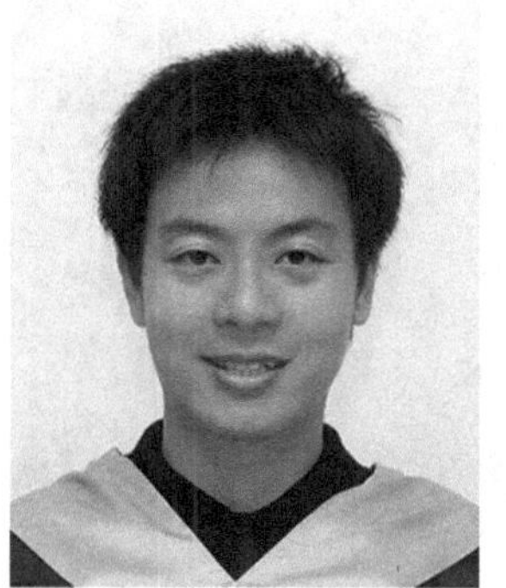

Dr. Li-Chih Yu is currently a project manager at IT division, Delta Electronics, Taiwan. His scope of work at Delta includes implementation of smart manufacturing systems and business intelligence in smart manufacturing.
Email: randy.lc.yu@gmail.com

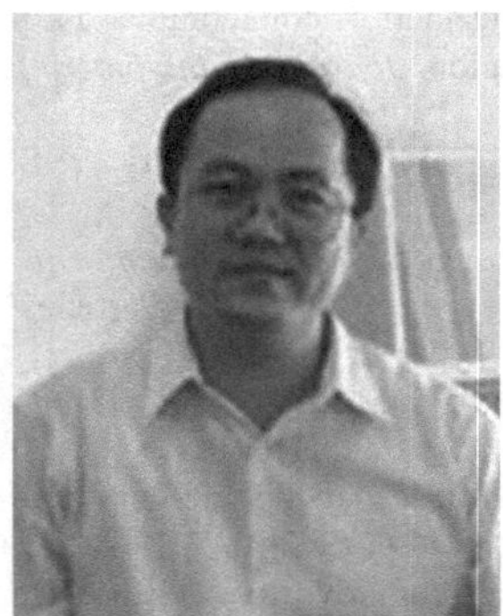

Jing-Chuan Lee is Professor of Graduate Institute of Educational Leadership and Evaluation in the Faculty of College of Humanities and Social Sciences at the Southern Taiwan University of Science and Technology. His research concerns school leadership, workplace stress management, workplace well-being, and data visualization in education.

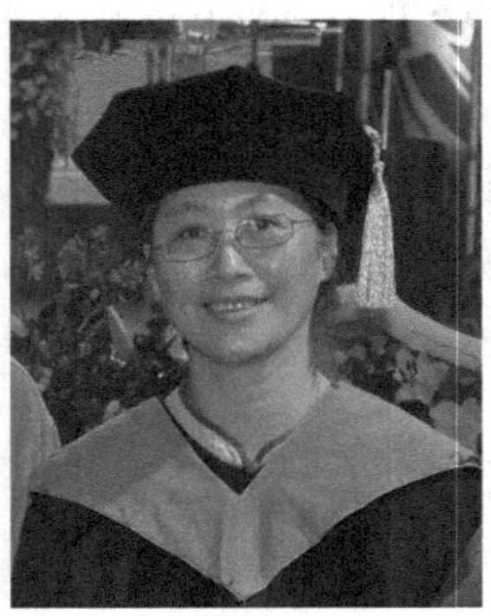

Chih-Lien Wang is an Assistant Professor of Graduate Institute of Educational Leadership and Evaluation, in the Faculty of College of Humanities and Social Sciences at the Southern Taiwan University of Science and Technology (STUST). Her research centers on the cultivation of liberal arts competencies in students, inspiring and awakening their innate desire to learn and their inherent potential to be creative/imaginative/innovative, and human resource development.

Science, Youth, and Integration: The Quest for Mindfulness Through Birding

Erin A. Ingle and Mike Mueller

1 Introduction

We start with a definition of what it means to be mindful in science education and then describe the unveiling of birding habitats across Alaska's public schools for cultivating mindfulness among youth and their teachers. Then we will tell the story of how the Youth Birding Alaska (YBA) project developed, and how it serves as an example of STEAM (that is, by explicitly recognizing how the visibility of the arts in science can enhance scientific inquiry). The story of this project will be told from the authors' conversations and perspectives of YBA.

YBA emerged from a small village in rural Alaska. Mike had been forging a relationship with the Lake and Peninsula School District for 3 years before, and that effort was rewarded when the district paid for his graduate students to fly to Newhalen or Naknek to participate in a biannual weeklong academic event organized around athletics. For this event, youth were flown from 12 villages to Newhalen, a village located within a Pennsylvania-sized region in Alaska. Newhalen has 200 residents and is located on the northern end of Lake Iliamna, near the Iliamna volcano. Naknek, a village on the southwestern corridor of Bristol Bay where the largest sockeye salmon migration in the world occurs, and its next-door neighbor, King Salmon, have a combined population of 850. This number swells to 10,000 in the summer due to the salmon migration. The event in Newhalen, dubbed AA (Academic and Athletic week), generally occurs in the spring, whereas the event in Naknek occurs in the fall. Up to 15 graduate students from the University of Alaska Anchorage (UAA) have the opportunity to work with village youth (more

E. A. Ingle · M. Mueller (✉)
University of Alaska Anchorage, Anchorage, AK, USA

© Springer Nature Switzerland AG 2019
A. J. Stewart et al. (eds.), *Converting STEM into STEAM Programs*,
Environmental Discourses in Science Education 5,
https://doi.org/10.1007/978-3-030-25101-7_9

than 50% of which are Alaska Native). Much insight about growing up in a village can be gained from listening to the youth. What might seem unusual to some readers, the perspectives can range from growing up in a tight-knit community, to craving the rare pizza that arrives home with someone heading back from Anchorage, to thinking nothing of the fact that a classmate commutes to school in a Super Cub airplane. At AA week, youth participate in school governance, a talent show, a science fair, speeches, and a prom (dance). Most importantly, these K-12 youth bond through athletics, such as Native Youth Olympics, which consist of various sports that mimic many of the skills needed by traditional hunters.

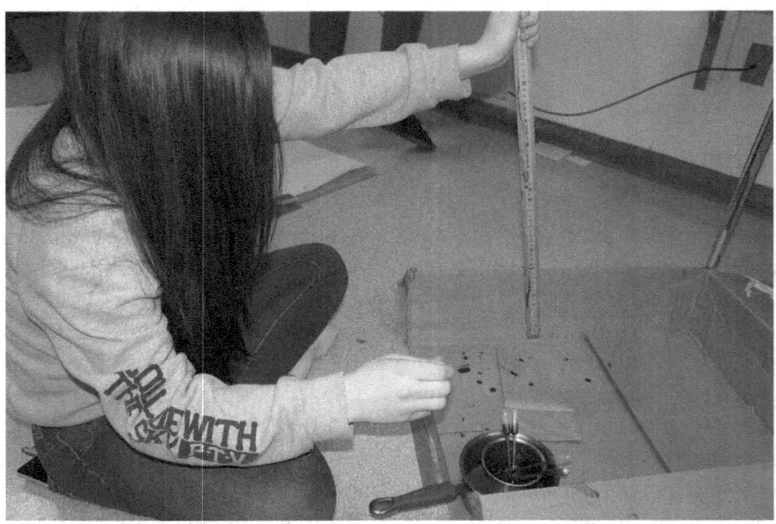

©2016 Mike Mueller, *Student participating in Blood Spatter Analysis Lab during the AA Week in Naknek*

The event involves about 120 youth and their teachers, crammed into one school in a rural village with 15 graduate students and a professor, and is exceedingly lively. Oftentimes, the sleeping bags in classrooms are pushed to the side of the room along with mats to make room for the learning experiences that have been prepared by UAA graduate students; these experiences focus on activities ranging from wildland survival skills to making cosmetics from the local plants and rocks. We also engage the students in STEM activities such as building bridges, egg-drop contraptions, and even a trebuchet. A nuance to this work involves showing students the value of the arts in STEM designs. This emphasis always becomes paramount to the success we have with these K-12 students.

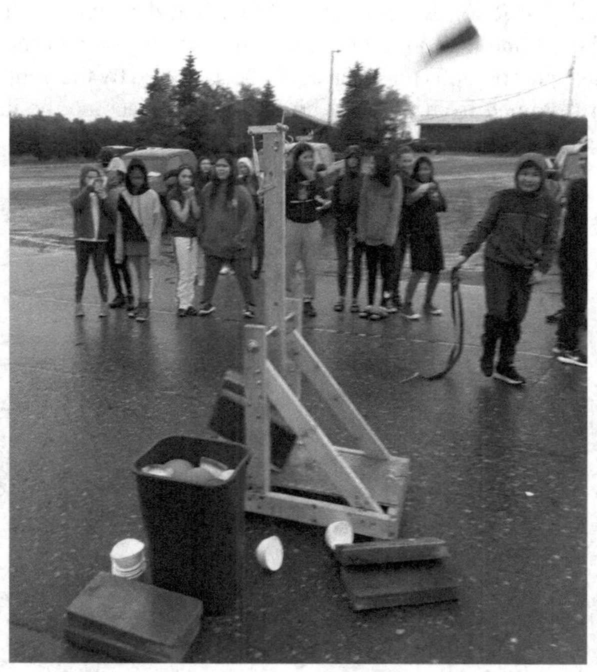

©2018 Mike Mueller, *Students observe a trebuchet in action! Naknek*

One spring, we decided to do a learning experience focused on birding in Alaska and building birding habitats at the Newhalen School. Students and their teachers showed a great deal of interest in this project, and we connected with several teachers at the Newhalen School that wanted to write a grant and build a birding habitat for their students. The birding habitat would help engage students in science inquiry, as they designed and conducted science investigations organized around teaching students about the nature of science. The grant was funded, and the first birding habitat was established at the Newhalen School in 2017. This habitat consisted of various types of bird feeders, bird seed, binoculars, a spotting scope, nest boxes, and a curriculum designed by UAA graduate students for engaging youth in the project. The first habitat took about a year to establish after a location had been selected. The following spring, UAA graduate students worked with the youth at Newhalen School to make observations of local birds visiting the habitat. Chickadees and redpolls then predominated. Youth made nature journal entries with elaborate drawings of the birds, including visible characteristics factors associated with nature journals—date, weather, temperature, and so forth. Newhalen youth also began observing a nearby Osprey nesting on a telephone post not far from the school, across the Newhalen River. Because the students and their teachers could only see the Osprey take off and land in the nest, they decided to use a drone to hover high above it and take photographs that then could be analyzed by the youth. The innovative ideas that stemmed from this first birding habitat become the impetus for the next phase of funding for others.

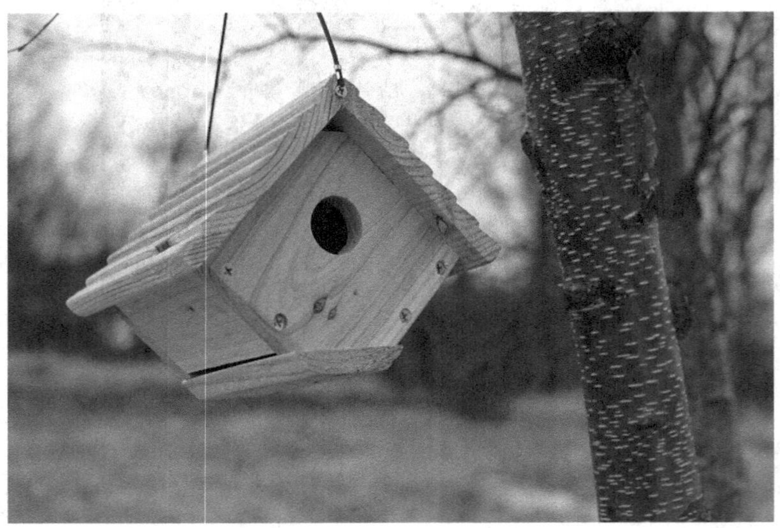

©2017 Mike Mueller, *Bird habitat at the Newhalen School on Lake Iliamna*

Before moving into the next stage of project development, we will first discuss how mindfulness is being used to frame the way we conceptualized and facilitated this work with the schools. In her work to develop an educational framework for cultivating mindfulness through citizen science, Erin uses singing. She uses singing as a way to metaphorically describe how engaging in science establishes a similar way of being that is paramount to the way we think about birds. Singing serves as a heuristic, where the "thing-you're-doing" cultivates the mindfulness that fosters the activity of engaging with the phenomena itself. In the same way, bird-watching is a heuristic (through photographic and arts-based data forms) that fosters the mindfulness around an understanding of the behaviors of birds, bird migrations, phenology, nesting, and so forth. As another example, in her book on citizen science, Caren Cooper (2018) notes how mindfulness creates positive behaviors through education:

> Advocating for dark skies is a battle against the human primordial fear of the dark. Our childhood cry for nightlights grows into demands for street lighting that functions to reduce feelings of fear—particularly fear for personal safety among women. Studies bear out of the idea that crime is reduced by street lighting, but not as one would expect. Lighting is a placebo that everyone, even perpetrators, swallow. Lights don't deter crime by increasing surveillance ability; when lighting is present, nighttime and daytime crime is reduced.

> Researchers speculate that lighting strengthens social pride, confidence, community cohesion, and social control of neighborhoods, and these social factors operate to reduce crime. (pp. 188–189)

In a similar vein, Mike has advocated using photographs of the northern lights to reduce light pollution within neighborhoods. When people in the community begin enjoying photographs of the aurora, and begin aspiring to see the lights themselves, or take photographs of them, they turn their house lights off during forecasted aurora events or when the lights are said to be out. With time, the entire neighborhoods can get darker as people feel more confident with the darker skies. In short, crime declines when people in the community agree to more social solidarity; floodlights, or other lights, decline in value as crime deterrents, as mentioned in Caren Cooper's example, above. In other words, the photographs serve as a way (or heuristic) for cultivating "social pride, confidence, community cohesion, and social control of neighborhoods," because people go out to look for the lights more frequently. These things can be described in terms of *mindfulness* because of the heightened awareness. Mindfulness then can be said to increase feelings of pride, confidence, cohesion, and control.

Singing, like photography, is an arts-based practice that can be used for engaging people in mindfulness. The arts are not separate from sciences, but rather are imperative to producing the most robust theories and laws in science. For example, Rachel Carson (2000) used literature to tell the story of her specimens and the natural history of ecosystems in her scientific work. There is an explicit acknowledgment of the arts through STEAM, whereas STEM tends to deemphasize or ignore the very thing (heuristics and more) that we are claiming vibrantly influences mindfulness. And mindfulness, as we note above, can lead to increased pride, confidence, cohesion, and control of our behaviors toward phenomena in the world. And here is where we take our ideas back to birds.

Effectively, heuristics and mindfulness, as revealed in our bird-habitats project, cultivate a scientifically literate populace that is more in tune with the changes of the world and that acknowledges the importance of engaging with these changes. Such is the nature of democracy, when youth and their teachers have unequivocal access to, and participate in, the local processes in decision-making in the local community. In our Newhalen School build-a-bird-habitat project, we began to see the diffuse but important benefits of building a network of schools to participate in the project, premised on the emerging ideas of citizen science as a heuristic for mindfulness. Thus, project, phase two.

The second phase of birding habitat construction led to another funded grant that provided supplies and equipment for Homestead Elementary School, some 30 min north of Anchorage, in Eagle River, Alaska. Homestead Elementary School has a history of engaging students in science even with an already overwhelmed curriculum focused on reading and mathematics. As we worked with the elementary teachers to think about how to also include science while teaching reading and mathematics, it became even more apparent that we could not separate the arts from sciences. Thus, a STEAM curricular approach would be necessary if we were going to move into the next stage of building birding habitats across Alaska's public schools. Why more than one habitat? Simply because we wanted students and their teachers to collaborate with students and teachers at other schools and share and communicate their findings, like scientists do professionally. Because birds migrate from all over the world to Alaska primarily for summer breeding, and since Alaska is expected to change faster than most other places on Earth, under various climate-change scenarios, we foresaw value in building a network of schools across the state that could monitor birding phenology. Phenology is the study of cyclical and seasonal natural phenomena in relation to climate. One of the most successful citizen science projects organized around phenology is the National Phenology Network (see https://www.usanpn.org/usa-national-phenology-network). We envisioned the Youth Birding Alaska project could "take the pulse" on changing bird migrations and behaviors in relation to climate changes on a statewide scale. Further, since Alaska is about 1/5 the size of the entire continental United States, Alaska alone could provide a wide-sweeping data collection opportunity for analyzing bird migrations and behaviors that could be of interest to scientists worldwide. We started to ask some simple but very important questions: Could teachers and their students become part of a meaningful project for measuring climate-change effects around the world? Could arts-based data and photography stimulate mindfulness in gauging Alaska's phenology? What types of partnerships and networks would be needed to establish this STEAM-based endeavor?

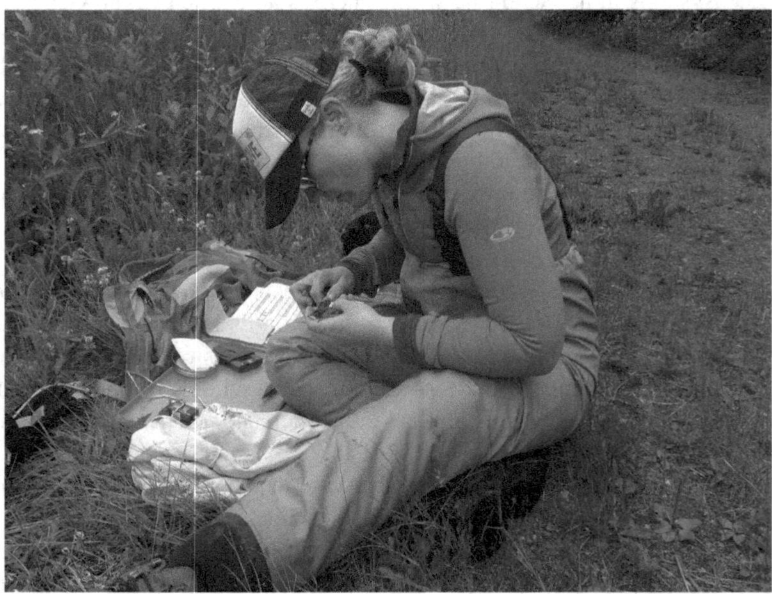

©2017 Erin Ingle, *Erin working with Tree Swallow chicks for the Alaska Swallow Monitoring Network Project*

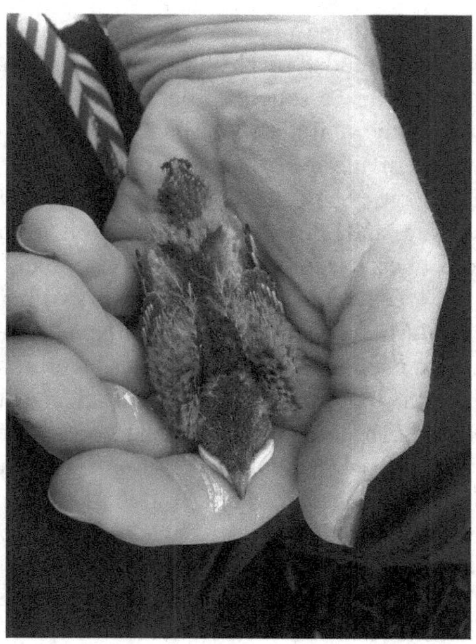

©2017 Erin Ingle, *Tree Swallow chick*

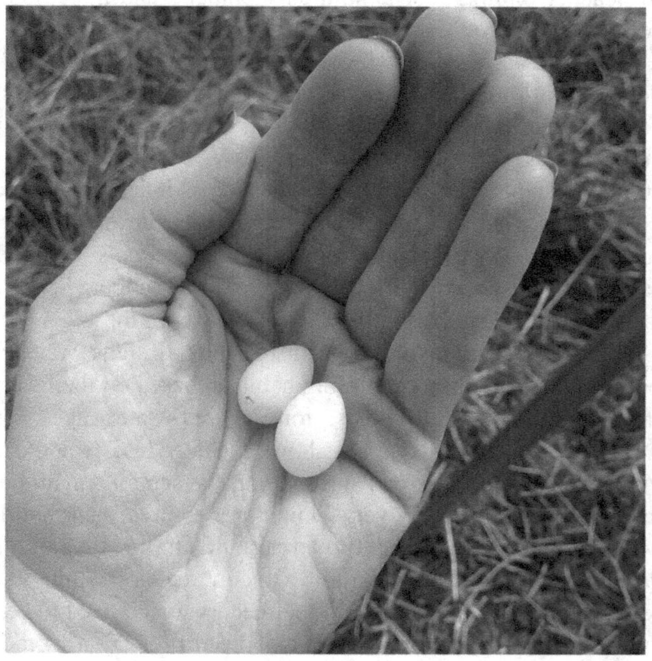

©2017 Erin Ingle, *Tree Swallow eggs*

The next stage in developing the YBA project involved a Resilient Communities Initiative. This initiative was funded by UAA's Center for Community Engagement and Learning, and through it, we sought to demonstrate the impact that the bird habitats would have on youth's ability to contribute to decision-making processes in relation to the resilience of the environment. At this stage, Erin and Mike co-conceptualized the next stage of YBA development: they sought to establish a social media presence, a website, and added four more schools to the network. Erin co-directed the YBA project because of her interdisciplinary thesis work and scientific background in studying Tree Swallow nesting behaviors with a wildlife biologist at UAA. Together, Erin and Mike conceived of focusing the YBA project on monitoring local birds as part of a phenology network in Alaska, orbiting around the educational standards of scientific inquiry and the nature of science in public schools. Our larger perspective focused conceptually on fostering heuristics and mindfulness. We also discovered that working with schools in the summer months was almost impossible: we had to wait until the academic year started again in the fall semester. So, during the summer, we created a Facebook page and website to align with our partnering schools. We also convened a meeting of community members to help craft the vision for YBA: this meeting included principals, teachers, university faculty, university graduate students, and educators from the Anchorage Museum and scientists from the National Park Service regional office in Alaska. The meeting participants decided the vision of YBA should focus on bird-related citizen science, with a mission of teacher training, curriculum development, and increasing awareness and science identity for youth. The science identity connected with nature and incorporated multiple ways of knowing (e.g., the arts and Indigenous knowledge) and phenology monitoring. Learning events were associated with habitat monitoring, related experiences, information sharing, and biodiversity tracking and observations.

Beginning in September 2018, the importance of maintaining and creating partnerships became paramount to YBA's success. Erin began developing the infrastructure for cultivating positive relationships with the schools by meetings with principals and teacher sponsors at each collaborating school. She worked with Lauren Casey, a graduate student in the Master of Arts in the UAA Teaching program, to foster these connections and coordinate habitat designs. We knew that each of the schools would have different birding habitats, based on the location and the campus structure in relation to fields, forests, and waterways. For example, Homestead Elementary School discussed above is located in the middle of a wooded area at the edges of the concrete pad surrounding the school's main building: this location lacked trees and had only very limited grassy areas near buildings. Thus, we knew we would have to work hard to attract birds to this type of a habitat. We put three posts with feeders and classroom window feeders within 30 feet of one another. We knew, too, that posts holding the feeders would need to be robust 4×4 treated lumber and that these posts would also need protection from hungry and ever-clever red squirrels. So, we worked with Lauren's husband to design a 4×4 post surrounded with metal stove-pipe tubing to protect the seed from squirrels. This design proved to be both ornamental and functional. For another school, like the

Machetanz Elementary School (located on the Palmer Hay Flats), the feeders fit right in with the school's mission of providing arts-based STEM activities for youth; at this school, we embedded the feeder posts among raised garden beds, weather station, and a chicken coop containing six chickens.

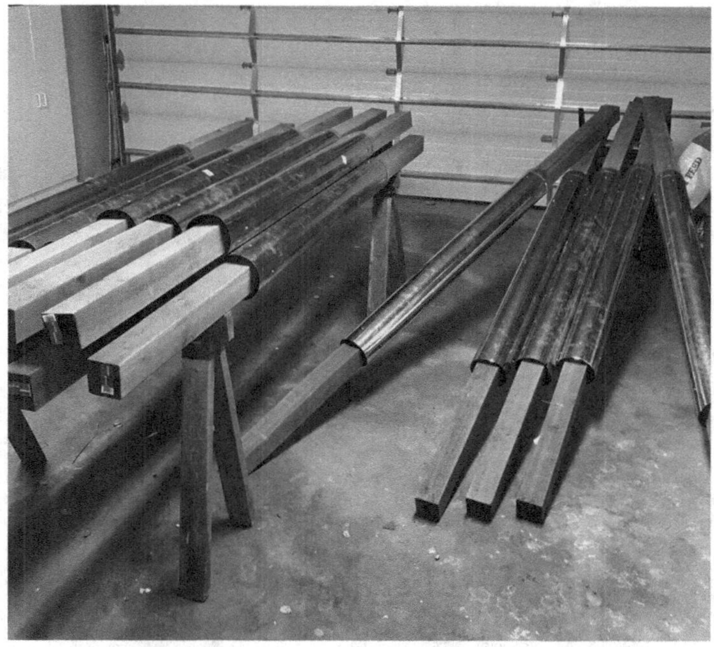

©2018 Lauren Casey, *Posts designed with a special squirrel guard before delivering them to the schools*

In Alaska, the Alaska Department of Fish and Game regulates a suggested timeline for putting out bird feeders, because of the possibility of bears eating the seed and becoming a nuisance and danger for local communities. The dates for putting out feeders used to be October 15; the time to take them down was April 15. More recently, however, these "open" and "close" dates have been moved to November 1 and March 1, respectively. As we began putting posts in the ground at schools and attaching hardware to the posts, preparing them for bird feeders, we had several situations in which the bears simply did not move into the mountains in response to warm fall weather throughout October. Finally, it snowed on October 31st. This weather-change brought some assurance that we might be able to put up feeders. The fact that bears would sometimes enter the schoolyard and eat the seed, and thus could be a possible danger to children, had us thinking hard about how not to lose the project entirely in response to fear of hungry bears. Some people in the community even noted how YBA might be creating a "bear baiting" situation, where bears would learn that the school grounds served as a place to get food before hibernation. Just after the first fully functioning birding habitat was established, on

November 1, a black bear climbed into a dumpster at the nearby middle school (less than a mile away): Alaska Department of Fish and Game officials were called, and scared the bear off with rubber bullets. We crossed our fingers, hoping that the completed birding habitat at Homestead would not complicate the situation.

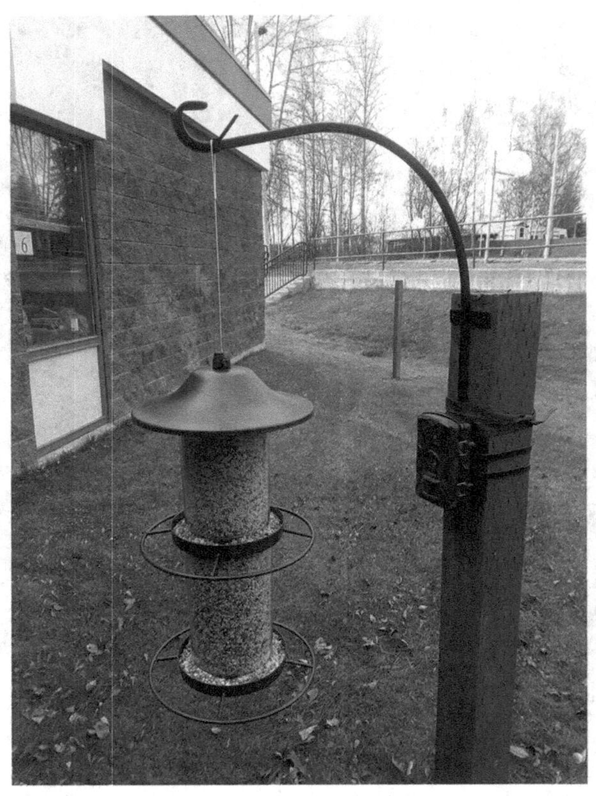

©2018 Mike Mueller, *Feeder and game camera setup at Homestead Bird Habitat*

After birding habitats had been established at the Homestead Elementary and Machetanz Elementary locations, we put fully functioning birding habitats at six other schools. Each of the birding habitats has our specially designed posts with 7-lb feeders, window feeders on the classrooms, and a game camera. In some cases, the birding habitats are positioned near tracts of forested land. We anticipate that the game cameras may capture images of more than birds! We look forward to learning how the teachers and their students end up using the game camera data and we are providing teacher workshops to help develop the process of uploading the photographs and other information. In addition to the data provided by the game cameras, we are providing arts-based methods for engaging teachers and their students in making good observations. Nature-journaling techniques, in addition to drawing skills, are used to help younger students develop skills in identifying birds visiting their feeders. Often, the skills associated with drawing and the arts are most memorable for youth.

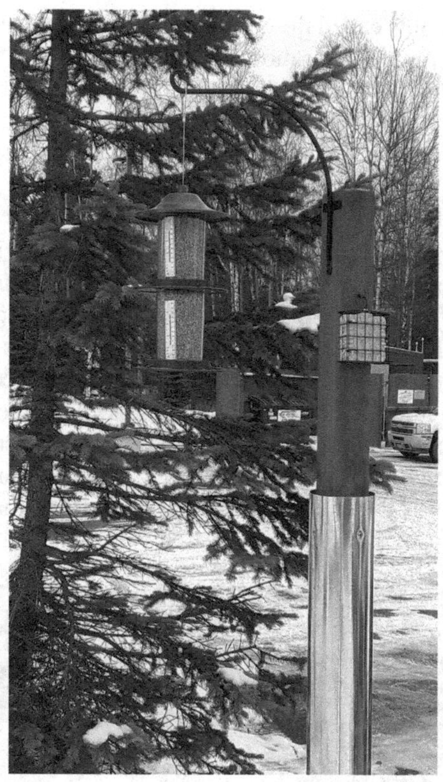

©2018 Mike Mueller, *Two types of feeders at a Birding Habitat*

We've also encountered several challenges with the YBA project. For example, we had to delay installing a habitat at one school, where the teachers have not yet committed to sponsoring full participation in the project, even though the principal and one teacher are interested in having feeders at the school for their students. Without teachers committed to doing the YBA, the project could essentially be stalled due to the heavy emphasis on non-science standards. Most teachers through 6th grade have limited time (often, no more than 30 min per week) for teaching science. The de-emphasis of science in the elementary grades could stall the project or prevent it altogether. Mike recognized this problem when we began working with the schools, so he established a time for UAA graduate students involved in a place-based education course to engage the students and encourage excitement with the teachers over the fall semester. Mike assigned two to three graduate students to each school, who helped teachers who sponsor the YBA project, which stimulated interest in the project. We recognized, too, the importance of providing continuous educational opportunities for the teachers.

STEM to STEAM is an advanced way of engaging youth in science and democratic participation more fully in their local community. One school we collaborated with for YBA was founded about a year ago, and is based entirely on this STEAM philosophy. This school is called the STrEaM Academy (r = research), and it offers a project-based, experiential education integrated with STEM for grades 6–8. We also collaborated with a K-12 school, Polaris, where the philosophy is similar but more focused on place-based education and interdisciplinary learning. One of the biggest issues at Polaris was getting the bird-habitat posts in the ground before the first big snow of the winter season. At the STrEaM Academy and at Polaris K-12, we don't know how the posts got into the ground—but the teachers made it happen, somehow, because the district refused to do the work until later in the spring. Finally, the last big challenge is that of having the birds appear. In some locations, such as Polaris, birds were practically landing on us as we put up the hardware and installed the feeders. But at other schools, where the feeders are not near shrubbery or trees, the jury is still out on how long it may take birds to find the feeders. At Homestead Elementary school, the posts and feeders were placed on the shadow side of the main building. So, these feeders may not receive any direct sunlight for the entire winter, because the sun is so low on the horizon in Alaska during the winter months. We are eager to learn if the teachers and their students at that school can design some sort of gadget to keep the feeders ice-free.

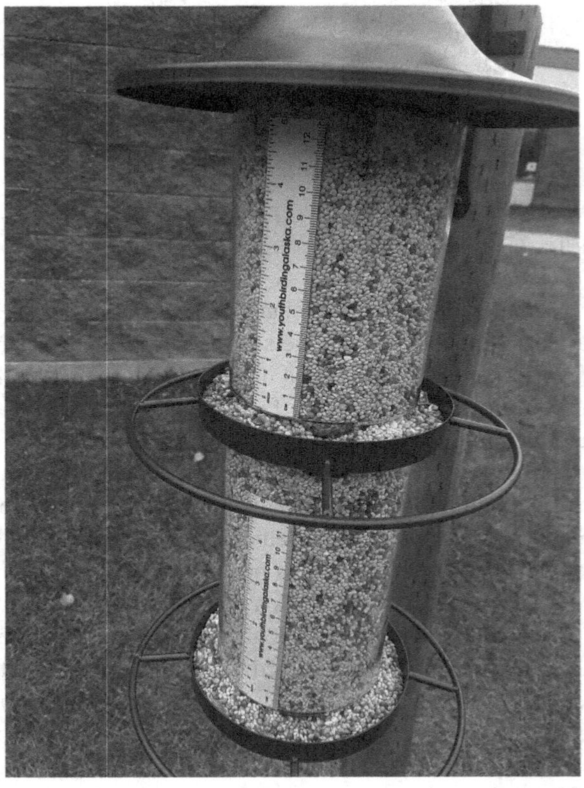

©2018 Mike Mueller, *These labels help students to ID birds by scale and the scale is also captured by the game camera for later analysis*

At the STrEaM Academy, students investigate real-world phenomena emphasizing research and the arts as part of science, technology, engineering, and mathematics. This school is small and meets in the basement of a Baptist university while their new building is under construction. STrEaM Academy students have access to teachers who are specifically interested in engaging them in the environment surrounding the school for all their classes. Teachers purposely select to work at the STrEaM Academy for its projects-based approach to learning, and we anticipate some interesting investigations by its students. Each school has a unique landscape for engaging students in the birding habitats. For example, Machetanz Elementary is in a very windy area of the Matanuska/Susitna Valley, and teachers there anticipate investigating which types of seed are more resilient than others, with respect to staying in the feeders and while still attracting birds. Because of their proximity to the Palmer Hay Flats, teachers and students at this school also see migratory birds, such as Sandhill Cranes and Snow Geese, that forage on the school's ball field. Another example of what students may observe (depending on their location) is at the Newhalen School. Students there have opportunities to investigate the behaviors of Osprey nesting nearby, because they are at a remote village location near the largest lake in Alaska. Where we originally thought different locations made it difficult to have similar protocol for investigating birds, we discovered, instead, that the different locations work to the benefit of the project as a whole. In this way, we count on finding locations across the state where the locations themselves are what make the difference. At the same time, some bird species, such as the red-breasted nuthatch and the black-capped chickadee, are more cosmopolitan: these species visit feeders at virtually every location, and since they don't migrate, we expect them to visit often. One challenge at schools in remote villages across Alaska is that we have to rely on distance technologies for communicating with the teachers and their students, and we have relatively few opportunities to send graduate students out to invigorate students and teachers at these schools about the project. Also, we don't know the extent to which they are using the equipment and actually engaging students in birding. However, we fortunately found a retired science teacher who has agreed to travel to school sites and help set up equipment and work with the schools when we get that grant funded.

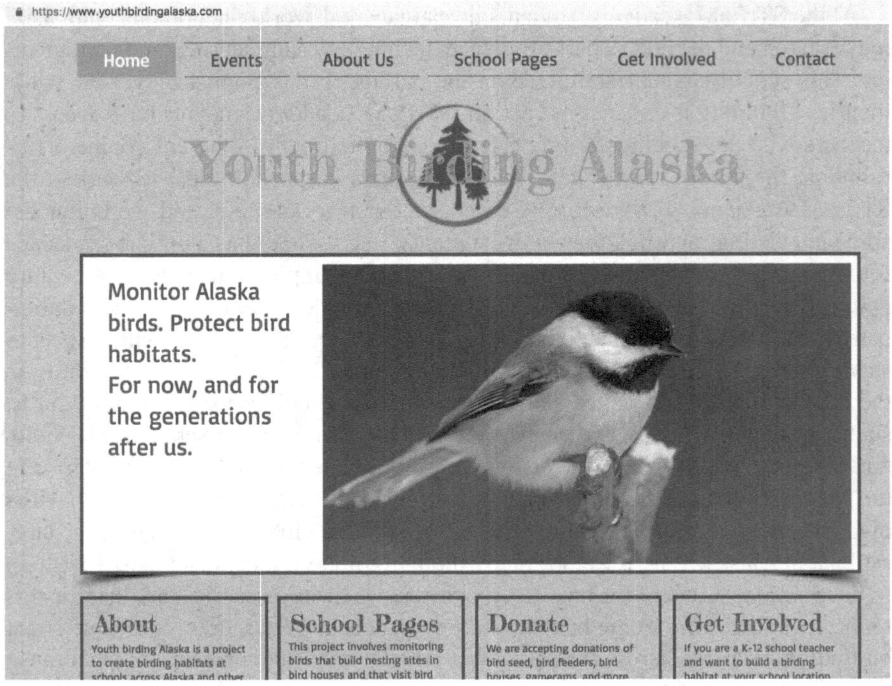

©2018 Mike Mueller, *Youth Birding Alaska website*

Some schools we're partnering with, such as Romig Middle School, offer the opportunity for an arts focus, specifically music education. In this case, the students have a chance to go on to the nationally competitive high school choir or orchestra. While we hope that all the schools in our network will embrace and cultivate the arts through science inquiry, the middle school model used at Romig offers very high-quality opportunities for this outcome. Since we weren't able to get posts in the ground at this school, we had to resort to hanging feeders from the trees and shrubbery next to the classrooms. Fortunately, Romig had more bird habitats already in place than we first anticipated, because it is nearest to the downtown area.

Bartlett High School is another school we are working with. Here, too, we were unable to get posts in the ground before the first heavy snowfall of the winter season. Further, Bartlett is surrounded by trees so that at this location, we may not need posts in the ground at all. In short, the habitats differ considerably from one location to another, and the equipment differs to some degree, as well, so the birds will differ as well. The larger purpose of engaging kids in birding is really at the heart of the project, and that is how we will finish this chapter.

With rapid climate changes occurring already in Alaska, the future of the region is not well known. Last year, for the first time ever, the Alaska Department of Fish and

Game prohibited all salmon fishing on the Kenai River, a popular destination for tourists and one of the most intensively fished rivers in the state. Other areas also received emergency orders that limited fishing. The problem was that the escapement goals for salmon were frighteningly low. In other words, very few salmon were migrating up the rivers, for the first time in a long time. Climate changes were thought to be responsible. Warmer ocean temperatures that decreased salmon food supply were said to have impacted the number of salmon actually migrating up the rivers. When escapement goals are very low, the Alaska Department of Fish and Game protects the viability of future stocks by limiting salmon fishing. In the same way, Alaska is experiencing warmer temperatures during the fall and winter months. Recall how the bear inadvertently became part of our project, in the story relayed earlier. This, too, could be a climate-change situation. Warmer temperatures are responsible for late snow accumulations and encourage precipitation in the form of rain, when normally it would be snowing. Now, it is not uncommon for the temperatures to be in the mid-40s and mid-50s during December and January. With these warmer winter temperatures, spring arrives earlier and insects, such as mosquitos, arrive on the scene much earlier than they did before. Birds will likely follow. And, we anticipate observing phenological mismatches, where insects arrive but birds have not yet appeared, or birds arrive but insects are depleted or have not emerged at all. The significance is that birds can help, as one of several factors, tell the story of climate changes, and that teachers and their students can be at the forefront of this effort, which may lead to some interesting discoveries that matter to policy. By involving youth this way, we expect the students to be better enabled to participate in local decisions. They also will develop an understanding of the baseline for the ecosystem they are a part of, which that makes all the difference when discussing whether something has been degraded or vulnerable.

Most importantly, the heuristic that is the birding itself, through photographs and arts-based methods of collecting scientific data, leads students to become more aware of their local environment and become astute about what is happening in the places where they live. For us, this mindfulness is the most significant part of the YBA project, and this part carries much of the scientific literacy, functionality, and appreciation for STEAM. Birding data in the form of photographs collected digitally through a camera or phone, and arts-based drawings, paintings, and so forth will help lead students to mindfulness. The heuristic and mindfulness may lead students to careers in STEM or even in the arts. With declining oil prices and revenues for the state of Alaska, ecotourism may become part of the next big wave of the economic base for Alaska. Students who are prepared to understand science inquiry, research communication, decision-making, and we hope, talking to youth in other places around the world about their birds, will be the innovators of this new economy. As we continue to build the YBA network and add more schools to the project, we hope to include science kits that teachers can use to teach their students about birds, and we would like to organize conferences that bring together student teams to present their work at a type of youth summit. We also hope to offer more exciting teacher workshops and conferences as the project grows, and bring the teachers together to help celebrate the ways they are engaging youth in birding. We hope these things when taken together will positively influence social pride, confidence, community cohesion, and social control of the places where project participants live, and ultimately, make youth birding a necessary endeavor.

Acknowledgments The authors thank the Center for Community Engagement and Learning and Northern Journeys for providing the funding for supplies and student assistants to build the birding habitats described in our chapter.

References

Carson, R. (2000). *Silent spring*. New York: Penguin.
Cooper, C. (2018). *Citizen science – How ordinary people are changing the face of discovery*. New York: Overlook.

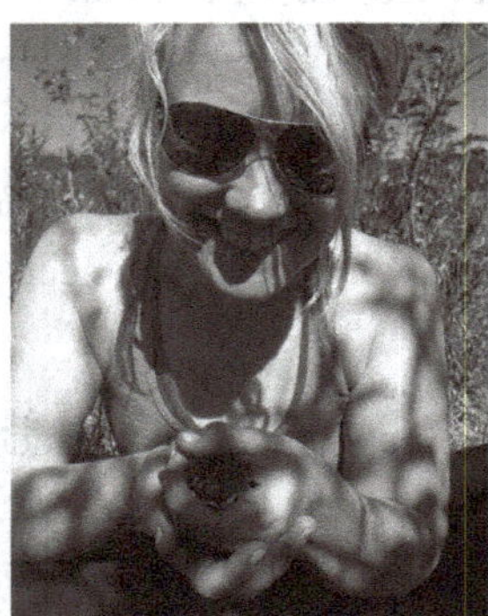

Erin Alexander Ingle is a graduate student working on Masters degrees in teaching and biology focused on citizen science. She co-directs the Youth Birding Alaska project. Her goal is to inspire and facilitate curiosity in the local environment with children.

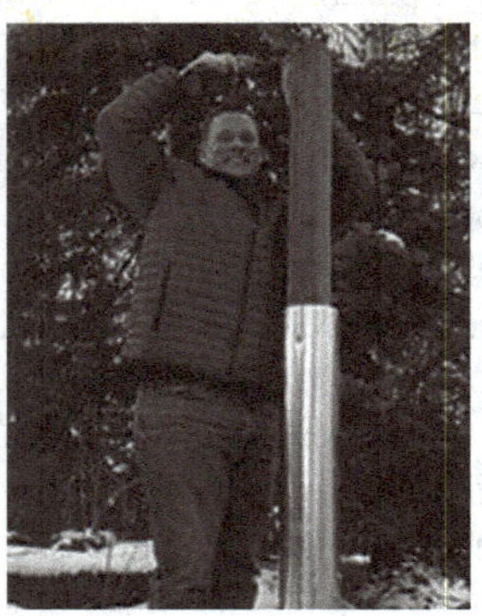

Mike Mueller is a Professor of Secondary Education (PhD) with expertise in environmental and science education at the University of Alaska Anchorage. He is the director of Youth Birding Alaska and fosters the value of biodiversity, cultural diversity, and nature with teachers.

The Role of STEAM in a Sustainable World

Shushman Choudhury, Sohn Cook, and Brittany Bennett

Professor Nichols enters the lecture hall for her Construction Engineering class. Today's discussion is on bridges. She describes a hypothetical city and mentions several potential locations. Then she asks the students for their opinions on the most important and relevant questions to ask in the design phase. Some queries involve expected technical issues—the nature of the soil, the anticipated amount of usage, the anticipated budget, and so on. However, she also gets several questions about the impact of the bridge on the surrounding flora and fauna, the opinions of people in the surrounding communities, and the long-term viability of the bridge.

1 Introduction

When we think of STEAM education programs, the idea of sustainability as an integral component often does not come to mind. We tend to view sustainability as a niche of environmental or energy engineering, or as an optional component of other engineering disciplines. However, the current and projected future state of climate and worldwide resources is such that sustainability needs to become an intrinsic part of an engineering education. Therefore, we need to incorporate sustainability-related ideas and concerns from the ground up, while planning and

S. Choudhury (✉)
Stanford University, Stanford, CA, USA
e-mail: schoudhury@eswusa.org

S. Cook
Engineers for a Sustainable World, Orange County, CA, USA

B. Bennett
Engineers for a Sustainable World, Denver, CO, USA

© Springer Nature Switzerland AG 2019

A. J. Stewart et al. (eds.), *Converting STEM into STEAM Programs*,
Environmental Discourses in Science Education 5,
https://doi.org/10.1007/978-3-030-25101-7_10

designing small- and large-scale projects and initiatives. Otherwise, we have little hope that humanity can persist in our increasingly fragile ecosystem.

At Engineers for a Sustainable World (ESW), we believe in building better engineers for a better, sustainable world. In this chapter, we show how incorporating the arts into an engineering approach to solving sustainability challenges leads to better solutions. The journey to a sustainable world must run on STEAM, and we outline how we foresee this can happen. To illustrate our values and principles in practice, we describe two projects that ESW has undertaken—a vertical garden at UC Berkeley and the Undocu-Wall at CSU Long Beach.

2 Sustainability, in Principle and Practice

The term "sustainability" is often tossed around in discussions and is understood more by lamenting its absence than by striving for its presence. Several definitions have been used since the term became popular, but the one with the broadest consensus is due to the 1987 Bruntland Report (World Commission 1987) from the UN World Commission on Environment and Development—*the ability of humanity to meet the needs of the present without compromising the ability of future generations to meet their own needs.*

The Bruntland report itself has elaborate details on motivating the need for sustainability, and the principles of sustainable development. More than the definition, however, it is the spirit behind the terminology that is important. The basic principle of sustainability is to evaluate the effects of our potential choices—on the environment, our natural resources, the local and global economy, society, and other large systems which we are a part of—and focus on choices that mitigate long-term harmful effects while still fulfilling current requirements.

Sustainability is a multi-faceted issue, occurring at multiple scales in several contexts of human civilization. It is widely considered to be one of the so-called wicked problems we must deal with as a species. Clearly, sustainability is far too complex and extensive to capture in a single chapter. Further, the impact that any individual can have on sustainability varies greatly depending on their personal resources and workplace. Here, we direct our focus to a more ubiquitous and familiar topic relevant to emerging professionals—namely, the role of education in inculcating the values and principles of sustainability in tomorrow's "movers and shakers," particularly engineers.

ESW is a US-based 501(c)3 nonprofit organization that empowers technically minded individuals to solve sustainability challenges in their communities. Our members include more than 50 collegiate chapters and several professional chapters across the United States and Canada. We are driven by a vision of a world of environmental, social, and economic prosperity created and sustained by local and global collective actions. Our mission is to help forge innovative, lasting solutions to local and global sustainability challenges by designing and implementing sustainability projects, educating and training individuals and organizations on sustainability policies and practices, and building a network of communities with a shared culture of sustainability.

We began as Engineers Without Frontiers in 2002 with the aim of promoting international sustainable development. The organization was restarted and incorporated in 2013 under the name Engineers for a Sustainable World to fill a gap between engineering education and sustainable thinking, and help build better engineers to build a better world. Higher education institutions have prepared students to tackle the technical aspects of real-world grand engineering challenges, but perhaps not so thoroughly the broader sustainability and longevity issues.

We advocate for education that works through hands-on projects and initiatives. Our members work in several areas, from solar irrigation pumps to community greenhouses to low-cost wind turbines. Since 2014, ESW members have completed over 250 projects dealing with topics from solar energy to food waste.

While we at ESW are ambitious in our goals and optimistic about the future, we realize the importance of practicality, because the idea of sustainability often tends to conjure up (misguided) notions of severe austerity or a minimalist lifestyle. Therefore, we focus on better equipping budding engineers with the tools, resources, and networks to design and create products and initiatives where sustainability is a key feature, not just an afterthought.

One way to achieve these objectives is through our chapter projects (two of which we will speak about later), where our members work together on something of tangible value toward the cause of sustainability. We are motivated not just by the environmental impact of the projects, but also by our members getting the opportunity and support to experience sustainability in practice, while getting a feeling of ownership about the project they have worked on. Their experience becomes much more influential this way. From working on ESW projects, our members are able to apply what they have learned to their actual professional careers. The connections that they make help create a sense of community, which allows them to overcome the skepticism that many other sustainability efforts engender. They all contribute in some way to ESW's vision for *a sustainable world*.

3   A Sustainable World

The phrase "Sustainable World" is a part of ESW's name, but it also has a broad definition. This phrase is inherent with certain embedded conflicts of interest. In relation to resources, the world we inhabit is materially a closed system (c.f. Bejan 2016 for an engaging discussion). Of course, technology and human innovation can transform raw materials into usable products, or create new materials, but these things in turn require energy and equipment. Therefore, it is not immediately clear that a sustainable world can allow for continuous economic growth over all strata of society.

This dilemma is captured vividly by the journalist, Thomas L. Friedman (2009), in his book *Hot, Flat and Crowded*, where he makes the case that three of the most influential and important factors for human civilization on Earth are global warming and climate change ("Hot"), extensive connectivity across the world thanks to the Internet ("Flat"), and rapid population growth ("Crowded"). Friedman explains how, with increasing globalization and consumerism, billions of people are aspiring to the quality of life representative of the middle class in the United States. Historically, people around the world have been deprived of the right to such a quality of life and the fair access to goods and services. However, on a planet with dwindling resources and a burgeoning population, we must balance this desire for a continuous and growing economic output with the ability of future generations to survive and provide for themselves.

This is where the importance of *envisioning* a sustainable world becomes clear. We need a shared conception of what a sustainable world looks like, at local, regional, national, and international levels. We also need a generally agreed-upon list of sustainability principles that we expect from our corporations, and hold them accountable for manifesting them. Those of us who care should engage in self-reflection, to help ourselves identify the aspects of our world in which we can embrace sustainability, whether they are social, cultural, economic, or technological.

We cannot simply jump into implementing individual initiatives that seem to be the most appropriate for the immediate future, even if those initiatives are driven by improving sustainability. Without a robust, well-articulated vision, we risk losing sight of the larger goal of a sustainable world; we risk overlooking the important questions we should ask about the effects of our decisions beyond immediate stakeholders; and we risk otherwise well-intentioned ideas becoming ineffective, or worse still, counterproductive. A common example is the poor handling of the displacement of local inhabitants near a proposed hydroelectricity project (Mathur 2006). A shared vision of a sustainable world would include the voices and concerns of the community, and an articulated plan to address displacement, allowing them to maintain their quality of life. As a byproduct, it becomes easier for a community to acknowledge and practice sustainability when they are not forcefully displaced.

A seminal thinker on the issue of sustainability was Donella Meadows, lead author of *The Limits to Growth* (Meadows et al. 1972) and co-founder of the Balaton Group. In her talk at the 1994 meeting of the International Society for Ecological Economics in San José, Costa Rica, she made an eloquent and convincing case for the importance of envisioning a sustainable world. According to her speech, an "alternative, sustainable world is, of course, where resource regeneration is at least as great as resource depletion. It is a world where emissions are no greater than the ability of the planet to absorb and process those emissions. Of course it is a world where the population is stable or maybe even decreasing; where prices internalize all costs; a place where no one is hungry or desperately poor; a place where there is true enduring democracy."

Any vision for a sustainable world is influenced greatly by the individual or organization. For us at ESW, this vision primarily concerns the work of like-minded engineers. In a sustainable world, engineers are like the students of Professor Nichols' class. They treat environmental impact and longevity not as an afterthought, but as a core value. When they make decisions relating to their projects, they care not only about the technical aspects but also those relating to the local community and the society at large. When they evaluate the success of their endeavor, it is based not just on the profit made or the targets met, but also on the resources saved.

A significant change that we want for engineers in a sustainable world must take place at the root level of their education. Sustainability concepts do not and cannot exist in a vacuum. Educators should be responsible for teaching not only the foundations of their discipline, but also real-word impacts and events in the classroom. We suggest how a well-rounded STEAM education program can help shape the future of engineering in a sustainable world.

4 STEAM Education for a Sustainable World

We will not dwell too much on the general benefits of STEAM over STEM. Much has been said on the topic already by STEAM advocates (Boy 2013; Jolly 2014; Feldman 2015). One of the most emphatic arguments for a broader STEAM approach was made by Professor Guy A. Boy, from the Florida Institute of Technology and NASA KSC (Boy 2013). In a world where the collective information obtained by humanity is available instantly at our fingertips, the importance of *understanding* equals and perhaps exceeds that of *knowledge*. The abundance of technical tools and resources enable us to better *contribute* solutions to society rather than just *consume* media generated by it. These traits are essential for moving toward a sustainable world, with complex challenges that will require innovation and critical thinking beyond conventional engineering approaches to solve and a sense of ownership and contribution to the cause of sustainability on the part of engineers to better motivate them toward that end.

The arts are a fundamental component of the kind of education system we need for a sustainable world. They inculcate a range of social, cultural, and moral experiences in children. They nurture and expand creativity and intellectual capacity, which is critical for taking a holistic approach to difficult real-world problems. They can also inculcate an appreciation for the beauty and fragility of nature and the desire to preserve it. For any large-scale endeavor to better achieve sustainability, in engineering or legislation or policy, both of the above aspects are crucial—the *ability* to foresee sustainability concerns and address them with ingenuity and the *motivation* to ensure that these measures are carried out as a key component of the endeavor.

Since their inception, the arts have been a potent tool for gathering an audience, cultivating widespread awareness, and inciting social change. This characteristic is prominent in sustainability, from local drives for renewables and energy-efficient homes at the community level to the ubiquitous documentary "An Inconvenient

Truth" increasing consciousness at the international level. Several artistic instalments and projects around the world have been conceived with the intent of incorporating green choices and making them aesthetically appealing (Waterglass at the Toronto Harbourfront Centre), preserving the traditions of peoples displaced or affected by climate change (Song of the Bird King), and expressing love for one's native land or natural environment (the Blue Cow, by Creative Carbon Scotland), all of which are very important in obtaining public support for endeavors. This issue is discussed in some depth in Garrett (2015).

To equip professionals of tomorrow with the vision, abilities, and resources to harness the power of the arts in achieving sustainability endpoints, we must take multiple related but distinct initiatives with students of today. First, we must motivate them to care about sustainability right from the K-12 stage, over and above the environmental education in school curricula. Key principles of sustainability—namely, respect for the environment, judicious usage of resources, and examining the long-term effects of our choices—should be integrated into every facet of their education. Arts-based activities can provide practical and lasting means of achieving this integration. A sustainability perspective can be introduced through school plays with an environmental message, or through field trip activities that appreciate nature and biodiversity, or through song-and-dance performances celebrating the heritage and culture of indigenous communities and their sustainable lifestyles, or through art-and-craft projects using trash and recyclables—or through any of a multitude of other routes.

A concrete instantiation of this point has been made by Professors Sybil S. Kelley and Dilafruz Williams from Portland State University, through their articles on Learning Gardens in the *Journal of Sustainability Education* (Kelley and Williams 2013) and *Clearing Magazine* (Kelley and Williams 2014). Their premise is that for most current K-12 programs, educators and students alike spend far too much time inside rigid school walls, devoid of any connection with nature and thereby sustainability. Kelley and Williams instead propose using the school grounds and gardens as a form of place-based education, a key component of sustainability education, and a means of understanding our niche in the delicate ecological system and developing connections with it. The school gardens and the larger principle that they embody allow for "experiential, integrated, and collaborative learning." In light of the adoption of the Next Generation Science Standards (2013), Kelley and Williams make several recommendations. First, plan early and start small. Gradually incorporate gardens as part of the learning pattern by encouraging students to spend as much time in nature as inside. Students can continue learning when they are outside: with teacher guidance, they need not treat the outdoor experience as recess; rather they can learn, share successes and offer challenge stories to other stakeholders to increase the feeling of community. Outside the often restrictive confines of a school classroom with its desk, blackboards, and projectors, the arts become a key medium of activities and expression in these learning gardens.

Second, sustainability must continue to remain a key issue even as engineering students begin learning the specialized topics of their major. Many universities now incorporate sustainability as an educational initiative. The Change Leadership for Sustainability Program at Stanford University's School of Earth, Energy and Environmental Sciences, for example, provides executive education programs, a Master of Science or, more significantly, a Master of Arts program to prepare individuals with the knowledge, mindsets, and practical skills required to understand and intervene in complex systems to advance sustainability. Unlike the K-12 education in which students may encounter environmental education as part of their curriculum, in the university environment, engineering majors can go through their college years without ever being required to consider the implications of their future career on the environment. This disconnect is problematic, especially once they enter the workforce.

Consider any large-scale architectural project, like a massive building or campus. Such an endeavor will have several stages, from planning to design to prototyping to construction, as well as several stakeholders who affect the outcome. There is considerable incentive for the building to meet sustainability benchmarks, in terms of reputation (LEED and other ratings), legislative requirements (Dernbach and Mintz 2011), and the financial savings from prudent resource usage. An entire sub-field of architecture, sustainable architecture, has thus become quite prominent now. However, as we have repeatedly stressed, the challenges of sustainability are quite daunting, with several interlinked aspects. To address these challenges at the various stages of the project, it is not sufficient to just have a subset of the engineers and architects who specialize in sustainability. For instance, it is not enough to design a building with good natural insulation to obviate the need for excess central heating, if the construction of the building itself is done inefficiently and wastefully. Different groups have different subtasks with different sustainability challenges. The most potent way to overcome these challenges in large-scale endeavors is to ensure that engineers across the board have an appreciation of sustainability being the need of the hour, as well as a more holistic and creative approach to the technical decisions they need to make. Potential employers like government and private agencies are actively paying attention to the above requirement.

As our previous example explains, in the current job market, sustainability-minded engineers are more suitable and, quite frankly, better, than conventional engineers who do not factor in broader impacts and aspects into their problem-solving. Therefore, it is necessary to alter the tone of how the various engineering disciplines are presented to the students. This should not require significant modifications to the core curriculum itself, because the fundamentals are always important. However, some changes to how the content is presented, as well as perhaps some changes to the practical and training components, are definitely warranted. The example of Professor Nichols' Construction Engineering Class is relevant here. Computer engineers should be familiar with the immense footprints of electronic waste and the best practices for e-waste disposal and recycling. Similarly, materials engineers should be conversant in the life cycle and durability of the various materials they work with, and focus on those that can be maximally reused. Mining engineers should have field projects centered on proper runoff management and maximizing equipment longevity. Overall, an objective in every engineering discipline should be to keep sustainability at the forefront, such that students do not lose sight of its importance. Better, more well-rounded engineers can formulate cheaper and more efficient solutions to difficult intermediate problems, making them more desirable to employers.

The arts and the broader humanities come into the picture here as well, for the conventional STEM approach has not been too successful. A working knowledge of the humanities can help engineers understand not just their technical responsibilities but also their social, civic, and moral ones (and there are few moral responsibilities more significant than that toward the environment). The wider issues that engineers need to grapple with, which come up in all the examples we have stated previously, and in engineering initiatives across the board, are best presented and motivated not in a lecture format, but through documentaries and other audiovisual presentations. Aesthetics and artistic design are an important component of an engineer's education. Often, sustainable alternatives are viewed as being inelegant and coarse (think of how eco-friendly toilet paper and hybrid cars were viewed when they first arrived on the scene), so a fairly potent way to make a sustainable design choice more palatable to the stakeholders is to make it aesthetically appealing (some of the best advertisements for sustainable choices now, such as ecotourism, LED lights, and even reusable tote bags and containers, appeal not just to our environmental consciousness but also appear to be more intrinsically attractive and enjoyable). Furthermore, incorporating artistic design principles can itself directly lead to a better, more sustainable engineering project, because of the creativity and lateral thinking required.

Third, as students get ready to graduate and enter the workforce, they must be given support and incentives for working in the sustainability domain. While inviting companies and agencies for engineering career fairs, universities must actively seek out smaller entities working on renewable energy, waste management, green products, and the like. In particular, agencies that focus on sustainable design (whether it be architecture or system planning) should be invited. Job fairs often emphasize just the giants in technology, conventional energy, and consumer products, to student disadvantage. Vocational training programs and career guidance counselors should also encourage students looking for work to consider nontraditional workplaces and job roles, and help them develop a communication vocabulary that inherently promotes their sustainable thinking, so that they can make a difference if they join a large existing company where sustainability is not a priority. The global markets are slowly realizing that sustainability is a valuable investment for the future; however, universities should stay ahead of that curve to make it easier for their largest resource, students, to contribute to this trend.

The importance of a well-rounded STEAM curriculum at all levels is clear. From imbibing an appreciation for the beauty and fragility of nature at an early stage to encouraging creative solutions to technical problems in college and to developing the tools of expression and advocacy at the workplace, incorporating the Arts into STEM domains to advance sustainability endpoints seems natural. In the following section, we will illustrate some of the principles we described above, manifested through authentic engineering initiatives. We will do so by discussing two prominent ESW chapter projects.

5 UC Berkeley Vertical Garden

At the University of California Berkeley, the Bechtel Engineering Center is used significantly throughout the year, in terms of student footfall and energy consumption. Therefore, it was an ideal location for the local ESW chapter's attempt to create a lasting representation of UC Berkeley's commitment to sustainability, through a beautiful vertical garden.

Pugh et al. (2012) showed that vertical gardens can reduce levels of NO_2 and fine particulate matter (PM_{10}) in the surrounding area by up to 15% and 23%, respectively. Reductions of this magnitude cannot generally be achieved just by planting trees. These vertical gardens can also reduce wall surface temperatures by up to 50 degrees Fahrenheit via evapotranspirative cooling and provide thermal insulation when outside temperatures drop. Collectively, these features reduce overall energy consumption, and based on the Bechtel garden's size, an estimated 256 kWh is saved per month.

The garden will also help boost pollinator populations. Native plants in California can attract and sustain wild pollinator populations. The vertical gardens contain several varieties of plants, which benefit pollinators such as native bees (e.g., bumblebee, sweat bee, and mining bee), butterflies (e.g., California tortoiseshell), and humming birds (Fig. 1).

Following ESW's holistic approach, the chapter plans to use sustainably sourced untreated redwood for the garden framework, and recycled materials for the mosaic of the famous UC Berkeley bear. To aid in improving the mental health and well-being of viewers, the team will emphasize aesthetically pleasing and uplifting environments for the garden. To increase engagement by visitors and to enhance their learning experience, a website will be created with educational information about the health and environmental benefits of the vertical garden, and information about the plants.

However, the real centerpiece is the team's incorporation of artistic aesthetic into the engineering design. Project leader Susanna Ming-Yu Dang revealed that she was inspired by other vertical garden designs, including a garden at the San Francisco Museum of Modern Art. One of the design requirements the team devised was to improve the mental health of students on campus. The team's solution was to consider the layout of the plant beds and arranging them to create the 28th official golden bear on the Berkeley campus. The California golden bear is the mascot of UCB, and as a symbol, it inspires strong feelings of attachment and loyalty among the students, faculty, and alumni. Having a golden bear on the vertical garden would

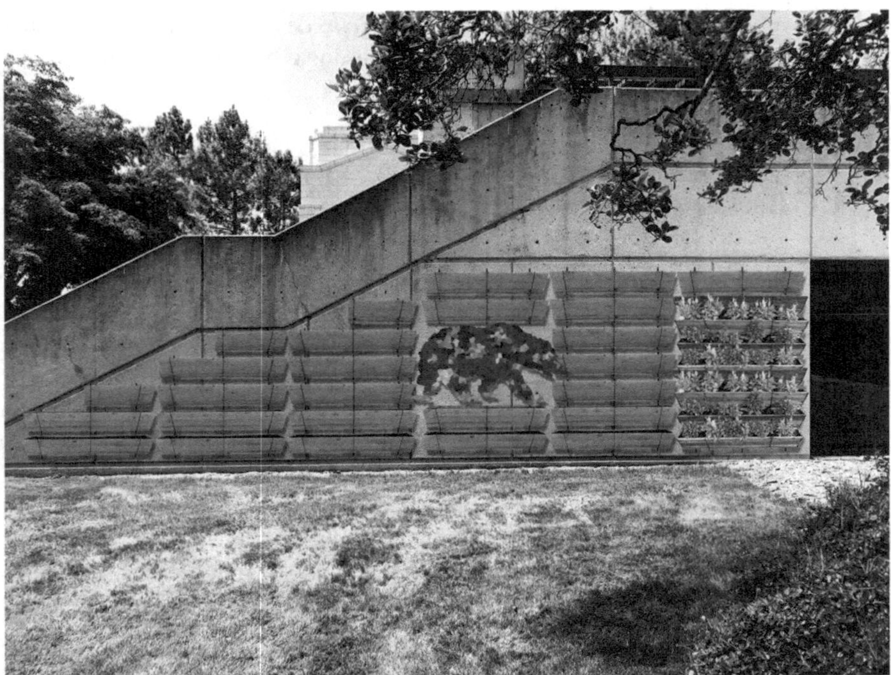

Fig. 1 Final design for the vertical garden at the University of California-Berkeley. The ESW chapter considered sustainable/recycled materials, colors, textures, and plant arrangement for our final design. Their wall will incorporate sustainably sourced redwood, recycled mosaic bear, and several species of drought-resistant plants that include California native plants, succulents, herbs, and flowering perennials. (© 2018 Sohn Cook)

be a potent way to engage the community. Although she never worked on a STEAM project in her classwork, Susanna and her peers at UC Berkeley recognized the importance of leveraging an artistic framework to help their project meet its design requirements.

6 Undocuversity

Undocumented students in the United States have recently had their daily lives and future plans put under considerable uncertainty. Given that they are a minority compared to the rest of the student population, their troubles and struggles often go unnoticed, which further enhances the feeling of isolation and lack of agency.

At the California State University Long Beach (CSULB), Jeff Ogle, a senior in international studies, felt that it was high time to raise awareness about the issues faced by undocumented students, their rights and protections, the services that CSULB offers them via the newly opened "Dream Success Center," and legislation, both now and in the future, which may affect their status. The aptly named "Undocuversity" is a long-term project he conceived to achieve this goal. Jeff became interested in immigration and undocumented students during his major, particularly due to study abroad in Spain and Costa Rica, and his Latin American studies.

Undocuversity had various suggested components—panel talks with students and community representatives about questions and concerns, screening related short films and documentaries, and so on. The local ESW chapter was involved with the Undocu-Wall, an interactive display on campus which students could use to share their experiences through writing and art. Verbal discussions at scheduled times are not the only means through which people wish to express themselves. The display allows them to use that with which they are most comfortable, to best get their message across.

Initially, the "Undocuversity" project was to be comprised of five pieces representing a solid wall between Mexico and the United States similar to an art installation in Downtown Los Angeles. However, after meeting with students and faculty, Ogle realized the political and psychological impact a wall could have on the undocumented community on campus. Most notably, vandalism and graffiti were the largest concerns by dissentients. The installation changed. The once-solid wall was separated into two larger pieces so individuals could freely walk through the installation. It had art. It was more inviting than isolated. It was a symbolic representation of "opening" the barriers that have blocked off the two countries for more than two decades.

There would be large structures placed in a public area with considerable foot traffic, so it had to be designed and created to be stable and secure, and of course in a sustainable manner. The original design consisted of five structures made of custom wooden pallets and a sturdy concrete foundation supported by a metal pole fixed to the ground (Fig. 2). A durable and smooth surface, such as Plexiglass, would be drilled and fastened onto both sides of the pallets. After Ogle's team went over their concept design, ESW project members examined every nook and cranny of the design and materials chosen. Sustainability wise, while the wooden pallets and concrete foundation were more sustainable than a prefabricated display metal and glass display board, building custom pallets and sourcing fresh concrete would produce more waste and carbon emissions. Plexiglass was another problematic material; it was expensive, heavy, and difficult to fabricate custom sizes. Reusing existing pallets and using recycled concrete would not only be more sustainable, but many of these materials can be sourced locally. The team obtained wooden reclaimed pallets donated by a local motorcycle dealer. Recycled concrete was cured in the College of Engineering Geotechnical Lab. The expensive Plexiglass was replaced with recycled thick card stock paper.

As the construction proceeded, the project team grew concerned that an overload of messages would overwhelm passers-by. They were swarming with ideas of how to incorporate these large visuals and images of real immigrants on to pieces of cardboard. One suggestion was to have the boards painted with large portraits or images. Contacting the campus art department, the team came across visiting artist, Narsiso Martinez. Martinez's artwork is a portrait collection of agricultural workers

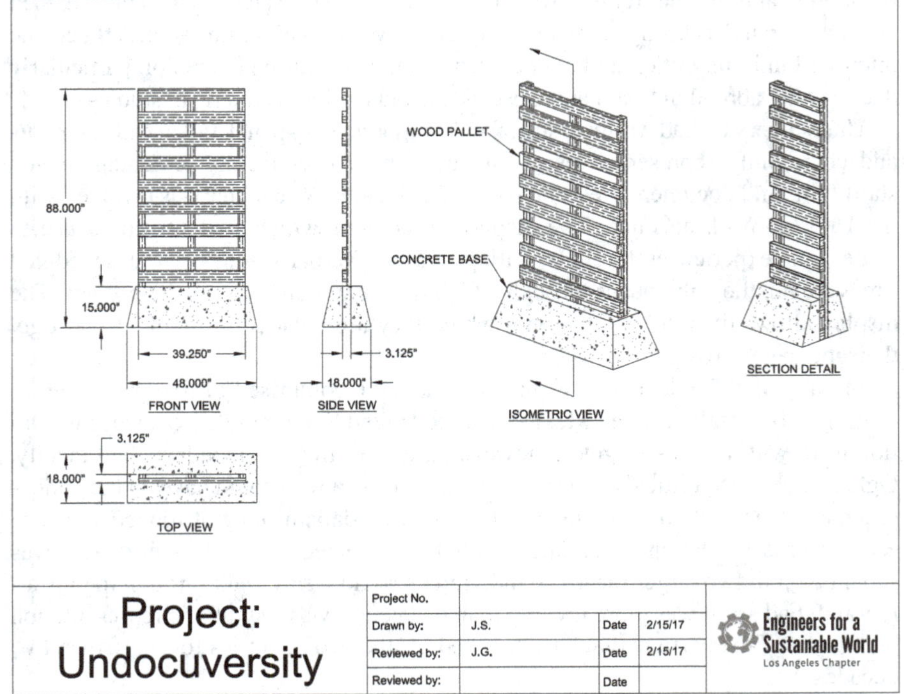

Fig. 2 Design sketch for the Undocu-Wall which was used for the construction on the CSULB campus. (© 2018 Sohn Cook)

painted or drawn on recycled produce boxes he collects from grocery stores. The design was remade incorporating Martinez's produce boxes as the backdrop and his haunting portraits featuring the struggles of immigrant field workers. With many revisions, the final product took two and a half months from conception to construction (Fig. 3).

The common takeaway from both of these projects, as it relates to the thesis of this chapter, is that two student teams independently decided that artistic concerns would improve their engineering design. In the case of the Berkeley team, the project lead repeatedly expressed that she could not envision designing a vertical garden without thinking about the visual aesthetic. She could not imagine divorcing the artistic element from the design. Both teams expressed that they had not been taught to think about incorporating art into their engineering in the classroom. In recent years, we have taken several steps to bring sustainability studies into engineering pedagogy, but we have a long way to go before STEAM is discussed significantly, and we can properly harness its potential to train engineers to build for a sustainable world. These two projects show that when students are given the freedom to define, design, and imple-

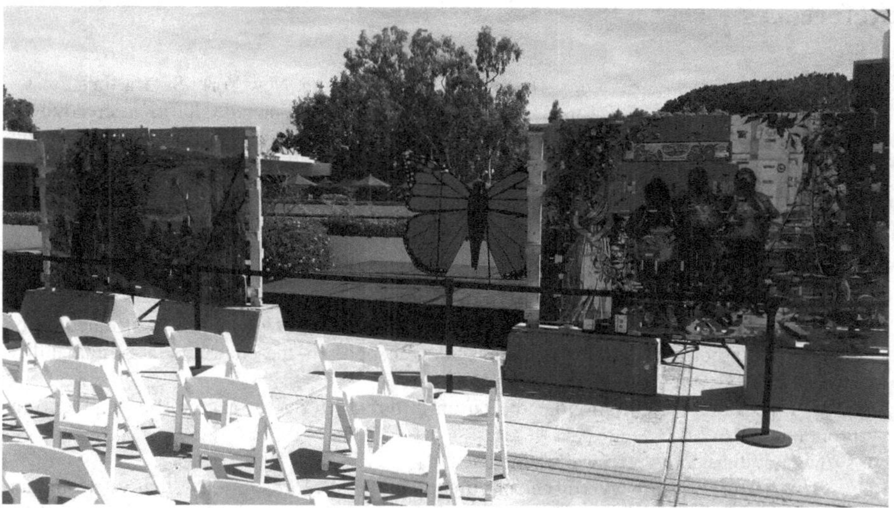

Fig. 3 The installed Undocu-Wall before the unveiling ceremony. (© 2018 Sohn Cook)

ment their own initiatives, they are able to take advantage of modern trends in engineering pedagogy that have not yet trickled into the classroom. It is time for educators to think about meeting them halfway, to everyone's mutual benefit.

7 Looking Ahead

Throughout this chapter, we have refrained from undue pessimism or from using apocalyptic language. However, there is essentially a universal consensus among environmental scientists that human civilization has an uncertain future, as a result of problems such as global warming, overpopulation, impeding food crises and resource exhaustion, and other large-scale problems. Whether these conditions will lead to an existential crisis or not is not yet known, but there is no doubt that the current generation of students will inherit a world with significantly many challenges to deal with.

Therefore, it is incumbent upon the educators of today to prepare them for these challenges. To that end, a business-as-usual approach of STEM education for engineers is no longer sufficient, as we have repeatedly advocated. Students must confront the issues of sustainability at every step in their education, and be aware of how to address them. Whatever their field of study or work, they must be familiar with not only the technical aspects of the applications, but also the larger social and cultural context in which their work exists. A well-rounded STEAM education is better positioned to achieve this future, to achieve a future where Professor Nichols' class is no longer hypothetical.

References

Bejan, A. (2016). *The physics of life: The evolution of everything*. New York: St. Martin's Press.

Boy, G. A. (2013). From STEM to STEAM: Toward a human-centered education, creativity & learning thinking. In *Proceedings of the 31st European conference on cognitive ergonomics* (p. 3). ACM.

Dernbach, J. C., & Mintz, J. A. (2011). Environmental laws and sustainability: An introduction.

Feldman, A. (2015). http://www.slate.com/articles/technology/future_tense/2015/06/steam_vs_stem_why_we_need_to_put_the_arts_into_stem_education.html. Accessed on 31 Dec 2017.

Friedman, T. L. (2009). *Hot, flat, and crowded 2.0: Why we need a green revolution–and how it can renew America*. New York: Picador. ISBN-13: 978-0-312-42892-1.

Garrett, I. (2015). Arts, the environment, & sustainability. http://www.americansforthearts.org/by-program/reports-and-data/legislation-policy/naappd/arts-the-environment-sustainability. Accessed on 31 Dec 2017.

http://www.abet.org/accreditation/accreditation-criteria/criteria-for-accrediting-engineering-programs-2016-2017/

https://www.ambius.com/green-walls/benefits/

Jolly, A. (2014). https://www.edweek.org/tm/articles/2014/11/18/ctq-jolly-stem-vs-steam.html Accessed on 28 Dec 2017.

Kelley, S. S., & Williams, D. R. (2013). Teacher professional learning communities for sustainability: Supporting STEM in learning gardens in low-income schools. *Journal of Sustainability Education*.

Kelley, S. S., & Williams, D. R. (2014). http://clearingmagazine.org/archives/10755. Accessed on 28 Dec 2017.

Mathur, H. M. (2006). Resettling people displaced by development projects: Some critical management issues. *Social Change, 36*(1), 36–86.

Meadows, D. H., et al. (1972). The limits to growth. *New York, 102*(1972), 27.

NGSS Lead States. (2013). *Next generation science standards: For states, by states*. Washington, DC: The National Academies Press.

Pugh, T. A., MacKenzie, A. R., Whyatt, J. D., & Hewitt, C. N. (2012). Effectiveness of green infrastructure for improvement of air quality in urban street canyons. *Environmental Science & Technology, 46*(14), 7692–7699.

Wicked Problems. https://www.wickedproblems.com/1_wicked_problems.php

World Commission on Environment and Development. (1987). *Our common future*. New York: Oxford University Press, 400 p.

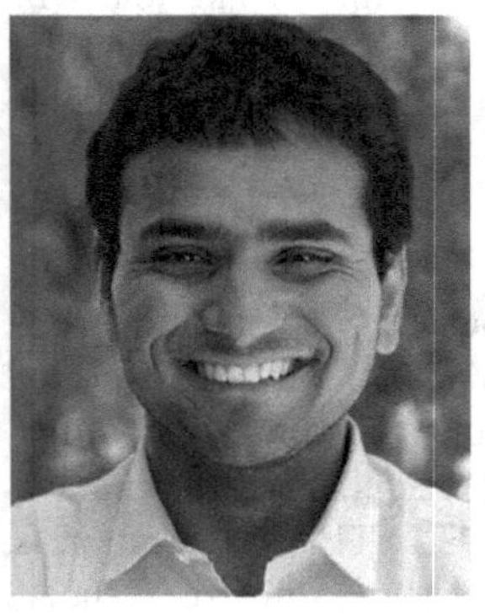

Shushman Choudhury is the Content Coordinator of Engineers for a Sustainable World. He is also a PhD student in Computer Science at Stanford University, researching decision-making for robotics. He is interested in social and policy issues of autonomous technology and how robotics and AI can be applied toward sustainability.

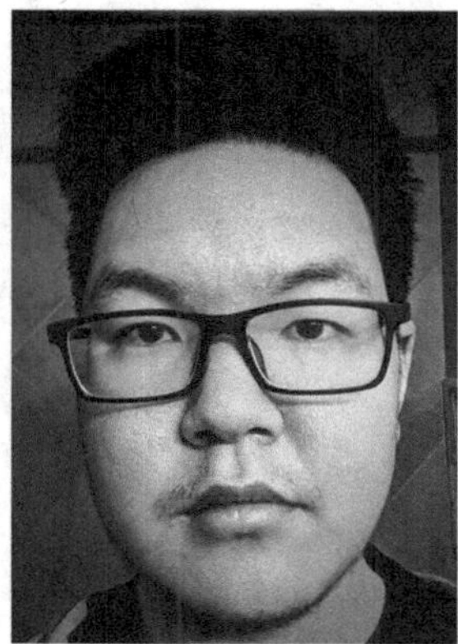

Sohn Cook is a professional and technical writer, wordsmith, and Oxford comma advocate with nearly a decade of experience from developing punchy and inspiring copy for nonprofits to ghostwriting technical manuals for the Homeland Security. Sohn currently freelances and writes pro-bono for NGOs serving disadvantaged, underrepresented communities worldwide.

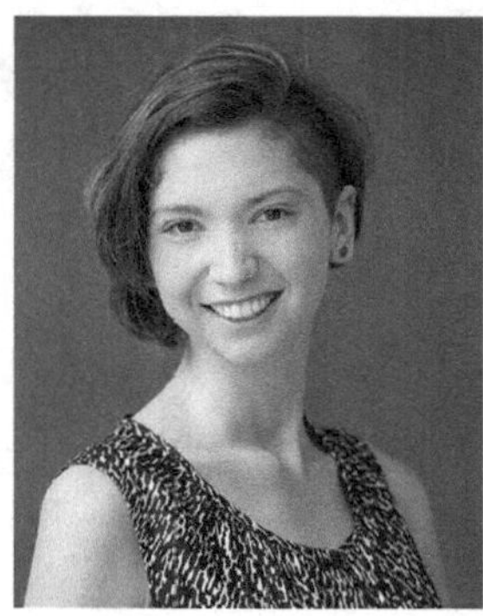

Brittany Bennett is the Executive Director of Engineers for a Sustainable World. After earning a degree in Engineering Science from Smith College, she decided to pursue a career in the nonprofit sector. She lives in Denver, Colorado.

Transgressing the Disciplines Using Science as a Meeting Place: The Science, Art and Writing Initiative

Anne Osbourn

1 The Origins of the SAW Initiative

The SAW Initiative was born in 2004 as a consequence of my frustrations as a scientist. I was established in my field and leading an internationally recognized research programme on plant natural products – how they are made, what they do and how new pathways arise. Plants are outstanding chemical engineers and make an enormous array of diverse compounds. The study of natural products provides an opportunity to address important questions about small-molecule-mediated dialogues between plants and their environment, and to discover new drugs and other high-value molecules. Although I enjoyed this area of research very much, I felt that the scientific culture that I was part of at the time was very inward looking, and that I was beginning to lose touch with the outside world.

Of course, scientists are specialists. They have become hardwired into using jargon and an exclusive vocabulary. To paraphrase Newton, they stand on the shoulders of giants, building on a mass of preordained knowledge. They are intimately familiar with the most detailed of details in the area of research in which they work. Specialization is important. Specialized communities, be they of scientists, writers, artists, builders, electricians, plumbers, doctors, lawyers, teachers or others, provide a forum for efficient shorthand communication – an environment in which groups of specialists can function effectively. But if, during the process of specialization, we lose the ability to communicate with those outside our fields of expertise, then our value to society declines. Even more importantly, fundamental breakdowns in communication may put the future of society in jeopardy.

My sister and I are both scientists. However, our parents both taught English, and we grew up in a house full of books on literature, history and art. I always enjoyed

A. Osbourn (✉)
John Innes Centre, Norwich, UK
e-mail: anne.osbourn@jic.ac.uk

© Springer Nature Switzerland AG 2019
A. J. Stewart et al. (eds.), *Converting STEM into STEAM Programs*,
Environmental Discourses in Science Education 5,
https://doi.org/10.1007/978-3-030-25101-7_11

creative writing, but the constraints of the school curriculum did not allow me to pursue interests in literature and writing alongside my other interests in science. I had to make a choice, and so I became a scientist. I went on to earn a degree in botany and a Ph.D. in genetics. I moved to Norwich to build my research career, acutely aware that in addition to being an international centre for biosciences, Norwich was also home to one of the best creative writing courses in the world – at the University of East Anglia. Thus it was in 2004 that I saw an advertisement in the science magazine *Nature* for a Dream Time Fellowship programme funded by the UK-based National Endowment for Science, Technology and the Arts (NESTA). The advertisement read 'Are you a high achiever with ten years' experience in the science and technology sector but looking for space to develop your ideas away from the demands of your professional life? Then you should apply for a NESTA Dream Time Fellowship'. I realized that this offered a perfect opportunity for me to combine my interests in creative writing with science and to see whether I could find new ways of making science accessible to all. Shortly after that, I became a NESTA-funded fellow in the School of Literature, Drama and Creative Writing at the University of East Anglia. I was on an interdisciplinary adventure, supported by my mentor Professor Clive Scott (then Head of the School). I had weekly meetings with Clive, but this was not a taught course. My only remit was to find ways of bringing science into everyday lives and language through creative writing.

To my surprise, I started exploring science and my own origins through poetry. I had never previously thought of myself as a poet, and was somewhat bemused when I found myself thrown into this crystalline world of words. In retrospect, I think I can pinpoint the trigger to my poetry writing. I was listening to the radio 1 day and heard a programme about Rebecca Elson, an astrophysicist and poet who had died of non-Hodgkin's lymphoma at the age of 39 and whose anthology *A Responsibility to Awe* had been published posthumously (Elson 2001). I was intrigued by Rebecca Elson and read this book, hungry to find out how and why she wrote poetry. Now, more than 10 years later, some of my own poems have been published in poetry magazines, and I hope to have my collection out soon (although in my view, publishing poetry is even harder than getting scientific papers into top international journals – perhaps that is some reflection on my poetry!). An account of my experience as a NESTA-funded experimental animal in the School of Literature and Creative Writing can be found in *Nature Reviews Microbiology* (Osbourn 2006).

While scientific terminology and concepts may be off putting for many, stunning scientific images are not. Scientific images intrigue, drawing in children and adults alike. They provide a meeting place, a starting point for scientific exploration. I learned this when trying to find different stimuli to trigger my own creative writing. This led me to establish the SAW concept, a cross-curricular initiative that uses science to fire imaginations. Images such as the earth from space, salt crystals under the microscope and thermograms of houses provide fascinating stimuli to promote discussion about science in schools. In a SAW project, children explore a scientific theme through practical science, supported by carefully chosen images that illustrate different facets of the science that they are investigating. The theme and the images are then used as a springboard for separate sessions involving creative writing and art. The children realize that science and the arts are interconnected, and they discover new and exciting ways of looking at the world around them. Most importantly, they become interested in wanting to learn, and they begin taking ownership of the learning process.

The path to specialization starts early. By the time children leave primary school, they have already been taught to view science and the arts as separate disciplines, rather than as interlocking pieces that together lay the foundation for a deeper understanding of the world. These divisions are reinforced in high schools, where the constraints of the rigorously compartmentalized curriculum further isolate subjects, stifle inquisitiveness and quell creativity. SAW offers a route to breaking down these barriers, or – if introduced early in the education process – a route for preventing them from becoming established in the first place.

2　Using SAW to Explore a Major Societal Challenge: Control of Infectious Diseases

SAW provides a different approach to teaching curriculum science topics in schools. It also is a highly effective means of introducing topics that would normally be regarded as too advanced for schoolchildren, even at the highest levels of school education. Research into topics such as the control of infectious diseases is relevant to all members of society. By making such research accessible through SAW projects, children can gain insights into how science disciplines are used in the real world.

Antibiotics were introduced into clinical medicine in the 1940s and revolutionized the treatment of infectious diseases. However, we now face the worrying prospect of a return to the pre-antibiotic era due to emerging antibiotic resistance (and frequently multidrug resistance) in various human pathogens. Antibiotics not only play crucial roles in the treatment of infectious diseases contracted by human individuals, but are also essential components of many aspects of modern medicine. For example, surgical procedures such as hip replacements, Caesarean sections, heart or bowel surgery, and cancer therapy all rely on antibiotics; without effective antibiotics, many of these procedures would become life-threatening and potentially impossible. The UK government's Chief Medical Officer, Professor Dame Sally Davies, recently highlighted the urgent need for global action to tackle the potentially catastrophic threat of antimicrobial resistance, commenting 'We need to encourage more innovation in the development of antibiotics'.

The development of new and improved anti-infective therapies and strategies requires a multidisciplinary approach, involving fundamental and clinical microbiology, molecular genetics and biology, bioinformatics, natural product and medicinal chemistry, rapid diagnostics and pre-clinical assessment. It must also encompass new ways of approaching the treatment of prevention of infectious diseases. In Norwich, the partner organizations of the Norwich Research Park (NRP) – the John Innes Centre, the Earlham Institute, the Quadram Institute Bioscience, the University of East Anglia, the Sainsbury Laboratory and the Norfolk and Norwich University

Hospital – have a diverse range of expertise spanning from genomics to the clinic. Collectively, these organizations are making major contributions in the mission to find innovative ways of combatting infectious diseases.

As mentioned above, intriguing scientific images are central in SAW projects: they make science accessible and help link the science, art and writing sessions in the classroom. To help disseminate and showcase NRP science through SAW and other diverse channels, we established a library of images based on our research. The NRP image library contains images submitted by scientists from across the Park. All images are freely available for download and reuse with attribution of the source from: http://images.norwichresearchpark.ac.uk. The images and legends are all carefully quality-controlled to ensure that the calibre and resolution of the images is high and the legends are accessible to non-specialists.

2.1 The Images

The images within the NRP image library encompass diverse swathes of science. However, for the purposes of the NRP antibiotics SAW projects, we selected images relating to different aspects of control of infectious diseases. These included 'bad' bacteria (e.g. *Escherichia coli* in the human intestine), microbes isolated from soil (the source of many antibiotic-producing bacteria), structures of engineered enzymes that make novel antibiotics, next-generation DNA sequencing technology for rapid diagnostics, DNA-based antibiotics and innovative methods for vaccine production (links can be found at the end of this article). These images provide a visual introduction to the spectrum of NRP research in this area. They also serve as a freely available resource for SAW projects in schools focused on the theme of tackling the problem of antibiotic resistance. Other resources are available, some without restriction or copyright. However, it is often difficult to find fee-free images that can be used in schools. Science is beautiful. Collections of images that are freely available represent extremely important resources for schools. Examples of image libraries that can be used for teaching science in schools (some of them without restriction, others requiring a fee) are listed in the "References" section.

2.2 The Projects

We conducted eight one-day SAW projects in schools on themes relating to different facets of control of infectious diseases. The projects were planned in close collaboration with the teachers and involved children of various age groups, ranging from 8 to 12 years. Fourteen scientists took part, working in teams with professional artists and writers. For each project, the scientist(s), artist and writer that comprised the delivery team were all in school for the whole day and participated in each other's sessions. In this way, they not only provided their expertise and imparted knowledge, but also enabled the children to understand what real-life scientists, artists and writers are like – i.e., just like any other person. Each project began with a game in which the children were invited to guess which of the visitors does what. This is a very good ice-breaker activity that challenges children's preconceived ideas of what people from these different professions look like. Unfortunately, the children often assume that the men are scientists, while women are commonly classified as artists, particularly if they are wearing colourful clothes. Writers, on the other hand, are often thought to look drab and boring! When the moment of revelation comes and true identities are revealed, the children's surprise is evident as their judgements are brought into question. We ask them to put their hands up if they think the various individuals in front of them are (a) a scientist; (b) an artist; or (c) a writer. Usually they will only put their hand up for one or possibly two of these. The children are encouraged to think that they can try their hand at anything that they want to be. The children are then invited to become scientists, artists and writers for the day.

The science sessions involve hands-on practical activities. In addition to any scientific equipment available in the school, the visiting scientists may bring other general laboratory equipment with them, such as gloves, pipettes, petri dishes and test tubes – specialized kit that the children are thrilled to use. During the planning phase, each class teacher gave a brief summary of the children's knowledge in the areas of microbes and health so that the scientist could find a sensible starting point for the session that enabled the children to go on a journey with them to extend their understanding. Often, children were familiar with the word 'germ' but not the word microbe, although some had heard of bacteria. Most children are very happy to tell visitors what they know, and so a discussion at the start of a workshop is always productive! The science was not oversimplified nor was the language dumbed down. Children are used to learning new words and they found the flood of new vocabulary exciting. The experiments carried out in various projects involved setting up experiments to isolate bacteria from soil; comparing the effectiveness of various household substances (e.g. lemon juice, vinegar, soap) and an antibiotic in inhibiting bacterial growth; learning about DNA by extracting DNA from strawberries and using a computer programme to compare different DNA sequences; using a handheld sequencing device known as a MinION to sequence DNA; and playing games and building molecules to learn about new diagnostic tests. One of the projects was led by a local medical doctor who showed the class what was in her doctor's bag. The children were then allowed to use the equipment to test their classmate's temperatures, listen to heartbeats with stethoscopes and pulse meters, measure oxygen content in the blood with a pulse oximeter, use peak flow for measuring lung capacity, examine inside each other's ears and throats, measure simulated urine samples for glucose and learn proper handwashing techniques using UV hand cream and a light box.

The science sessions were followed by a poetry session and then finally an art session, each of which drew on the scientific theme, supported by the images, as a starting point for creative exploration. Each of these sessions was carefully planned and the children were given direction and encouragement, but the outputs were very diverse because they reflected personalized responses to the science. Some examples of the children's creative responses can be found below, along with their comments. These responses illustrate how this multidisciplinary approach to teaching and learning enables children to explore and grasp scientific concepts and encourages them to consider the importance of science in addressing major societal issues. More about this set of projects can be found in the recently published book '*SAW antibiotics: Science from the Norwich Research Park*' (Osbourn et al. 2017). The UK government Chief Medical Officer, Professor Dame Sally Davies, very kindly provided a quote for the cover of this book, which reads as follows:

> *I thoroughly enjoyed this stunning book. Educating children in such an interactive way is so important. I am passionate about addressing AMR (antimicrobial resistance), as are the researchers, and I hope that this can help inspire future generations to engage in the efforts to save modern medicine.*

The importance of scientific intrigue, serendipity and creativity in solving the major challenges faced by humanity is further illustrated by this quote from Dr. Jane Osbourn, Vice President of Research and Development at the drug company, MedImmune:

> *Antibiotics are a pillar of modern health care which cannot, and must not, be taken for granted. They were discovered through curiosity, careful observation and creative thinking, and these are the same skills we need to nurture as we respond to the evolutionary pressures of microbial resistance. SAW is an innovation tool to do just that; encourage our next generation of problem solvers to understand, consider and respond to scientific challenges in creative ways. The work presented here is an inspiring celebration of the beauty and complexity of scientific endeavour and its ability to fascinate generation after generation.*

3 The Children's Responses

3.1 Poetry

Invasion of the plague

The first body of the victim fell to the floor.
Invading the next,
unleashing a wave of unstoppable minions,
tirelessly attacking,
infinitely resisting.
Too tiny to see,
too much pain to handle
the feel of fright.
Occupying the first cell,
changing its will,
cloning itself,
sending an inferno at the body's defenses,
too tiny to see,
no trace of goodness.

With one destiny,
only one desire,
growing infinitely,
taking more lives,
outrageously sucking life gone.
Too tiny to see,
to panic.
They're coming.

The body's defences are weakened.
its might grows,
destroying the last,
striking down the heart,
its body fell to the floor.
The job is done.
They're moving on.

Thales (Age group 10–11)
Monkfield Park Primary School

Poem

These little bacteria look like
cupcakes that sit in a tray
all day, ready to rise to
a person's mouth, but
when you chew they make
your teeth go from white to black, then
bacteria will lie in your gum forever
until you're older
than ever.

Maddison (Age group 9–10)
Colman Junior School

Enemies

Evil force being kept captive.
Just by an innocent marshmallow heart.
Surrounded and trapped.

Spark, crack, pow!
Evil force getting stronger.
Pink, pretty, poisonous.
Good force overlapping.
Trying to defeat.

Fighting off its every move.
But saving each other at the same time.
Rays of darkness.
Poison spreads.
Intensity fills the air.

Battling,
defeating,
saving,
overtaking,

enemies.

Amber (Age group 10–11)
Monkfield Park Primary School

Cells

Although deadly,
it carries on
growing and
getting bigger it
spreads knowing
nothing it goes
through life blinded
but although blind
it grows happily
spreading its deadly genes and making
more to replace those
gone although knows
nothing it carries on
and on just like
a large waterfall
and it don't matter
what we do it will
never stop fighting.

Layla (Age group 10–11)
West Earlham Junior School

War of bacteria

Spreading the message
risking hope
anger, cunning
protecting their home.
Evil flying
time ticking
heart burning
losing all their mercy
unstoppable warfare
gut taking
taking over
brain splitting
tactical planning
infected hearts
live
losing

spirits escaping
tears dipping
sorrowful eyes
happiness spreads
new plots begin
life restarting
guarding the house
guarding us.

Hamza (Age group 10–11)
Monkfield Park Primary School

The invasion of bacteria

The bacteria moved slowly upwards like beautiful hot air balloons
gliding towards the moon.
Knowing that they could soon cause an infection.
Ready to make trouble and spread.
But suddenly out from nowhere came a brave soldier who fought like a wild bear,
protecting its kingdom and others.
Spoiling the villain's plan to infect.

Esme (Age group 10–11)
St Francis of Assisi Primary School

Soil bacteria

At first glance, soil may look nothing but boring.
But the compost has a twist and the dirt has a story.
The way to find the hidden tale is to look slightly deeper,
for your eyes are the key and the soil is the keeper

once you're past the layers of mud and of grime,
discovering the secret is only a matter of time.
Try and get through the insect stage, however gross it may be,
you're nearly there now, you already have the key!

Just twist the knob on the microscope and perhaps adjust the lens,
and while you're at it, give the eyepiece a good cleanse.

Finally, you're there now, the treasure is revealed.
The bacteria are flowers and the soil is a field.
Fluffy, bubbly, a variety of microbes,
billions, trillions, there are absolutely loads!

So far, it's been fantastic, an eye-opening trip.
A whole universe of bacteria underneath your fingertips.

Mia (Age group 11–12)
City of Norwich School

Ants

The ants that farm hurriedly,
the ants that scurry alongside each other,
who shield themselves,
who protect their food,
who leak the smell of friendship,
extending their family,
extending their home.

Protecting their nest,
they work hastily,
farming non-stop,
astoundingly clever,
farming even before us.

Pat (Age group 11–12)
Monkfield Park Primary School

We are here!!

We are here,
everything you
touch,
see,
eat.

We are here.
No matter what, we are here.

Your everyday life,
is thanks to us,
your living
is thanks to us.

We are here.
No matter what, we are here.

We keep you safe.
We provide medicine.
We are inside you.
We are what you eat.

We are here.
No matter what, we are here.

David (Age group 11–12)
City of Norwich School

Bacteria

Unnatural bubbles drifting, slowing.
Fluorescent blue bacteria,

glowing, dazzling little dots.
Masterpieces of the world,
so small yet so bright.
Stars that shine all night.
So vibrant to miss,
yet we do.
Millions upon millions,
their own little world.
Some big, some small,
but so different among them all.

Abigail (Age group 11–12)
City of Norwich School

DNA

DNA is a twisted world with more of itself in between.
This wonderful thing makes us who we are.
It's like life without choices, but choosing depends on our DNA.
This majestic strand creates with others and won't grow on its own.
It is us, how we feel, like we are, soft pillows.

Max (Age group 10–11)
St Francis of Assisi Primary School

The microscope

The glass laid by a big metal pillow.
Aroma disinfected of pine or lemon.
Made of metal towers.

Brandon (Age group 10–11)
West Earlham Junior School

Bacteria

Bacteria comes in many different places
and even gets into secret bases.
When people get these horrible diseases,
they start to cough and do lots of horrid sneezes.
Of course, these people try to stop it,
and will probably have to mop it.
When people recover from their illness,
they will have no more illness!

Nathan (Age group 9–10)
Colman Junior School

Untitled

Here are the coins rattling in the coin box,
rusty and dirty.
Here are the tablets curing the illness,

red, brown and purple
tablets fighting to make me better,
working hard, to cure the pain.

Here are the skulls watching me with every step I take,
cracking and banging against the sides.
I hear tiny noises murmuring and telling me that I'm okay.

Shawnie (Age group 10–11)
West Earlham Junior School

Untitled

Our search for a cure
is failing
but we can't give up
we are at the edge of the earth
maybe we will go to another
but alas we've grown old
so now it is up to you to continue
our fight for I cannot
finish it. So
succeed, for I have failed.

Declan (Age group 11–12)
City of Norwich School

3.2 Artwork

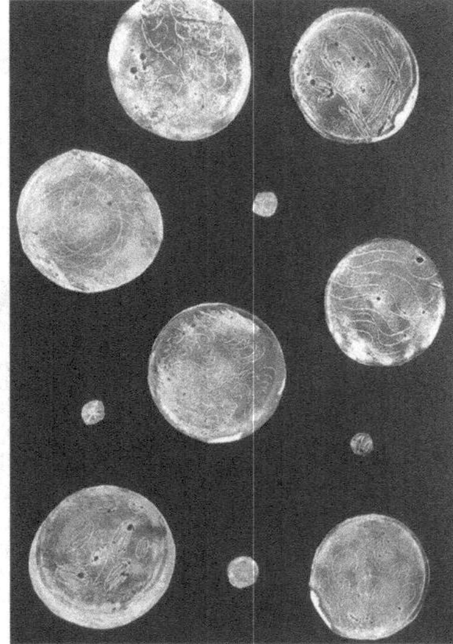

Bacterial colonies on agar

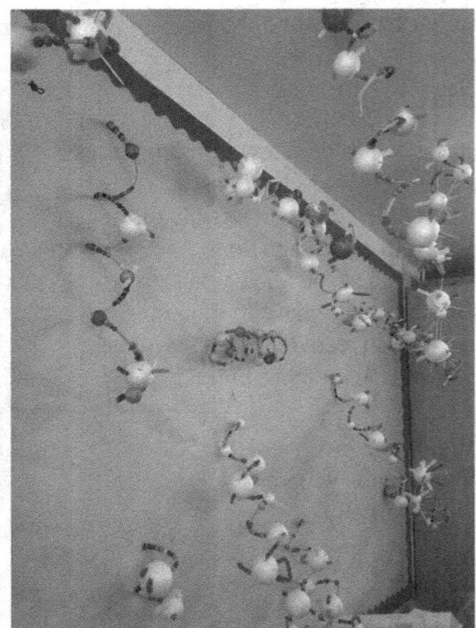

DNA

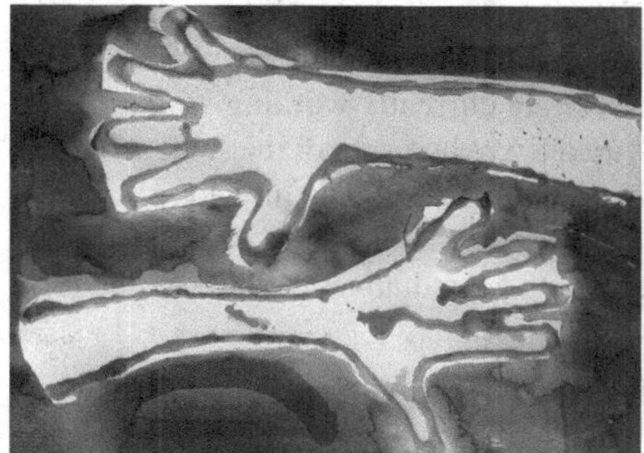
Hands transmitting bacteria

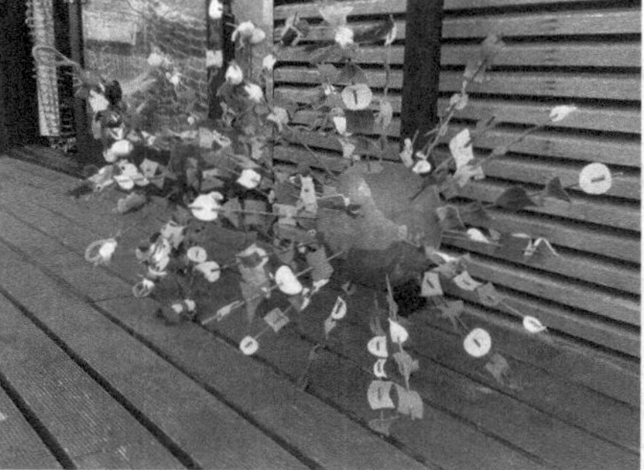
Vaccines

3.3   *What the Children Said*

Let your mind go free!
Whatever you do wrong you can turn into something else.
Everything we did was as if we were real scientists/writers/artists.
I learned that not all the poems have to be about what we do, it could be about science and art.
I learned how to use more of my imagination.
Almost everything has bacteria in it.
Antibiotics kill bacteria not viruses.
If you let your brain go wild it makes a good poem.
I learnt that there are so many different types of painting, wood cut and even bubble wrap.
All the science surprised me today.
I learnt about the process of how the good bacteria work.
I liked science best because it was fun to learn about microbes and do an experiment.
When you are doing poems you can do it in your own style.
I learnt how to use a microscope.
I didn't know that scientists can be writers or artists.
I was surprised that you could get DNA out of a strawberry.
I learnt how good my body is.
It is not just about painting or drawing it is also about experimenting and creating.

4   The Impact of SAW

Based on the above, it seems that taking part in SAW projects certainly has an impact on the children. The children's creative outputs and comments provide one means of capturing outputs. Establishment of robust methods for the evaluation of the impact of educational programmes can, however, be challenging. How can we assess the effects of SAW on children, teachers and teaching practice? What impact does taking part in SAW projects have on the other team members – the scientists, artists and writers? These are difficult questions to answer. Although we gather feedback from all of those involved in SAW projects (teachers, children, scientists, artists, writers) at the end of each project, quantification of the impact of SAW in the immediate and longer term is problematic. We are currently working with the local education authority here in Norfolk to set up a SAW hub school that will provide a central point for dissemination of the initiative to other schools in the region and a framework for longer-term evaluation.

We do, however, have a considerable amount of qualitative feedback in the form of comments from those involved:

4.1 What the Teachers Say

The comments from teachers have provided important insights, as illustrated by these words from a teacher of children aged 7–8 years:

> I was asked to take on a SAW project by our head teacher and was very excited, as I had come across previous SAW work in other school and had been impressed by the high quality work produced by the children. I was keen that the children should experience activities that we would not normally be able to provide in the classroom, especially in science. I also wanted the children to start to make links between the work we do in the classroom and things that happen in the 'real world.
>
> The children were really enthusiastic about the day and loved having lots of adults to work with. It was interesting to see how some children who normally find literacy or art challenging found it more interesting in this context, and others who didn't think they enjoyed science were actively engaged. It certainly gave us an opportunity to work in new ways and try things we would not normally experience. I noticed increased levels of engagement and enthusiasm from some children. In science lessons there is always excitement about practical work and I think it has really embedded the theoretical work we had already done. There were several who took real pride in their writing and creative skills and we have been able to draw on this success in other lessons.
>
> The most noticeable impact has been in confidence. Having their work on display at the Norfolk Show was a great boost and the children loved explaining it to the public. We have since had the work displayed at school and I have noticed most of the class pointing out their work to family and friends. Knowing that work will be displayed and celebrated gives the children an incentive to improve and refine their work. We capitalised on the enthusiasm generated on the day and they happily continued to work on their poems and develop their writing skills after our SAW day.
>
> The children enjoyed all aspects of the work they did and came away with lots of new knowledge and enthusiasm. Most importantly for me they came away with lots of questions, which shows heightened curiosity about the world and forms a platform for future learning. Thank you so much for the opportunity. The SAW project has been an experience that has captured the imagination of my class and something they will never forget. I have learned a lot about practical exploration of the science curriculum and cross-curricular links which I will be using in the future. We would love to do it all again.

Comments from other teachers include:

> Today was a tremendous success. Everyone enjoyed it immensely. The children's response was 'Epic, epic, epic! The best day ever!' One little boy who is normally a huge problem gave me a hug at the end of the day and thanked me for arranging the best day that he had ever had in school. That means a lot, coming from him. We still have some work to do on

our poetry and artwork, but hope to have it completed by the end of next week. Our SAW display should be in place by Friday.

And:

Antibiotics and bacteria may not seem the first choice for a primary science, art and writing day but the expertise of the SAW project team and their choice of activities made it a day to remember. The science was presented in an interesting and accessible manner with the task, extracting DNA from strawberries, being many of the children's favourite part of the day. As a teacher, I was impressed with the creativity and quality of the writing and poetry produced by the children. The art resulted in another favourite part of the day – to quote a pupil 'I loved making the lethal bacteria mobiles – it was great fun.' *To sum up, in the words of another pupil* 'The day was epic!' *If you can get a SAW project team into your school, jump at the opportunity.*

Clearly the word 'epic' is in vogue amongst the younger generation at present! The importance of quality, pride and presentation comes over in the teachers' comments; and perhaps most critically, they notice increased enthusiasm, curiosity and confidence in the children.

4.2   What the Scientists Say

SAW came into being because of my frustrations about feeling as though, as a scientist, I was losing touch with the outside world. My adventures since my NESTA fellowship have been very rewarding personally. I have discovered poetry; worked with other scientists, writers and artists on interdisciplinary SAW projects; and have spent a lot of time in schools learning from teachers and children alike. We routinely ask all those involved in SAW projects to fill in feedback forms about their experiences at the end of each SAW project, including the scientists. Below are some comments from scientists who took part in the antibiotics projects (including students, postdoctoral researchers and professors):

I love seeing the boundaries between disciplines dissolve, and have as much fun helping with and discussing their poetry as I do I teaching them about science.

The children were naturally enthusiastic about science, highly creative and great fun to work with. If scientists and teachers can foster children's enthusiasm for science into adulthood, the future will be bright.

I went on a SAW trust workshop expecting to tell people about science but came away amazed by the creative potential of the poets I met. Perhaps if we can mix that creativity with scientific insights we can achieve amazing new ways to solve problems.

We were impressed with the enthusiasm the students displayed learning about the importance of why antibiotics are prescribed, the importance of taking your full prescription and the effect they can have on our overall gut microbiota. Their desire to learn new concepts of science through experiments, poetry and art filled the day with laughter and fun for all.

Extremely impressed by the ability of children to turn our research into poetry and art.

A fantastic and engaging way to communicate current research to the next generation.

I love the unexpected questions that children ask – it makes me look at my research from a very different perspective.

I find these comments heart-warming. The scientists are connected, feel that their research is important, are fulfilled by their engagement experiences. They enjoy the stimulation of working with open-minded young people who ask questions, as part of an interdisciplinary team. There are hints that in some cases they perhaps even feel as though the door to the creative arts is ajar and beckoning, should they choose to explore further.

4.3   Science as Inspiration for Experience Writers and Artists

The school projects involved experienced writers and artists working as part of the SAW teams. These experts were then given the opportunity to respond to the theme of antibiotics through their own creative explorations. The poetry and artwork from these adult workshops are included in the *SAW antibiotics* book, along with narratives that they provided to accompany each piece of work. Some examples are shown below. These provide a very powerful means of engagement on antibiotic resistance and bring new dimensions to the multifaceted impacts of SAW.

Process – by Rebecca Thompson
A study of the impact of bacterial elimination and growth after the use of antibiotics, created in fused glass and wood.

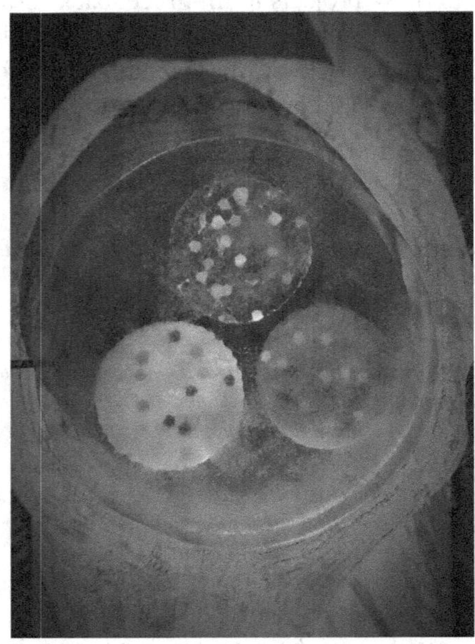

The Greater Good

Every day resistance increases –
old wounds that won't heal
common infections come back to plague us
indifferent and deadly.

It seems there's something in us
that's incurable –
    nightly our living rooms flood
with the faces of mothers
begging their terrible questions.

In the lab the leafcutters hoist
the bright green sails
of what might be our future,
streaming their live data 24/7.

They've been ceaseless,
tending their nebulae of fungus,
for 50,000,000 years

which is why it always takes him by surprise
each time he goes beyond the observable –.
    lifting one clear of the water with his fingertip.

Esther Morgan

'The poem brought together several different ideas which had struck me during the course of the project – the concept of resistance, the importance of data and numbers in scientific research, and the huge disparities in scale implicit in the project. I found it very intriguing that the future of human health, perhaps even our existence, may hinge on something as tiny as the leafcutter ants helping us to counter the threat from microscopic infection. Set against these more abstract themes, I kept returning to an image which had struck me when I was taking a tour of the lab, of one of the research scientists bothering to rescue an ant which had fallen into the water around its colony. We often use the metaphor of ants to describe humanity in its teeming masses, but here was a sudden, vivid reminder of the value of a single life.'

References

Elson, R. (2001). *A responsibility to awe*. Manchester: Carcanet Press Ltd.

Osbourn, A. (2006). The poetry of science. *Nature Reviews. Microbiology, 4*, 77–80.

Osbourn, A., Hutchings, M., McClean, T., & Rant, J. (2017). *SAW antibiotics: Science from the Norwich Research Park*. Norwich: The SAW Press. ISBN 978-0-9550180-3-9.

To find out more about SAW, please visit the SAW trust website – www.sawtrust.org

Links to Scientific Images Illustrating Research on Control of Infectious Disease at the NRP Image Library

A novel vaccine design strategy: http://images.norwichresearchpark.ac.uk/imagedetails. aspx?imgid=168

Bacterial biofilm coating: http://images.norwichresearchpark.ac.uk/imagedetails.aspx?imgid=95

DNA gyrase: http://images.norwichresearchpark.ac.uk/imagedetails.aspx?imgid=78

Engineering antibiotics: http://images.norwichresearchpark.ac.uk/imagedetails.aspx?imgid=76

Green vaccine machine: http://images.norwichresearchpark.ac.uk/imagedetails.aspx?imgid=135

All images are freely available for download and reuse with attribution of the source from: http:// images.norwichresearchpark.ac.uk

Transforming MRSA: http://images.norwichresearchpark.ac.uk/imagedetails.aspx?imgid=102

Death's-head hawkmoth: http://images.norwichresearchpark.ac.uk/imagedetails.aspx?imgid=105

Life in earth: http://images.norwichresearchpark.ac.uk/imagedetails.aspx?imgid=173

Leafcutter ant closeup: http://images.norwichresearchpark.ac.uk/imagedetails.aspx?imgid=174

Leafcutter ant with rose petal: http://images.norwichresearchpark.ac.uk/imagedetails. aspx?imgid=177

Actinomycetes isolated from leafcutter ants: http://images.norwichresearchpark.ac.uk/imagedetails.aspx?imgid=178

Streptomyces coelicolor gusArt: http://images.norwichresearchpark.ac.uk/imagedetails. aspx?imgid=180

Postgraduate student dedication: http://images.norwichresearchpark.ac.uk/imagedetails. aspx?imgid=183

Escherichia coli in the human intestine: http://images.norwichresearchpark.ac.uk/imagedetails. aspx?imgid=192

Analysis of antibiotic production: http://images.norwichresearchpark.ac.uk/imagedetails. aspx?imgid=199

Streptomyces coelicolor: http://images.norwichresearchpark.ac.uk/imagedetails.aspx?imgid=222

Mass spectrometry on a leafcutter ant: http://images.norwichresearchpark.ac.uk/imagedetails. aspx?imgid=232

Comparison of biosynthetic gene clusters: http://images.norwichresearchpark.ac.uk/imagedetails. aspx?imgid=234

Genome of a *Streptomyces* bacterium: http://images.norwichresearchpark.ac.uk/imagedetails. aspx?imgid=236

Sequence alignment of bacterial peptides: http://images.norwichresearchpark.ac.uk/imagedetails. aspx?imgid=237

Microbisporin production: http://images.norwichresearchpark.ac.uk/imagedetails.aspx?imgid=69

Streptomyces cinnamoneus: http://images.norwichresearchpark.ac.uk/imagedetails. aspx?imgid=72

Streptomyces glaucescens: http://images.norwichresearchpark.ac.uk/imagedetails.aspx?imgid=73

3D leafcutter ant: http://images.norwichresearchpark.ac.uk/imagedetails.aspx?imgid=82

Mycobacterium tuberculosis: http://images.norwichresearchpark.ac.uk/imagedetails. aspx?imgid=89

Fluorescence microscopy: http://images.norwichresearchpark.ac.uk/imagedetails.aspx?imgid=90

Pseudomonas fluorescens: http://images.norwichresearchpark.ac.uk/imagedetails.aspx?imgid=93

MinION channel: http://images.norwichresearchpark.ac.uk/imagedetails.aspx?imgid=96

SMRT cell technology: http://images.norwichresearchpark.ac.uk/imagedetails.aspx?imgid=98

Virus-like particles produced in plants: http://images.norwichresearchpark.ac.uk/imagedetails. aspx?imgid=169

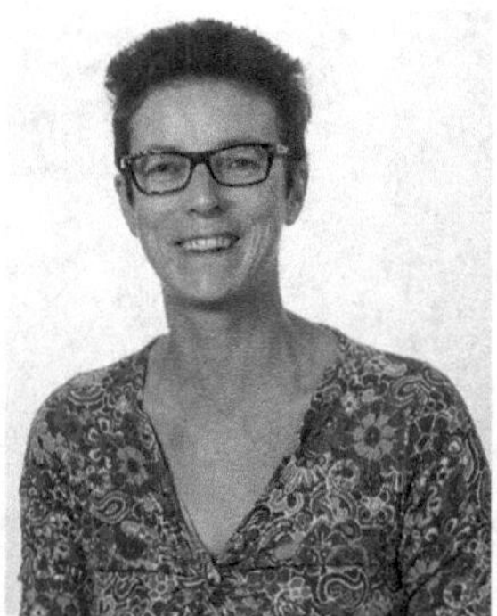

Professor Anne Osbourn leads a research group at the John Innes Centre in Norwich, UK, and is also a professor at the University of East Anglia. Her research focuses on plant natural products – function, synthesis, mechanisms of metabolic diversification and metabolic engineering. She is also a poet and has developed and co-ordinates the Science, Art and Writing (SAW) Initiative, a cross-curricular science education outreach programme (www.sawtrust.org).

Artistic Ways of Knowing: Thinking Like an Artist in the STEAM Classroom

Joanne Haroutounian

1 Introduction

The third graders are learning about the water cycle. Colorful posters display the different stages, from evaporation, transpiration, and condensation to precipitation and run-off. Tables are strewn with art projects depicting these stages and water-filled plastic bags are taped to the window showing signs of evaporation to measure each day.

Instead of sitting at tables, students begin today's class lying on the floor with their eyes closed, listening to *Clouds (Nuages)* by Debussy. After a reflective pause, questions arise. Which part of the water cycle does this music describe? How can we show this through movement? After students gently show the rise of evaporation to the music, they continue realizing the cycle through interpretive movement to *Beethoven's Pastoral Symphony (Storm)* and Smetana's *The Moldau*. The lesson closes as the rapids of the river widen to the ocean, with students swirling around the room personally expressing the water cycle through interpretive movement (Haroutounian 2017).

We can imagine what discussions may arise following this science/arts lesson. How did the music describe a cloud – a storm – a river? How did your movements change from one stage to the next in describing the water cycle? The students were actively engaged in the arts in this lesson while reinforcing their understanding of the water cycle. This STEAM lesson encouraged them to interpret creatively, through movement, how water evolves from gentle evaporation to exciting precipitation and fluid run-off. And it was fun!

The lesson also encouraged students to "think like artists" by listening with focused attention to the music (perceptual awareness and discrimination), internalizing how they would move (metaperception), and creatively moving to interpret

J. Haroutounian (✉)
George Mason University, Fairfax, VA, USA

© Springer Nature Switzerland AG 2019

A. J. Stewart et al. (eds.), *Converting STEM into STEAM Programs*,
Environmental Discourses in Science Education 5,
https://doi.org/10.1007/978-3-030-25101-7_12

what the music expresses (creative interpretation – dynamic of performance). As the students critique the activity, they realize how the movements they chose mirrored the concepts they were learning in science.

The educational community is excited about the possibilities that arise when we integrate the arts with STEM fields to create STEAM (Science, Technology, Engineering, Arts, Mathematics). The rationale for including the arts rests on the importance of creative innovation during the problem-solving process within the STEM fields (Quigley and Herro 2016; Wynn and Harris 2012; Guyotte et al. 2015). This fresh approach incorporates creative collaboration across disciplines that challenges students to broaden their perspectives to include somewhat novel answers arising from working through the arts.

Educators have the opportunity to examine *how* children learn through this transdisciplinary process. Students engrossed in the sciences can learn the value of thinking like an artist when working through complex problems. Students in the arts can realize how their unique way of "knowing" can contribute to scientific results.

> *What most of us lack in order to be artists is not the inceptive emotion, nor yet merely technical skill in execution. It is the capacity to work a vague idea and emotion over into terms of some definite medium.* (John Dewey 1934, p. 75)

2 The Arts: An Overview and Rationale for the "A" in STEAM

The role of the arts in education has always been influenced by political fluctuation and pragmatic challenges. During the 1980s to 1990s, most schools offered music and art classes at the elementary level with several classes a week. Secondary school options included orchestra, band, and chorus performance classes with some dance and drama options rounding out the arts presence in schools. Nonperformance options often included humanities classes, which offered an interdisciplinary approach merging history, culture, and the arts. For people educated in the arts during that era, those were decidedly "the good old days."

We now must face the stark reality of the role that arts education holds in the twenty-first century, where only 27 states consider the arts as a core academic subject to even include in the curriculum. Most of these states refer to the arts in general terms rather than delineating the specific artistic domains of music, visual arts, dance, and theater. *A Snapshot of State Policies for Arts Education* (Arts Education Partnership 2014) notes that most states include policies that relate to arts instruction: however, these policies vary widely by discipline and grade level, content, frequency, duration, and qualification for delivery of instruction.

Establishing the voluntary *National Standards for Arts Education* in 1994 provided some guidance for arts educators by creating content standards in the four art domains (music, visual arts, dance, and theater). As the educational climate steered toward "core standards," the arts developed the *National Core Arts Standards* in

2014 (National Coalition for Core Arts Standards 2014). This conceptual framework describes the process of creating, performing/presenting/producing, responding, and connecting. Figure 1 illustrates the National Core Arts Standards Matrix. It is significant that the arts chose to organize their standards by *process*, reflecting the artistic-creative process that is used in artistic ways of knowing, described later in this chapter (Haroutounian 2015, 2016, 2017).

Always on the defensive in proving the value of the arts in American education, multiple studies have linked arts education with achievement in academic areas. Such studies include traditional arts-specific curriculum, as well as arts integration. The Arts Education Partnership (http://www.aep-arts.com) lists 40 studies on their website linking the arts with academic achievement and thinking skills in various areas. A sampling of these studies:

- Music students do better in math, with arts-integrated math instruction facilitating mastery of computation and estimation skills (Courey et al. 2012; Harris 2007; Kinney and Forsythe 2005; Smithrim and Upitis 2005).
- Arts education develops students' critical thinking skills – comparing, hypothesizing, critiquing, and exploring alternative viewpoints (Heath et al. 1998; Montgomerie and Ferguson 1999).
- Students studying the arts score higher than their peers on tests measuring the ability to analyze information and solve complex problems (Costa-Giomi 1999; Korn 2010).

Winner and Hetland's meta-analysis (2000) showed minimal causal relationships between studying the arts and non-arts cognition. However, they explained that "we do not justify the presence of mathematics education by whether such

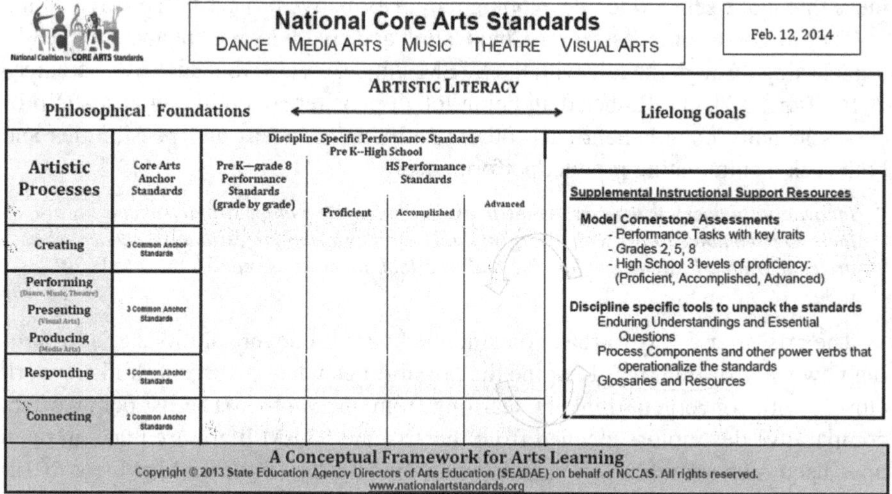

Fig. 1 National Core Arts Standards Matrix. Source: National Coalition for Core Arts Standards, http://www.nationalartsstandards.org/. Used with permission

study leads to stronger skills in English or Latin; nor should we justify the presence of arts education by whether such study leads to stronger skills in traditional academic areas" (Winner and Hetland 2000, p. 7). They conclude that there is a glaring oversight because "there are no studies to document the kinds of thinking that are developed through study of the arts" (Hetland et al. 2007, p. 4). This oversight may be addressed as STEAM research develops, recognizing how artistic ways of knowing play a part in the creative process for artist/student interdisciplinary experiences.

> *Whether work in the arts has consequences that extend to all aspects of the world cannot now be determined with any degree of confidence. What can be determined with a high degree of confidence is that work in the arts evokes, refines, and develops thinking in the arts. We might cautiously reason that meaningful experience in the arts might have some carryover to domains related to the sensory qualities in which the arts participate.* (Elliot Eisner 2002, p. xii)

Influenced again by political/educational priorities, the era of STEM immediately impacted the presence of the arts in American schools. With curricular emphasis on STEM subjects, arts programs were greatly marginalized. Schools devoted more time to STEM subjects by decreasing or eliminating the arts (Wynn and Harris 2012). Music and art classes, the mainstay of elementary arts education, met once a week, if at all. Budgetary limits in low-income schools eased out the arts to include more STEM for this population of students (Sousa and Pilecki 2013).

Used to ongoing pragmatic challenges from those beyond the arts education arena, the arts community forged ahead, offering ideas that are the basics of learning through the arts ("A") to those working in the academic/sciences of STEM. How can the arts make a difference in STEM? How can these "opposites" attract to create a more dynamic education for students? The STEAM literature offers concrete ideas that make sense when developing a transformative project-based curriculum.

Their rationale emphasizes how innovation and creativity are enhanced by meshing learning through the arts with the STEM subject areas (Duncan 2010; Saraniero 2016; Tarnoff 2011). Rather than being total opposites, scientific and artistic processes actually complement each other. Both explore ideas and possibilities and both seek a culminating product/performance.

> *Innovation happens when convergent thinkers, those who march straight ahead toward their goal, combine forces with divergent thinkers – those who professionally wander, who are comfortable being uncomfortable and who look for what is real.* (John Maeda 2012, p. 1)

The arts offer a way to reach outside the box with novel solutions and intriguing new ways of thinking. Imagine the possibilities when divergent thinkers work closely with convergent thinkers, learning from one another. The list below offers comparative descriptors gleaned from the arts and STEM literature that can open up a discussion about the rationale for including the "A" in STEAM (Maeda 2012; Sousa and Pilecki 2013; Thurley 2016; Haroutounian 2015, 2016):

STEAM – opposites attract

Arts	STEM
Subjective, intuitive, perceptive, novel	Objective, logical, analytical, useful
Divergent thinking	Convergent thinking
Creative interpretation and collaboration	Scientific process and research
Problem-finding, critical thinking/making	Problem-solving, critical thinking

Reviewing the vocabulary of thinking skills and processes on the list above, I see a challenge in future discussions that would be exciting for students who wander, enjoy being uncomfortable, and invigorated by the creative process. It's time to hold hands with your science colleague and search together for what is *real* – shall we dance?

3   STEAM Is Rising

3.1   *Early Childhood: A Natural!*

What better way to present the idea of STEAM education than with *Elmo the Musical*, appearing in 11-min episodic segments throughout the 43rd season of Sesame Street! Elmo sings and dances while using his imagination to explain math skills such as enumeration, relational concepts, addition/subtraction, and geometric shapes to solve problems. Sesame Street reinforces at-home activities through STEAM-based games and learning tools offered through their website. Rosemarie Truglio, Senior Vice President of Education and Research at Sesame Workshop, explains that "incorporating the arts into our STEM curriculum was an exciting and natural addition, as Sesame Street has always used music, visual and performing arts as tools to education and entertain children" (stemtosteam.org).

Hedda Sharapan worked with Fred Rogers of Mister Rogers' Neighborhood for decades and explains that many early childhood educators feel uncomfortable working with science-related topics. However, the arts are naturally part of early childhood education. "This new term, STEAM, can help early childhood educators to build the foundation of science-related knowledge, using the arts to encourage children to express their ideas in a wide variety of creative ways" (Sharapan 2012, p. 36).

Sharapan shared a STEAM everyday experience between an early childhood teacher and her class of 3-year-olds. The class came into the school after an exceptionally hot playtime outside. She lined them up in the hallway and asked if they felt a difference between the air outside and the air inside. It was a simple question that required a hesitant pause and the answer "It's colder." The everyday experience lesson followed with a discussion of air conditioning, what makes things hot, cold, and how engineers devised a way to create air conditioning. Students created other ways to stay cool making their own fans, feeling the air created through their own "engineering and art." These youngsters experienced authentic learning by paying attention to detail in the world around them.

3.2 K-12 Schools and Programs

The Rhode Island School of Design championed the STEAM movement early on, aligning it with their interest in furthering art and design as an element merged with STEM. Their objectives were to transform research policy to place art and design at the center of STEM, to encourage integration of art and design in K-20 education, and to influence employers to hire artists and designers to drive innovation. The school has connections with many STEAM projects and hosts STEAM conferences to help train teachers. Their website includes a number of case studies of STEAM school programs worth investigating (stemtosteam.org).

Not surprisingly, fresh new STEAM program initiatives in schools are beginning to grow nationwide, sponsored by national and local funding. Here is a sampling of projects emerging in different parts of the country:

Fort Garrison Elementary School (MD) is enjoying a dance-integrated program called "Teaching Science with Dance in Mind" with a goal to show how "dynamically dance can bring deep and complex learning to children" (Robelen 2010, p. 1). The program is sponsored by the nonprofit organization, Hands on Science Outreach. Program Director Rima Faber explains, "The more we teach through dance integration, the more we realize how dynamically it brings deep and complex learning to children" (Robelen 2010, p. 14).

Philadelphia Arts in Education Partnership is working with city schools to help elementary students better understand abstract concepts in science and mathematics (fractions, geometric shapes) through art-making projects, including a "fraction mural" displayed at one school. Education Director Raye Cohen notes that "visual arts just seems to be able to hone in on the concept, taking it from the abstract to the concrete so students are really able to understand it" (Robelen 2010, p. 8). The project has an intensive research component whose findings will be welcome in this new STEAM field.

The Wolf Trap Foundation for the Performing Arts, based in Vienna, VA, has developed early childhood initiatives that blend STEM with the arts. Funded by a 2010 federal Education Department grant and Northrop Grumman, it includes performing artists in theater, music, dance, and puppetry working alongside kindergarten and preschool teachers. Findings from the American Institute for Research study show that arts-integrated methods in early childhood education can increase students' math achievement by providing the equivalent of more than a month of additional learning. Wolf Trap participants outperformed their peers in the Early Math Diagnostic Assessment (EDMA) in 2 consecutive years (American Institute for Research, wolftrap.org, February 16, 2016).

The ArtScience Prize is built around the ideas of Harvard University Professor David Edwards, and the competition has expanded to Boston, Minneapolis, and Oklahoma City as well as international locations. The contest merges abstract concepts in the arts, design, and the sciences. Students work with abstract real-world concepts such as The Future of Water or Virtual Worlds bringing in the field of synthetic biology (Robelen 2011). Carrie Fitzsimmons, Executive Director of the ArtScience Labs in Cambridge, MA, states that "we are empowering young people to come up with their own ideas while exploring and playing in the arts and sciences. It's all fun, experiential learning, but we're teaching them to be critical thinkers and problem-solvers" (Robelen 2010, p. 8).

Hampton University Museum and School of Science, Engineering and Technology (VA) initiated "Jam Session: Jazz and Visual Art in Engineering" in 2010. The project applies jazz jams and improvisation principles to engineering students' design processes. Participants enjoyed John Sims' exhibit "Rhythm of Structure: Mathematics, Art and Poetic Reflections" (Wynn and Harris 2012).

Robious Middle School (VA) included "Keep Our Watershed Together…Be a Part of the Whole" project during the 2011–2012 school year which involved creating a mosaic depicting the natural processes within the ecosystem, the James River, and the food web. 380 sixth graders developed project sketchbooks creatively explaining what they learned from fieldwork at the river. Students sat with their sketchbooks on the riverbank painting botanicals with watercolors, using water from the river (Wynn and Harris 2012).

3.3   *STEAM University Research*

Texas A&M STEAM Camp   The study focused on student perception about using creativity in STEM projects during a 2-week STEAM camp experience for 104 students from seventh to twelfth grades. The camp used problem-based learning activities that built bridges with popsicle sticks, made lip gloss from organic materials, prepared a video to explain created products, created an app for a cell phone, built robots with Legos, and designed an object with 3D modeling software. Student surveys indicated students believed that problem-solving requires artistic solutions.

Three implications for education emerged from this study:

- The arts should preserve or regain their prominence in the educational system.
- Opportunities should be provided in formal school settings for students to use both creativity and logical thought processes in solving problems.
- Engagement in the arts has benefits emotionally, giving the arts an importance on their own, outside of STEM (Oner et al. 2016).

Transdisciplinary Design Studio This college course incorporated collaborative creativity using a visual-verbal methodological approach. It included 11 undergraduate and graduate students from disciplines of environmental and civil engineering, landscape architecture, and art education. Design challenges included exploring waste reduction and a water ethic, with local and global implications. Students created sculptures while working together to develop individual visual-verbal narratives.

STEAM approaches engage students in interdisciplinary explorations of complex social issues and offer collaborative engagements through the arts that nurture holistic, authentic, and dialogic perspectives. A transdisciplinary STEAM curriculum fosters reflection of and understanding of an individual's creative process. "While many focus on what the arts bring to the STEM conversation, we are also interested in what STEM might bring to the arts." (Guyotte et al. 2015, p. 31).

4 Artistic Ways of Knowing and STEAM

Artistic Ways of Knowing describes the way students who are fully engaged in the arts perceive and create through these experiences. They describe the perceptual/cognitive processes inherent while learning and working through the arts (Haroutounian 1995, 2002, 2014, 2015, 2016). This artistic process includes perceptual awareness and discrimination, metaperception, creative interpretation, the dynamic of performance/product, and critiquing (summarized in Fig. 2).

There is a broad consensus that the arts are important to include in STEM because they bring creativity to the problem-solving process. However, creativity is a very generalized term, so we lose sight of exactly how students think as they create. If we examine each step of this creative-artistic process, we more fully understand how students "think as artists." Teachers and researchers can document and study what they witness in STEAM classes for future research of artistic thinking (Hetland et al. 2007).

The explanation of each element in the artistic process will include brief descriptions of how one can observe this element of artistic knowing in a STEAM classroom. Enjoy realizing these activities vicariously, and imagine how students can expand conceptual understanding as they work together to solve problems.

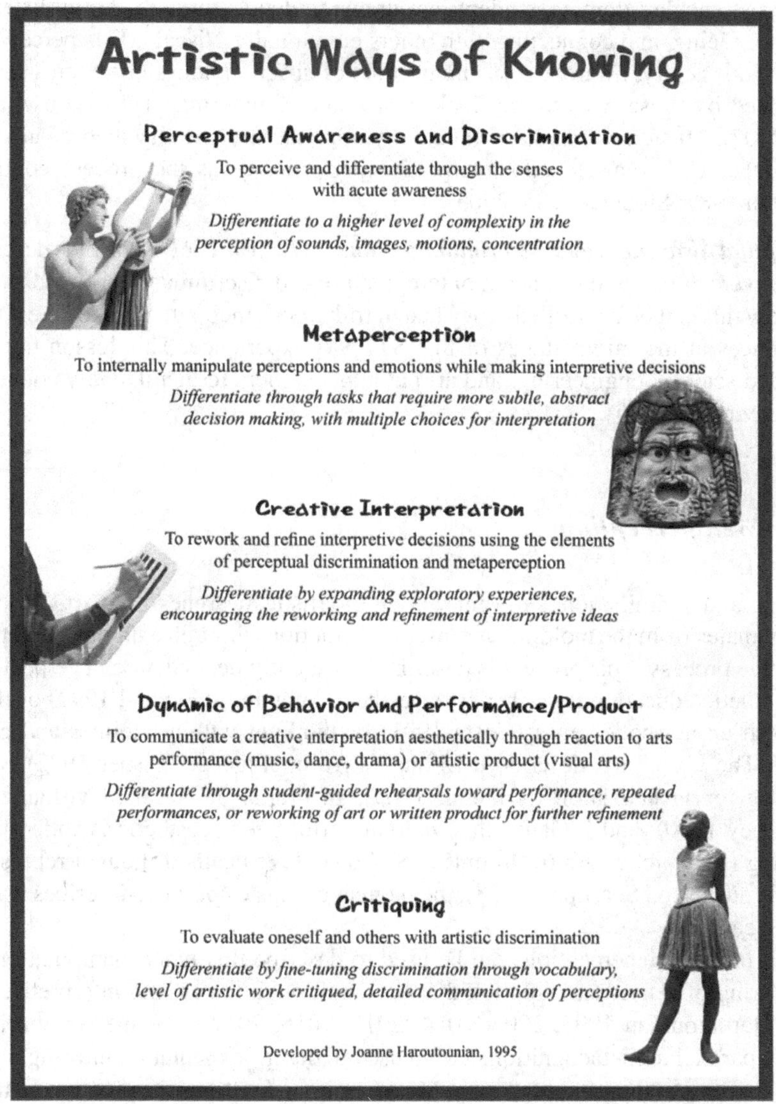

Fig. 2 Artistic Ways of Knowing. (Haroutounian 2015, p. xi)

4.1 Perceptual Awareness and Discrimination

Artistic knowing begins with fine-tuned sensory awareness and discrimination. Eisner (1986, p. 8) describes qualitative awareness as "critical abilities to differentiate, distinguish, to recognize and to make distinctions between many qualities that constitute our world."

Across each art domain, students are drawn to details through careful listening, moving, seeing, and connecting with others emotionally. Visual artists perceive the world with acuity, awareness of dimensions of space, color, and textures that go unnoticed by those who simply look (Clark and Zimmerman 1984; Hurwitz and Day 2007). Musicians can discriminate rhythm patterns, melodic shapes, and tonal colors that will translate into interpretive performance as the process continues (Gordon 1987; Sloboda 1985, 2005).

Perceptual Awareness and Discrimination and STEAM The 3-year-olds described earlier were drawn to the concept of temperature and discrimination of cold and hot and how this can be controlled. They had to truly focus their attention to "feel" these differences at the initial stages of this STEAM experience. This lesson not only included science, engineering, and art but allowed them to qualitatively connect to their environment.

4.2   Metaperception

Perceptual discrimination is the entry point of artistic awareness, but artistic knowing emanates from the molding of senses and emotions through a unique perceptual/cognitive process. This process is described differently dependent on perspectives.

Aesthetic education describes it as aesthetic knowing (Goodlad 1992) or thinking with an aesthetic sense (Costa 1991) or thinking within (Reimer and Smith 1992). The visual arts describe it as qualitative intelligence (Eisner 1972), visual thinking, or visualization (Arnheim 1969). In drama, it is called virtualization (Courtney 1990), and in dance this internal performance aesthetic is called body-thinking (McCutchen 2006). In music, Sloboda (1985) calls it thought representation of music, and Seashore (1938), the pioneer of music aptitude, describes it as the mind's ear.

The term, metaperception, can be used to describe this inner manipulation and monitoring of senses and emotion that occurs through the artistic interpretive process (Haroutounian 1995, 2002, 2014, 2015, 2016, 2017). Metaperception is the artistic parallel to metacognition, a term used to describe mental monitoring in cognitive thinking and problem-solving. Metacognition describes the process of "thinking about thinking" while metaperception describes the process of "perceiving/thinking about artistic intent." Artists filter and manipulate sensory perceptions combined with cognitive and expressive decision-making in order to create artistic solutions (Haroutounian 2002, p. xvi). The term is useful because it is understandable to both the artist and the scientist.

Metaperception and STEAM The students engaged in the Philadelphia Arts in Education group developed a Fraction Mural that was displayed at the school. The process began with students individually manipulating abstract concepts underlying fractions internally through metaperception. They shared their ideas collaboratively to transform these abstract concepts into a concrete artistic interpretation as a mural.

4.3 Creative Interpretation

As students work metaperceptively through an arts medium, the expressive reworking of ideas becomes an artistic-interpretive process that results in a creative interpretation. This interpretation grows through experimenting and adjusting ideas that will eventually become a performance, a product, or a critique. The arts are a perfect way to blend the invention of thought with perceptive/expression manipulation of ideas. Here is where the arts working with STEM content can produce products that uniquely extend outside the box.

Creative Interpretation and STEAM The Hampton University Museum and School of Science, Engineering and Technology merged the engineering design process with improvisation principles in jazz. Hearing jazz improvisation with all of its creative combinations working together can ignite ideas in engineering design through experimenting with ideas across both disciplines. Viewing John Sims' art exhibit adds visual inspiration to the mix. The collaborative result of all this creative interpretation can be innovative.

4.4 Dynamic of Behavior and Performance/Product

The musician, dancer, or actor communicates an interpretation to an audience through a performance. The audience experiences the performance sharing in the interpretive process. This mutual aesthetic experience of audience and performer creates the dynamic of performance. When experiencing an artwork, we find details that personally draw us to the work, creating this same dynamic between the artwork and viewer. The more aware the perception by the audience/viewer, the more dynamic the aesthetic connection.

Dynamic of Behavior and Performance/Product and STEAM The Transdisciplinary Design Studio went through the artistic-creative process using visual-verbal narratives to create interpretive ideas as they collaborated to design sculptures related to exploring waste and water real-world problems. One group placed empty water jugs on the campus lawn to represent the amount of water used in a normal 10-min shower, and fewer jugs demonstrating water used in an energy-efficient showerhead. This design challenge involved creating a visual artwork that engaged the community and established social significance in the dynamic between the artwork and the viewer.

4.5 Critiquing

The cyclical artistic process requires self-assessment of one's developing work in an arts medium as well as the astute critique of the artwork of others. This critique involves examination beyond performance through perception and reflection to add depth to the artistic process. Affording opportunities for students to reflect and critique their own work fosters artistic reasoning (Haroutounian 1995, 2002, 2013; Winner et al. 1992).

Critiquing and STEAM In the Robious Middle School watershed project, students prepared ongoing self-assessment of their individual project sketchbooks, as well as comments and ideas for fellow student sketchbooks. These sketchbook/journals functioned as works-in-progress and reference books. During their expedition to the river, and while testing the river water, the students used its water as a medium for watercolor interpretive drawings, thus expanding on this reflective self-assessment. This STEAM experience provided ongoing critique opportunities reflecting artistic critique methods.

Incorporating artistic ways of knowing in STEAM classrooms will ensure that arts experiences are adding depth and breadth to the STEM curriculum while developing the specific thinking skills that define the artistic process. STEM educators will observe artists at work in their classrooms and arts educators will realize the opportunities afforded to their students through creative collaboration with the math and sciences.

STEAM – What are we learning? Even though concepts for broadening STEM to STEAM are still emerging, we can already see the impact of bringing the arts into STEM classrooms.

- When the arts are integrated with STEM concepts, creative collaboration between different ways of thinking works positively in solving problems (Sousa and Pilecki 2013).
- The arts draw attention to the idea of broadening approaches to learning and seeking novel solutions that provide room for students to explore and question as well as solve problems (Robelen 2011).
- Students are engaged in solving real-world problems, realizing how the arts play a role in effectively communicating their ideas, utilizing an appropriate arts field that works for that project (Piro 2010).
- When the arts are added to STEM disciplines, students are less intimidated and become more engaged in learning. STEAM classes are stimulating and inspiring, with creativity allowing students to express themselves as they solve complex problems (Wynn and Harris 2012).
- The STEAM approach to education should not be the only arts experience available to students. Each domain of the arts – music, visual arts, theater, dance – must be recognized as an essential component of the basic curriculum (Oner et al. 2016).

It is exciting to see how quickly the concept of STEAM has taken hold across the country – basically starting less than 5 years ago, it is already accepted as a valuable transdisciplinary approach to education. As someone dedicated to the arts in education, I was initially hesitant to embrace STEAM, worried about the possibility of lessons that may use the arts peripherally rather than as an integral part of the collaborative learning process. The examination of a handful of STEAM projects showed exciting possibilities of student exploration across disciplines that can also provide opportunities for teachers to see artistic ways of knowing in action in their classrooms.

References

American Institute for Research. (2016, February 16). *Wolf Trap's early STEM/arts program translates into one to two months of additional math learning, AIR study finds.* American Institute for Research. Retrieved from http://www.air.org/news/press-release/wolf-trap-s-early-stem-arts-program-translates-one-to-two-months-additional-math

Arnheim, R. (1969). *Visual thinking.* Berkeley: Berkeley University of California Press.

Arts Education Partnership. 2014. *A snapshot of state policies for arts education.* Retrieved from http://www.aep-arts.com

Clark, G., & Zimmerman, E. (1984). *Educating artistically talented students.* Syracuse: Syracuse University Press.

Costa, A. (Ed.). (1991). *Developing minds: A resource for teaching thinking.* Alexandria: Association for Supervisor and Curriculum Development.

Costa-Giomi, E. (1999). The effects of three years of piano instruction on children's cognitive development. *Journal of Research in Music Education, 47,* 198–212.

Courey, S. J., Balogh, E., Siker, J. R., & Paik, J. (2012). Academic music: Music instruction to engage third-grade students in learning basic fraction concepts. *Educational Studies in Mathematics, 81,* 251–278.

Courtney, R. (1990). *Drama and intelligence: A cognitive theory.* Montreal: McGill – Queen's University Press.

Dewey, J. (1934). *Art as experience.* New York: Perigree Books, Xxx p.

Duncan, E. (2010April 9). *Well-rounded education a necessity, not a luxury.* Education Week's blog. Retrieved from http://blogs.edweek.org/edweek/curriculum/2010/04/heres_something_yo u_probably_d.html

Eisner, E. (1972). *Educating artistic vision.* New York: Macmillan.

Eisner, E. (1986). Appreciating children's aesthetic ways of knowing: An interview with Elliot Eisner by T. Buescher. *Journal for the Education of the Gifted, 10*(1), 7–15.

Eisner, E. (2002). *The arts and the creation of mind.* New Haven: Yale University Press.

Goodlad, J. (1992). Toward a place in the curriculum for the arts. In B. Reimer & R. Smith (Eds.), *The arts, education and aesthetic knowing: Ninety first yearbook of the national society for the study of education. Part III.* Chicago: The University of Chicago Press.

Gordon, E. (1987). *The nature, description, measurement, and evaluation of music aptitudes.* Chicago: G. I. A.

Guyotte, K. W., Sochacka, N. W., Costantino, T. E., Kellam, N., & Walther, J. (2015). Collaborative creativity in STEAM: Narratives of art education students' experiences in transdisciplinary spaces. *International Journal of Education & the Arts, 16*(15). Retrieved from http://www.ijea.org/v16n15/

Haroutounian, J. (1995). Talent identification and development in the arts: An artistic/educational dialogue. *Roeper Review, 18*(2), 112–117.

Haroutounian, J. (2002). *Kindling the spark: Recognizing and developing musical talent.* New York: Oxford University Press.

Haroutounian, J. (2013). *Musician's performance portfolio: A process for musical self-assessment.* San Diego: Neil A. Kjos Music.

Haroutounian, J. (2014). *Arts talent id: A framework for the identification of talented students in the arts.* Unionville NY: Royal Fireworks Press.

Haroutounian, J. (2015). *Artistic ways of knowing: How to think like an artist.* Unionville: Royal Fireworks Press.

Haroutounian, J. (2016). Artistic ways of knowing in gifted education: Encouraging every student to think like an artist. *Roeper Review, 39*(1), 44–48. https://doi.org/10.1080/02783193.2016.1 247397.

Haroutounian, J. (2017). *Artistic ways of knowing: A curriculum for the artist in every student.* Unionville: Royal Fireworks Press.

Harris, M. A. (2007). Differences in mathematics scores between students who receive traditional Montessori instruction and students who receive music enriched Montessori instruction. *Journal for Learning Through the Arts, 3*(1). Retrieved from http://www.escholarship.org/uc/item/07h5f866

Heath, S. B., Soep, E., & Roach, A. (1998). Living the arts through language-learning: A report on community-based youth organizations. *American for the Arts Monographs, 2*(7).

Hetland, L., Winner, E., Veenema, S., & Sheridan, K. (2007). *Studio thinking: The real benefits of visual arts education.* New York: Teachers College Press.

Hurwitz, A., & Day, M. (2007). *Children and their art* (8th ed.). Orlando: Harcourt Brace Jovanovich.

Kinney, D. W., & Forsythe, J. L. (2005). The effects of the arts IMPACT curriculum upon student performance on the Ohio fourth-grade proficiency test. *Bulletin of the Council for Research in Music Education, 164,* 35–48.

Korn, R. (2010). *Educational research: The art of problem solving.* New York: Solomon R. Guggenheim Museum Visitor Studies. Evaluation and Audience Research.

Maeda, J. (2012). STEM to STEAM: Arts in K-12 is key to building a strong economy. *Edutopi*a, Oct. 2, 2012. www.edutopia.org/blog/stem-to-steam-strengthens-economy-john-maeda

McCutchen, B. P. (2006). *Teaching dance as art in education.* Champaign: Human Kinetics.

Montgomerie, D., & Ferguson, J. (1999). Bears don't need phonics: An examination of the role of drama in laying the foundations for critical thinking in the reading process. *Research in Drama Education: The Journal of Applied Theatre and Performance, 4,* 11–20.

National Coalition for Core Arts Standards. (2014). *National core arts standards.* Retrieved from http://www.nationalartsstandards.org

Oner, A. T., Nite, S. B., Capraro, R. M., & Capraro, M. M. (2016). From STEM to STEAM: Students' beliefs about the use of their creativity. *e STEAM Journal, 2*(2): Article 6. https://doi.org/10.5642/steam.20160202.06. Available at: http://scholarship.claremont.edu/steam/vol2/iss2/6

Piro, J. (2010). Going from STEM to STEAM: The arts have a role in America's future, too. *Education Week, 29*(24), 28–29.

Quigley, C., & Herro, D. (2016). "Finding the joy in the unknown": Implementation of STEAM teaching practices in middle school science and math classrooms. *Journal of Science Education Technology, 25,* 410–426. https://doi.org/10.1007/s10956-016-9602-z.

Reimer, B., & Smith, R. (Eds.). (1992). *The arts, education, and aesthetic knowing: Ninety-first yearbook of the National Society for the study of education, part II.* Chicago: The University of Chicago Press.

Robelen, E. W. (2010). Schools integrate dance into core academics. *Education Week, 30*(12)., 1, 14–15 online Nov. 16.

Robelen, E. W. (2011). STEAM: Experts make case for adding arts to STEM. *Education Week, 31*(13), 8.

Saraniero, P. (2016). *Growing from STEM to STEAM: Tips to team up the arts and sciences in your classroom*. The Kennedy Center's ArtsEdge. Retrieved from https//artsedge.kennedycenter. org/educators/how-to/growing-from-stem-to-steam

Seashore, C. (1938). *Psychology of music*. New York: McGraw Hill.

Steamtostem.org./case studies. Sesame street + STEAM. Retrieved from http:/www.steamtostem. org. September 10, 2017.

Sharapan, H. (2012). From STEM to STEAM: How early childhood educators can apply Fred Roger's approach. *Young Children*, 36–40. National Association for the Education of Young Children.

Sloboda, J. (1985). *The musical mind: The cognitive psychology of music*. Oxford: Oxford University Press.

Sloboda, J. (2005). *Exploring the musical mind: Cognition, emotion, ability, function*. Oxford: Oxford University Press.

Smithrim, K., & Upitis, R. (2005). Learning through the arts: Lessons of engagement. *Canadian Journal of Education, 28*, 109–127.

Sousa, D. A., & Pilecki, T. (2013). *From STEM to STEAM: Using brain-compatible strategies to integrate the arts*. Thousand Oaks: Corwin.

Tarnoff, J. (2011). STEM to STEAM – Recognizing the value of creative skills in the competitiveness debate. *Huffpost Education*. Retrieved from http://huffingtonpost.com

Thurley, C. W. (2016). Infusing the arts into science and the sciences into the arts: An argument for interdisciplinary STEAM in higher education pathways. *e STEAM Journal, 2*(2): Article 18. 10.5642/steam.20160202.18. Available at: http://scholarship.claremont.edu/steam/vol2/iss2/18

Winner, E., & Hetland, L. (2000). The arts in education: Evaluating the evidence for a causal link. *The Journal of Aesthetic Education, 34*(3–4), 3–10.

Winner, E., Davidson, L., & Scripp, L. (Eds.). (1992). *Arts PROPEL: A handbook for music*. Cambridge, MA: Harvard Project Zero and Educational Testing Service, Xxx p.

Wolftrap.org. (2006February 16). New research analysis shows arts-integrated teaching translates to additional month of math learning. *Wolf Trap Media and Newsroom*. Retrieved from http:// wolftrap.org. September 20, 2017.

Wynn, T., & Harris, J. (2012). Toward a STEM and arts curriculum: Creating the teaching team. *Art Education*, 42–47.

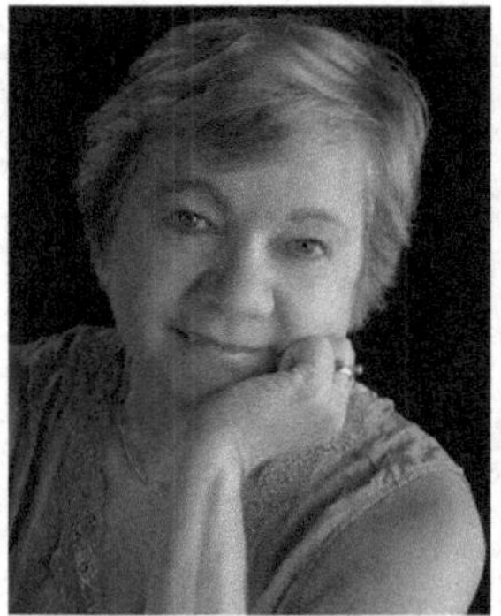

Joanne Haroutounian serves on the music faculty of George Mason University and is active internationally as a consultant in the areas of piano pedagogy, music, artistic thinking, creativity, and gifted/arts talent identification. She has 30 publications across gifted and musical fields, including *Artistic Ways of Knowing: How to Think like an Artist*.

Putting the STEAM in the River: Potential Transformative Roles of Science, Technology, Engineering, Arts, and Mathematics in School District Culture, Organization, Systems, and Learning Environments

John Puglisi and Beth V. Yeager

1 Introduction

The Rio School District's STEM-to-STEAM story was an effort to change a vision of learning and teaching. In 2012, Rio began building a culture of inquiry among key stakeholders, including teachers and students, to transform the District-wide vision, and teaching and learning, toward a twenty-first-century inquiry model. These efforts inevitably came to include the movement, in 2013, from STEM to STEAM (and beyond-STEAM) as an integral part of that model.

We present a case study, based on research from an ethnographic perspective (e.g., Castanheira et al. 2001), of a 5-year journey within Rio School District. In presenting this study, we make visible the importance of a leader with a clear vision and an investment in a distributed leadership model. The District's new vision included twenty-first-century practices and inquiry at its core and led to early decisions to focus on STEAM and other integrated disciplines, using a transdisciplinary approach rather than the more commonly used STEM approach.

Rio School District is a small- to mid-size public-school district on the West Coast of California. The District borders the state's Santa Clara River. Its population includes 90% Latinx students and 80% low-income students. Many of the District's students are English Language Learners. Rio (which means river in Spanish) currently has eight schools (six K-5 schools, two middle schools, and one K-8 Dual Immersion Academy). A new K-8 STEAM Academy will be the ninth school.

Dr. John Puglisi, one of the authors of this chapter, became the superintendent of Rio District beginning with the 2012–2013 school year. In the following excerpt

J. Puglisi (✉) · B. V. Yeager
Rio School District, Oxnard, CA, USA
e-mail: jpuglisi@rioschools.org; byeager@rioschools.org

© Springer Nature Switzerland AG 2019
A. J. Stewart et al. (eds.), *Converting STEM into STEAM Programs*,
Environmental Discourses in Science Education 5,
https://doi.org/10.1007/978-3-030-25101-7_13

from an interview with him, he provides an overview, context, and rationale for the journey we present, which began as he began his tenure as superintendent:

> Some five years ago… I took an initial assessment of the existing learning environments, teaching and learning practices, infrastructure and other aspects of the school district as part of my leadership plan. What I saw was not uncommon in many districts I had worked in or observed over the last thirty plus years. In fact, it was not unlike many school districts across America and especially in California. The Rio School District's vision of teaching and learning in 2012 was an amalgamation of a variety of pedagogical approaches. This amalgamation was loosely tied to a theory of standardization aimed at increasing the number of students achieving basic academic skills as they related to the California standards and as measured by the California Standards Tests (CSTs) which were multiple-choice assessments. There were many practices and activities that were attempting standardized and rote learning schemes and few opportunities for students to talk, make, create, and deeply think and problem solve. For this reason, I engaged staff and community in exploring new paradigms, practices, and mindsets. Many of these were well communicated and contextualized by the EdLeader21 network of Districts and schools (http://www.edleader21.com) and its focus on 21st Century learning practices, dubbed the '4 C's' (Collaboration, Communication, Creativity, Critical thinking). Along with participation in EdLeader21, we began to give permission for and encouragement of many learning activities that were less prevalent or non-existent in Rio's recent past. Since then, we have been evolving and clarifying a vision of learning and teaching that aims at developing 21st century practices through inquiry based learning activities that are more student driven and call upon students' abilities to think, create, and express their understandings by solving problems. Together with a broader set of organizational change processes, these catalytic endeavors helped forge a move towards a vision of inquiry-based 21st century learning environments and activities that naturally began to envelope the movement from STEM to STEAM. All of these efforts were aiming to transform learning environments such that children of all ages would be more engaged, more motivated, more included, and have more opportunities to collaborate, communicate in diverse ways, think critically about challenging academic work, be creative, and generally care for themselves, each other, their community and their learning and schooling. (September 2017)

In the following sections, we describe how this view unfolded. We share the framework that has guided our district's inquiry efforts. We also address some of the common constructs we have taken up as part of our approach to transforming teaching and learning by placing inquiry and the "5 C's" (the 5th C, "Caring," added by Rio in 2015) at the center. We do this to better explain what the 5-year inquiry efforts "looked and sounded like" and how these efforts informed the District's approach to "STEAM" rather than "STEM."

2   Using an Ethnographic Perspective in Order to Take an Inquiry Stance

We use Interactional Ethnography (IE) (Green et al. 2003), as a logic of inquiry, providing an orienting theoretical/methodological framework for our research and for our overarching inquiry as a District. In taking up this perspective (Green and Bloome 1997), we draw on Anderson-Levitt's (2006) argument to view ethnography as epistemology, a way of knowing, and therefore, as a logic-in-use rather than as simply a set of non-theoretically-based tools. From this perspective, we view classrooms and other educational settings, such as a school or a school district, as cultures-in-the-making, in which members co-construct patterned ways of being, knowing, and doing through their actions and interactions (e.g., Santa Barbara Classroom Discourse Group 1992).

By drawing on an IE perspective, complemented by microethnography (Bloome et al. 2005), we were able to conduct contrastive analyses both within and across time and events, including moment to moment interactions as reflected in oral and written discourse. This complementary approach further enabled us to examine, at multiple levels, the ways in which particular visions and perspectives, goals, processes and practices were initiated and (re)formulated over time. We identify, through analyses presented in our case study, processes and practices that were especially relevant at the district level over time.

3   A 5-Year Inquiry Process: A Case Study

In this case study, we describe how particular changes occurred over time, including changes in public discourse. We also address three factors that supported particular kinds of change – including the decision to incorporate STEAM as a central piece of the District's transformative efforts. The three factors are: (1) the role of particular kinds of leadership, (2) the importance of an inquiry stance (cf., Cochran-Smith and Lytle 2009), and (3) the importance of catalytic overlapping initiatives, which became patterns of practice/processes across time (e.g., Green et al. 2003).

Our consideration of leadership includes discussion of a leader who may also serve as an ideas/vision catalyst (sensu Brafman and Beckstrom 2008). We also consider the importance of a shared-expertise approach to distributed leadership (e.g., Gronn 2010). We found all of these factors were essential while building a culture centering on inquiry.

3.1   Key Leader as Catalyst

The "story" begins in 2012, when Dr. Puglisi entered as superintendent. He brought a clear personal and professional vision that, while continually evolving, was grounded in his 30 years as a teacher and administrator, including several years as superintendent of two smaller districts. He also brought a perspective on inquiring, an openness to observing and asking questions, on thinking in an integrated way about disciplines, and on thinking out of the box. These perspectives were grounded in both his professional and personal experience as an artist and musician, with interests in mathematics, science, and multiple other disciplines. His work as a graduate student in ethnography and technology also contributed to his personal and professional vision. This vision and experience would affect Rio's own vision and actions as a District.

From an ethnographic perspective, entering an existing community/culture required a period of stepping outside of his own experience in order to study that community. It required time to transition. The District's goals, adopted for 2012–2013, reflect that transition period, as well as what the superintendent was learning about the Rio school system's "vision of teaching and learning." The following excerpts from Rio's goals that year under the heading of "Student Achievement" serve as an example of what Dr. Puglisi was learning about the District in this area:

(a) *Improve % of students scoring proficient or advanced on CSTs...*(CA State Standards, California Standardized Tests)
(b) *Improve % of students with a positive level of changes on CSTs....*
(c) *Improve all API scores...* (CA Dept of Education Academic Performance Index, still in use in 2012–2013).

Clearly, Rio's orientation to teaching and learning at that time was tied strongly to student achievement, as measured on standardized, multiple-choice tests. However, the superintendent started inserting his vision, placing ideas into public space early in the academic year, via the superintendent's blog on the Rio website. In the following blog entry (Table 1), the language of "world-class learning" appears in the Rio District public discourse for the first time. Through this written discourse,

Table 1 Dr. John Puglisi blog entry – world-class learning – 12/4/12

World-class (wûrldkls)
adj.
1. Ranking among the foremost in the world; of an international standard of excellence; of the highest order: a world-class figure skater.
What does it mean to be a world-class learning organization?
Students have access to world-class educational opportunities and achieve world-class results?
Employees have access to world-class educational opportunities and achieve world-class results?
Partners have access to world-class educational opportunities and achieve world-class results?
The organization itself achieves world class results in its own learning processes?
The organization incorporates and develops integrating the best classical educational elements with the most modern and innovative twenty-first century tools and contexts.
It emphasizes access.
It emphasizes equity.
It emphasizes innovation and creativity.
It emphasizes excellence and striving.
It emphasizes critical thinking and collaboration.

http://rioschools.org/blog/world-class-learning-organization-12-4-12/

the superintendent proposed a potential change in Rio's vision and in its perception of itself as a District (i.e., how Rio might position itself in others' eyes).

An analysis of the discourse in the blog text makes visible the potential work being accomplished here. The superintendent shared, first, a dictionary definition of "world-class," rather than offering his own personal definition. He proposed this definition at the outset, inviting his readers (staff, families, the public beyond the District) to begin thinking about an educational organization within this context. The notion of "ranking among the foremost in the world" was new to the discourse by and about the Rio School District at this time. Rather than simply telling his readers, or positioning himself as the expert on world-class, he instead posed questions and invited readers to think about what these terms meant. Finally, he placed language into the public space that later, in various forms, became part of the District common language: examples included "21st century tools and contexts," "innovation and creativity," and "critical thinking and collaboration."

Fichtner (blog entry, retrieved 10/11/17) describes certain key leaders as catalysts when discussing Brafman and Beckstrom's (2008) review on theories of leaderless organizations. She describes catalysts as "visionaries that develop amazing ideas. But instead of holding onto those ideas for themselves, they share their ideas with others. And they *inspire* others to take action on them. And then, the Catalyst steps out of the way and lets the community carry the idea to incredible results." We don't suggest that Rio District is without leaders or a key leader. Rather, we contend that Dr. Puglisi was a visionary catalyst, in the context of a form of distributed leadership (Gronn 2010). In this particular distributed leadership arena, ideas are proposed and taken up, or not, by various stakeholders. We also suggest that the superintendent uses a particular approach to leadership, distributing it both vertically and horizontally. This leadership style focuses on notions of shared expertise, which is a critical factor in supporting change efforts in the area of STEAM. In other words, Dr. Puglisi has described, and our ethnographic evidence supports, that, as the number of participating stakeholders and their contributions grows, the collective result becomes increasingly co-constructed, rather than the product of one individual. Illustrative of this argument are the practices Dr. Puglisi identified in a presentation he gave at a STEAM Consortium in Northern California in 2015. These practices (Table 2) embody support for teachers in moving toward "do[ing] STEAM learning":

Table 2 Superintendent practices: "STEAM learning – do it!"

Setting goals/vision
Giving permission
Providing resources
Supporting professionals developing
Networking and partnerships with the broader community

3.2 Vision, Mission, and Goals Statements

The symbiotic process between superintendent vision/ideas and stakeholder ideas developed and became evident through adjustments in the publicly adopted District vision and mission statements and the Rio District goals. We offer evidence of such changes in vision and mission statements (Table 3). Information in Table 3, and information offered later in a discussion of Goals, establishes context for the subsequent discussion of corresponding changes in practice.

Although changes occurred in vision and mission statements between 2013–2014 and 2015–2018 (Table 3), references to empowering students were retained as a central aspect both times. The focus on empowering students appeared later in discussions of the work of Inquiry-Based Instructional Designers (IBID), a teacher-driven inquiry group. Empowering students also became a goal for the new STEAM Academy, emphasizing student-driven learning and student-driven/teacher facilitated investigations.

The focus on student-driven learning is even more explicit in Rio District's revised mission statement for 2015–2018. In addition, twenty-first-century "curriculum and tools" as well as the "5 C's," discussed below, are now central to what the District wants students to be "self-driven" to do. This change in mission statements, adopted by the Rio Governing Board, makes visible the efforts to coalesce and codify the transformative efforts summarized below.

Table 3 Rio vision and mission statements

2013–2014	2015–present
Vision statement	**Vision statement**
The Rio School District seeks to reflect a world and nation where society understands and values the **interdependency**∗ between nation, state, community, family, the democratic process, and the role of public schools in educating for the future. Within this vision, **students are at the center** of our commitment to stimulate empowerment and achieve the greatest possible potential as part of living in a diverse and changing world.	The Rio School District and community **empower** students to **achieve their full potential** in our community, our American democracy and our diverse and changing world.
Mission statement	**Mission statement**
The Rio School District nurtures the increasingly challenging **learning and development** of children from our pre-school through 8th grade utilizing the kind of curriculum and programs that can serve as a **hub for community development** as a whole.	The Rio School District and community nurture **learners to be self-driven** to fully engage with the twenty-first century curriculum and tools aimed to develop the **5C practices**: Communication, Collaboration, Critical Thinking, Creativity, and Caring.

∗Bolded words, inserted in originals by Rio District, represent intent to place emphasis

Examining District goal statements at three different points (2012–2013, 2013–2014, and 2015–2018) renders changes resulting from transformative efforts that occurred within the District even more explicit (Table 4).

For example, STEAM was first included in the District's goals in 2013–2014. This inclusion codified efforts to strengthen STEAM approaches through a Rio-led county network. STEAM Colloquiums were started, too (discussed below). In 2015–2018, Rio District adopted a goal to become "a 5C Focused, Digital Learning, S.T.E.A.M. (Science, Technology, Engineering, Arts, and Mathematics) community hub."

Table 4 District goals across the 5-year case study period

2012–1013 (adopted 1/16/13) *	2013–2014	2015–present
1. Student Achievement	Create a love for learning, engage in creativity, and value the process of inquiry and investigation.	Develop proficient and engaged readers, writers, and mathematicians.
(a) Improve % of students scoring proficient or advanced on CSTs District wide, at each school, and at each grade level	Provide world class learning opportunities for our entire educational community.	Improve the rate of English language development for all learners.
(b) Improve % of students with positive level changes on CSTs District wide, at each school, and at each grade level	Achieve and document achievement results based upon world class learning.	Develop teacher capacity as reading instructors and facilitators of the 5 Cs: Communication, Collaboration, Critical Thinking, Creativity, and Caring.
(c) Improve API scores District wide, and at each school, for all students, and for all sub groups of students	Be a role model for twenty-first century education in California and nationally, focusing on the 4 C's: Communication, Critical Thinking, Collaboration, and Creativity.	Develop the District as a 5C Focused, Digital Learning, S.T.E.A.M. (Science, Technology, Engineering, Arts, and Mathematics) community hub.
2. Student Health & Well-being	Develop our STEAM Education Center in Ventura County linked to broader efforts across the country.	
3. District Fiscal Well-being	Ensure the Financial Stability of the Rio School District.	
4. Dist/Schools short & long-term planning	Develop and implement the Master Plan for facilities growth and maintenance.	
5. Dist/Schools/Community Climate		
6. Tech Integration/ Innovation		
7. Facilities Development & Maintenance		
8. Partnerships		

*2012–2013 had eight major goal areas and multiple sub-goals. Only sub-goals for Student Achievement, with its focus on students, are used here for the contrastive analysis

3.3 Catalytic Patterns of Practice/Processes

Changes in vision, mission, and goal statements reflect the evolution of the District and its efforts toward building a STEAM-embedded culture of inquiry. We present a time-map of identified, interrelated major events for the Rio District, including those already discussed, to show what we refer to as transformative change efforts (Table 5).

Table 5 shows the flow of activities in the District, within and across years. It is also possible to see not only shifts in what has occurred in the District, but also the kinds of events or initiatives that appear to (re)occur across years, associated with those changes.

In 2012, Rio District was one of only a few, if not the only, districts in the Edleader21 PLC serving predominantly students of color and/or low-income or state-defined "low-performing" students, many of them second language learners. Still, Rio's vision of becoming an organization for "world-class learning" and of engaging in twenty-first-century learning was in synchrony with the vision of EdLeader21. Shortly, thereafter, "4 C's" (*C*ollaboration, *C*ommunication, *C*reativity, and *C*ritical Thinking) became part of the common language. Later, the team incorporated the 4 C's into the public language and actions of the District, and, as shown in Table 5, a fifth C, "*C*aring," was added in 2015. The 5 C's, in documents, in language and in classroom practice, are combined with, and as important as, the CA State Standards (e.g., Common Core and NGSS) in Rio District (see Table 3). We have identified and highlighted several additional catalytic processes/tracks in Table 5. The relationships among these processes are exemplified by EdLeader21, the 5 C's, and the evolving development of inquiry and inquiry design as central cores for teaching and learning. For example, we see inquiry as integral to the 5 C's. Inquiry practices such as asking questions, observing, making an argument, and collaborating with others, and processes such as innovating, and imagining, all flow through the actions of *c*ollaborating, thinking *c*ritically, *c*ommunicating, being *c*reative, and *c*aring (empathy). As a result, a focus on inquiry has been compatible with the efforts to implement the 5 C's, and the 5 C's are catalysts for movements in multiple directions.

Central to our STEAM effort has been the growth of a teacher-driven inquiry group, IBID (Inquiry-Based Instructional Designers). IBID is a grassroots inquiry group, evidence of the shared expertise approach to leadership. IBID began with seven teachers (representing five grade levels) and two facilitators. The facilitators were part of the superintendent's efforts to secure resources to support teachers in implementing an inquiry stance for teaching and learning. The efforts of IBID are deeply grounded in looking at their own practice, developing capacity to design inquiry-based instruction, guiding students in empowering themselves to drive their own inquiries/learning, and actively integrating disciplines using a practice-based approach. Table 6 contains an excerpt from a written reflection on IBID by a partici-

Table 5 Timeline of selected key activities driving the Rio district's inquiry process toward transformative teaching and learning and selected key patterns of practice/tracks

Year 1	Year 2	Year 3	Year 4	Year 5	Year 6
2012–2013	2013–2014	2014–2015	2015–2016	2016–2017	Fall, 2017 (Present)
New Super. starts 1st Super. "World Class Learning" statement	Super. active member of EdLeader21 (4 C's)	EdLeader21 teams begin meeting 'Caring' added – 5 C's	EdLeader21 Team meetings ongoing	EDLeader21 continues	EdLeader21 continues
Transitional goals adopted	New goals adopted, include "inquiry & investigation"	New goals, vision, mission adopted for 2015–2018			
IBID (Inquiry-Based Designers) formed from Summer PD Institute – Year 1	IBID Summer Institute & academic year meetings Year 2	IBID Inquiry group academic year meetings & Summer Institute Year 3	IBID Summer Institute & Inquiry group meetings. Year 4 (membership grows)	IBID Inquiry Group & Summer Institute (new members) – Year 5	IBID Inquiry Grp continues Year 6
VCSTEAM'N initiated - #s 1&2	VC STEAM'N Colloquium #3			STEAM Colloquiums #4 & 5	
	Summer Science Academy – Year 1	Summer Science Academy – Year 2	Summer Science Academy – Year 3	Summer Science Academy – Year 4	Plans for Summer Science Academy – Year 5
	Facilities Master Plan adopted (incl Science Academy, STEAM School)	STEAM School Bond passage	STEAM school stakeholder design continues	Recruitment for STEAM faculty & hiring	STEAM Academy official groundbreaking ceremony
		Stakeholder design for STEAM school begins		Orientation & first Summer Institute for STEAM Academy faculty (Curriculum & instruction design begins)	STEAM faculty testing & trying out – ongoing curriculum & instruction development

(continued)

Table 5 (continued)

Year 1	Year 2	Year 3	Year 4	Year 5	Year 6
2012–2013	2013–2014	2014–2015	2015–2016	2016–2017	Fall, 2017 (Present)
Inquiry-oriented consultants hired	Inquiry-based consultants as thinking partners begins	Inquiry-based partners as thinking partners (invited by teachers)	Inquiry-based partners as thinking partners	Inquiry-based partners as thinking partners (invited by teachers)	Inquiry-based consultants continue as thinking partners
CA Common Core State Standards adopted (PD)	CC & inquiry centered PD/Curriculum & assess development	PD – Project-based Learning and inquiry (CC)	PBL & inquiry-based PD (CC & NGSS)	Additional emphasis on literacies & mathematics 'Transdisciplinary' enters	Focus on literacies/ student-driven learning

Table 6 Teacher reflection on inquiry-based instructional design

Now I see IBID can take on many forms throughout the disciplines. Using literacy-based instruction, using animals, using plants, art. It is more about the <u>learning journey</u> that you allow your students to experience.
These empowering experiences allow students to unfold their thinking, their questions, and skills. When a student travels through this discovery and investigative process, they gain so much internally and are more confident in showing, sharing, telling, trying and collaborating. It is truly a transformative process that takes place and it will stay with them forever. (IBID Teacher, Summer, 2017)

pating teacher in the 2017 Summer Institute. This excerpt succinctly supports our view that the "IBID" track is a key catalytic process in moving to STEAM in our inquiry-based model.

With these overlapping efforts, several questions emerge. Why STEAM, for example? What is the relationship between STEAM and STEM, and why has STEM not been a part of the "official" language of the Rio District since 2012, even as it remains prevalent in other local, state, and national settings?

3.4   Why STEAM in the Context of a Culture of Inquiry?

To explore this question, we looked further at the events and patterns of practice in Table 5. We also examined the ways in which our catalytic leader and other partners have come to think about STEM and STEAM in the context of inquiry.

As noted earlier, the Rio superintendent began using STEAM rather than STEM in his personal and professional conversations in late 2012/early 2013. He did so because it made sense to him. At the time, STEM was a term used by teachers, and used in literature and in other settings. For the most part, the acronym STEM was invoked primarily by teachers and others in Rio (as it was in many districts and other public settings) as an acronym for four separate disciplines – as an easy way to think about these disciplines and encourage a focus on them in schools.

On one hand, while focusing on inquiry, as Dr. Puglisi described in an interview, "STEM (science, technology, engineering, and mathematics) learning activities provided an easy platform for many teachers and community to relate to in terms of doing things differently in schools. The District began these efforts by moving towards becoming a one-to-one District or one-to-the-world District, which provided each child a networkable computer device. Simultaneously, we looked at what we were doing in the name of science and math and for opportunities to engage in more hands on, minds on doing activities and projects that connected the STEM subject areas and practices."

As shown in Table 5, a key catalytic track, the Rio Summer Science Academy, began in 2013–2014 in the spirit of Dr. Puglisi's description above. This Academy was started in partnership with Dr. Jerome Clifford, a Physics Lecturer at California

State University, Channel Islands. The Academy has increased teacher and student capacity, and it allows teachers to pilot hands-on, minds-on activities for classroom use later.

But the issue of STEM versus STEAM, in public naming, was not really an issue. This was because Dr. Puglisi, first, and then the Rio District later, did not see it as an issue. As Dr. Puglisi described, his decision to focus initially on the STEM disciplines – as STEM – was pragmatic. In his mind, the arts were always present, as were the Humanities, Social Science, Literature, and so on. In addition, the superintendent and other partners also had inquiry and the 5 C's at the forefront of their thinking. Finally, multiple, sometimes arbitrary and confusing, definitions of STEM complicated matters (Seikmann 2016). Was it inclusive or exclusive? Was it compatible with our views of multidisciplinary approaches, for our view of twenty-first-century learning and inquiry? Some definitions hinted at yes, while others suggested no. So, in fact, STEM as a construct for District public discourse (and therefore overarching approach) was not particularly relevant or useful in Rio, despite particular activities in practice in the district in 2012–2013 and beyond. It just made more sense to include the arts and use STEAM.

In February 2013, Rio held its first VC STEAMN' Colloquium. This event centered on Symmetry (across disciplines) and was open to teachers throughout Ventura County. Dr. Puglisi named this widely popular event. By 2014, two additional Colloquiums had followed. As shown in Table 5, this key catalytic track, along with the later mission statement and goals, began codifying STEAM in the common Rio language and approach. STEAM became part of a system of professional development, too, that included a focus on Project Based Learning.

STEAM made sense to us in the context of our other transformative efforts. First, we saw inquiry and inquiry practices at the center of both STEM and STEAM. Second, as an artist and a musician, the superintendent understood how these disciplines fit with science, technology, engineering, and mathematics. Further, engaging in creativity (5 C's) and artistic thinking in multiple disciplines, and collectively bringing these multiple perspectives to solve complex problems, made sense to him, and to us. We took these transdisciplinary ideas and proposed them to teachers in various existing inquiry-oriented contexts. But key to this process was the leader as Catalyst, who proposed ideas and perspectives and then let go of them as members of our larger community (e.g., teachers, students, administrators, etc.) were inspired to assimilate, (re)formulate, and act on them. In doing so, they drew on the common language of inquiry, transdisciplinarity, and STEAM that we've co-constructed.

STEAM also makes sense in the context of our evolving notion of transdisciplinarity, which in turn has its roots in the context of the open nature of inquiry-based practices and design and of twenty-first-century learning practices. Transdisciplinary approaches are used primarily in higher education, but we see potential for their use in K-8 schools, as well. Transdisciplinarity involves (metaphorically speaking) looking through multiple disciplinary lenses (Klein 2000), making something new. We see value in working with students to draw on multiple perspectives, from multiple disciplines, to develop new approaches for solving problems. In other words, it takes multiple perspectives, whether they're included and named as STEAM disciplines or in addition to those disciplines, to address "wicked problems." Therefore, it makes sense to us to bring this to bear in problem-based learning approaches. At the same time, we value a practice-based approach. For this reason, we also consider artistic thinking and practice (Marshall 2014) as essential to transdisciplinary approaches. STEAM allows problems to be reviewed through different perspectives, but it is also a starting point for bringing additional perspectives to bear from other disciplines.

Why STEAM? Because the arts and artistic/musical thinking are essential and because all of our transformative efforts to date to build a culture of inquiry in the context of twenty-first-century learning, from a transdisciplinary perspective, cannot help but include STEAM as a central component.

All of this comes together (Table 5) in what has been a three-year journey to design, build, and open (in Fall, 2018) a new K-8 Rio STEAM Academy. While space does not permit a full description of this journey, we offer an event map to make it visible (Table 7).

As shown in Table 5, a new STEAM school was first envisioned in a Rio Master Plan and adopted in September 2014. There is no evidence in the Plan that there was ever any discussion of a "STEM" School.

Multiple interrelated processes occurred over time, including stakeholder Design Team meetings and conversations with a Chumash elder (a group spiritually and historically tied to the school site). All of these activities affected architectural plans and decisions about the STEAM Academy. As part of this same process, recruiting and selecting faculty for the school from existing teachers occurred in May 2017, more than a year and a half before the school was expected to open. Early hiring was unique, but also important, so that teachers could begin building community, develop a common language, and design guiding principles for planning curriculum and instruction.

Overarching goals (Table 8) for the STEAM Academy currently guide faculty work. Importantly, the Academy goals are grounded in the superintendent's vision, the broader Rio District vision, mission, and goals, and in what was co-constructed, over time, by the Design Teams and other stakeholders.

STEAM teachers as of this writing had participated in a 2-day orientation and a 3-week Summer Institute. Table 9 provides excerpts from reflective writings by faculty at the end of the Institute.

In these excerpts, we find references to key constructs, to community, and to STEAM, as well as reflections of the larger vision for the school that, in turn, was grounded in earlier and simultaneous transformative efforts over time.

Table 7 Event map/timeline: Overview of STEAM Academy development process to date

Year 1 (Aug 2014–July 2015)			Year 2 (Aug 2015–July 2016)			Year 3 (Aug 2016–July/Aug 2017)		
Aug–Dec	Jan–Apr	May–July	Aug–Dec	Jan–Apr	May–July	Aug–Dec	Jan–Apr	May–Jun
Facilities Master Plan Overarching goals for STEAM School	STEAM School sites tour (small group)	Citizens Oversight Committee meeting.			Draft initial Environ Impact Analysis report completed	Larger extended group development partners meet.		
Citizen Oversight Committee established (Board)	Initial meeting COC							
Measure "G" Bond Passage								
Design Team meetings 1&2 (sites and overarching hopes and vision for school)	Day-long Design Team Sessions and general meeting 3	Design Team general session 4 First meeting with Chumash elder and architect (and other partners) re: site spiritual and physical history, relationship to school design.	Design Team meeting 5	Ongoing meetings re: indigenous ways of knowing related to nearby river, to school design	Design Team meeting 6 (architect and Chumash elder present)			STEAM info meeting open to all Rio teachers Hiring process – positions opened to current Rio School faculty members.

					Core team leadership meetings begin (toward staffing, PD, etc.)	Preliminary draft framework for teamwork (professional work)	Core team leadership meetings	Applications, interviews, hiring (completed by April 6)
						Process for determining PD/ curriculum and instruction development facilitation	Finalize overarching goals description and application materials for staffing	
							Facilitator (PD) confirmed	

Table 8 STEAM Academy initial overarching goals

Twenty-first-century learning practices
Transdisciplinary, multi-age, student-driven/teacher-facilitated, STEAM-oriented content learning
Commitment to designing and planning inquiry-based curriculum and instruction in the contexts of place, culture, and "real-world" problems
Commitment to creating a STEAM school and team that will be a fluid and collaborative effort, welcoming of input and expertise from multiple sources, whether it be from students, other teachers, family members, or Chumash elders, Mixtec community groups, and other community members supportive of culturally responsive ways of learning, as well as other community partners

Table 9 Excerpts from STEAM foundational faculty end-of-summer-institute reflections

A	I used to think transdisciplinarity was a complicated concept and applying it will require a group of mindful people who can embrace chaos. Now I'm realizing that this group of educator/learners is up to the task and will be patient, empathizing practitioners as we navigate this new river.
B	I used to think the STEAM Academy was simply a school site that we would fill with students, teachers, and administrators. I'm now realizing that we are building a culture that will catapult the curriculum and our community into places that we've never been to. And that uncomfortable uncertainty is a good thing.
C	I used to think that we were designing a STEAM School. Now, I think we are building so much more. … a new way of being.
D	I used to think the STEAM Academy would be similar to a traditional school, just with more STEAM aspects to it and maker spaces. Now I'm realizing the STEAM Academy will be much different. I learned what transdisciplinary was and how everything will connect. I also realized our grades won't be siloed and we will be teaching multi-age.
E	We have a dynamic group of educators and leaders, where we all have individual strengths and personalities. What we have in common is the goal of becoming better teachers and leading our students to become stronger learners and future leaders. We are all willing to take chances and understand that we need to "go with the flow," regarding uncertainties/ challenges that will arise with a brand new school. I'm proud to be a member of this group… .

4 Conclusions

In this chapter, we reviewed a transformative journey toward change in teaching and learning over time in one school district – a journey toward building twenty-first-century learning within a culture of inquiry, which in turn encompasses a STEAM approach orientation rather than a STEM approach. A key leader who also served as a catalyst was critical in this journey, proposing ideas and visions, then stepping back to let the community consider them and reformulate them to create something new.

1. Rather than a "leaderless organization," the key leader supported a shared-expertise model of distributed leadership that facilitated change efforts.
2. Inquiry was a central aspect of both the transformative journey (inquiry as stance) and a central process (inquiry-based design and learning); it led to ways

of seeing inquiry at the center of STEAM, providing a rationale for including artistic thinking and practice as important components.

3. A series of through-time patterns of practice could be identified that served as catalytic processes. These processes were key in how STEAM efforts were named, codified within the public Rio space, and then (re)formulated.

The transition to new approaches to "doing school" has taken time to develop. It is not a top-down process of prescription and transmission. Rather, it is a process of envisioning and exploring and of dialogue and negotiation and a process of proposal and take up – or not. For some teachers, change has come slowly, particularly at the middle-school level. For others, especially in K-5, change has come more rapidly. The opportunity to really hear students' questions, to empower them to investigate them, or the opportunity to explore exciting possibilities through art and music together with science, technology, engineering, and math – it's all very exciting. Some of the teachers can take it up more fully; others are taking small steps. Over time, however, we see that STEAM activities and learning environment transformations are becoming more common each year.

Rio's classrooms are now more inquiry-based, transdisciplinary, engaging, differentiated, student-driven, blended, passion-based, and culturally situated. As a result, student and employee attendance has improved; student citizenship has improved; standardized literacy assessment results have improved; student, teacher, and parent perceptions of the District have improved; and a state-of-the-art K-8 STEAM Academy is (as of this writing) well on the path to opening. Rio is moving toward a more effective twenty-first-century learning organization that continues to aim to be "world class" by various measures or perceptions.

References

Anderson-Levitt, K. (2006). Ethnography. In J. L. Green, G. Camilli, & P. B. Elmore (Eds.), *Handbook of complementary methods in education research* (pp. 279–296). Mahwah: Lawrence Erlbaum & Associates for AERA.

Bloome, D., Carter, S. P., Otto, S., & Shuart-Faris, N. (2005). *Discourse analysis and the study of language and literacy events: A microethnographic perspective*. Mahwah: Lawrence Erlbaum.

Brafman, O., & Beckstrom, R. (2008). *The starfish and the spider: The unstoppable power of leaderless organizations*. New York: Portfolio. (Reprint Edition, July, 2008) (Penguin Group).

Castanheira, M., Crawford, T., Dixon, C., & Green, J. (2001). Interactional ethnography: An approach to studying the social construction of literate practices. In: Cumming, J.J. & Wyatt-Smith, C.M. (Eds), Special issue of *Linguistics and Education: Analyzing the Discourse Demands of the Curriculum,* Vol. 11(4), pp. 353–400.

Cochran-Smith, M., & Lytle, S. (2009). *Inquiry as stance: Practitioner research in the next generation*. New York: Teachers College Press.

EDLeader21. http://www.edleader21.com/

Fichtner, A. *The unstoppable power of leaderless organizations*. Hacker Chick Blog entry. Retrieved from https://hackerchick.com/the-unstoppable-power-of-leaderless-organizations/, 10/11/2017.

Green, J., & Bloome, D. (1997). Ethnography and ethnographers of and in education: A situated perspective. In J. Flood, S. B. Heath, & D. Lapp (Eds.), *Handbook for literacy educators: Research in the communicative and visual arts* (pp. 181–202). New York: Macmillan.

Green, J., Dixon, C., & Zaharlick, A. (2003). Ethnography as a logic of inquiry. In J. Flood, S. B. Heath, & D. Lapp (Eds.), *The handbook for research in the English language arts* (pp. 201–224). Mahwah: Erlbaum & Associates.

Gronn, P. (2010). Hybrid configurations of leadership. In A. Bryman, D. Collinson, K. Grint, B. Jackson, & M. Uhl-Bien (Eds.), *Sage handbook of leadership* (pp. 435–452). London: Sage.

Klein, J. T. (2000). Voices of Royaumont. In M. Somerville & D. Rapport (Eds.), *Transdisciplinarity: Recreating integrated knowledge* (pp. 3–13). Oxford: EOLSS.

Marshall, J. (2014). Transforming education through integrated art-centered learning. *Visual Inquiry: Learning and Teaching Art, 3*(3), 361–376.

Puglisi, J. (2012). *World-class learning organization 12/2/12*. District website Superintendent's blog entry. Rio School District. Retrieved from author personal archives, 10/7/17.

Santa Barbara Classroom Discourse Group. (1992). Do you see what we see? The referential and intertextual nature of classroom life. *Journal of Classroom Interaction, 27*(2), 29–36.

Seikmann, G. (2016). *What is STEM?: The Need for Unpacking Its Definitions and Applications*. Adelaide: National Centre for Vocational Education Research (NVER). Retrieved 10/11/17 from https://www.ncver.edu.au/__data/assets/pdf_file/002

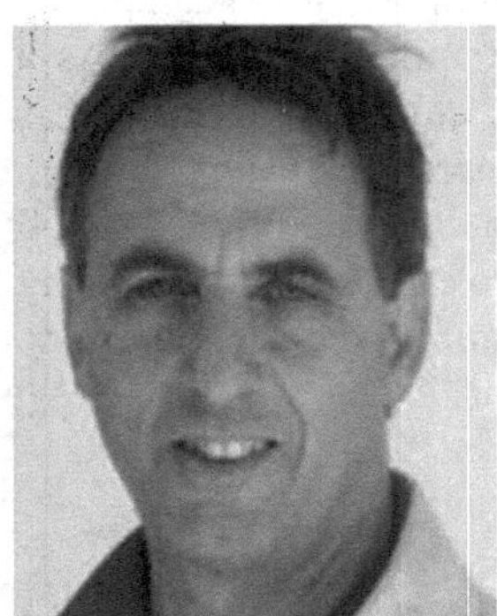

John Puglisi, PhD, is the Superintendent of the Rio School District and an Adjunct Professor in the Graduate School of Education at California State University, Channel Islands. His research concerns educational leadership, transdisciplinary, and the intersections of technology, teaching, and educational systems.

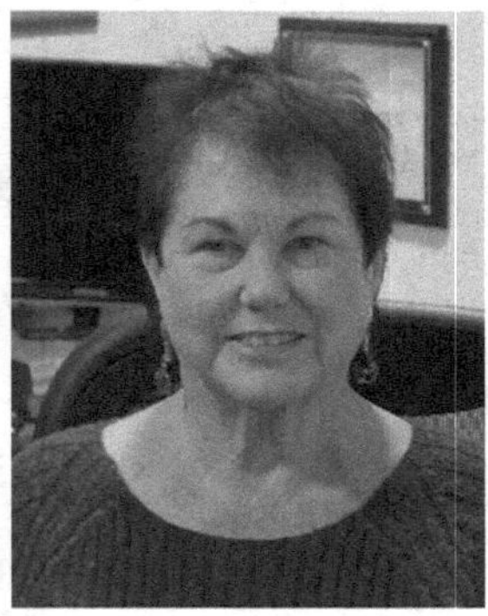

Beth V. Yeager, PhD, is a consultant in the areas of literacy, inquiry-based instruction, and practitioner research with Rio School District, as well as "embedded" ethnographer in the initiation of the STEAM school process in the District. A former elementary teacher, she is a researcher retired from California State University, East Bay (Institute for STEM Education) and Gevirtz Graduate School of Education at University of California, Santa Barbara. Her research interests include ethnographic and discourse-based studies in classroom and professional learning settings, with a focus on academic identities, language and social processes, and issues of equity of access (particularly for linguistically diverse students).

Emerging Scenarios to Enhance Creativity in Smart Cities Through STEAM Education and the Gradual Immersion Method

Jorge Sanabria-Z and Margarida Romero

1 Introduction

"Smart cities" are now envisioned as a way to improve civic life in the twenty-first century. In this context, smart cities need innovative industries, and innovative industries require creative people with profiles and skills, capable of solving problems through multidisciplinary means, with an acute vision for identifying opportunities and threats. These trends are driving a boom in creative spaces to learn and produce – things that break with the paradigm of traditional education, even in formal institutions. The projects that occur in these spaces are therefore disruptive: They combine, for example, the tools of digital fabrication and efforts to bring the community together, with the aim of looking for solutions to the immediate environment. The speed with which new technologies emerge stimulates a constant demand for guided learning in future scenarios of smart cities. We used the Gradual Immersion Method (GIM) to promote creativity and collaboration, taking advantage of the benefits of augmented reality in consideration of STEAM. In this chapter, we present two scenarios for applying GIM in a STEAM context, one of which addresses a proposal for co-creating an interactive and sustainable "smart cities" environment.

J. Sanabria-Z (✉)
Innovation Generation and Management Program, University of Guadalajara,
Guadalajara, Mexico
e-mail: sanabria@suv.udg.mx

M. Romero
Laboratoire d'Innovation et Numérique pour l'Éducation,
Université de Nice Sophia Antipolis, Nice, France
e-mail: Margarida.Romero@unice.fr

© Springer Nature Switzerland AG 2019
A. J. Stewart et al. (eds.), *Converting STEM into STEAM Programs*,
Environmental Discourses in Science Education 5,
https://doi.org/10.1007/978-3-030-25101-7_14

1.1   Challenges for the Twenty-First-Century Citizens

According to the World Economic Forum (2016), we have entered the fourth industrial revolution. In this context, new skills are needed to address societal challenges created by artificial intelligence (AI) and robotics technologies. Technologies that can replace some traditional human jobs are challenging the way we should consider education (Robinson and Aronica 2016; Romero et al. 2017), the rise of the creative class, and its impacts on cities (Florida 2014). According to Florida (op. cit.), there are three types of creative individuals: creative professionals, super-creative professionals, and bohemians. The creative class now comprises about 30% of the US workforce, including a 12% "super-creative core" of innovative professionals and a surrounding suite of about 18% of "creative professionals." Creative professionals add value to the (complex) solving task: These innovative individuals create new products and services. Moreover, the super-creative core goes beyond an existing problem and can set new ways of understanding or managing an existing reality. The super-creative core includes workers in science, engineering, education, computer programming, and research, as well as designers, artists, and media workers who are "fully engaged in the creative process" and "along with problem solving, their work may entail problem finding" (Florida 2002, p. 69). Creative professionals, in comparison, work primarily in healthcare, business and finance, and legal and educational sectors. They "draw on complex bodies of knowledge to solve specific problems," which requires a certain degree of education and lifelong learning. Florida (op. cit.) also describes a small group of Bohemians in the creative class, who develop creative activities without professional constraints. These different creative agents within Florida's concept of "creative class" can strongly affect cities. Studies in the field of innovation and economy of the last decade have identified interactions between the cities and the creative class. For instance, Hutton (2017) describes different types of creative activities that affect city organization and explains how cities shape the ways in which creative organizations, collectives, and individuals adapt their activities according to a city's constraints.

The city also is an important ecosystem for technologically based start-up companies which require a great concentration of talent, but do not need much for space for production, stock, or sales due to the high dematerialization of their creative activities. But even as city concentrates part of the creative class, another important phenomenon emerges: the smart city, which uses technologies to improve citizens' life in terms of "…livability, workability and sustainability. In simplest terms, there are three parts to that job: collecting, communicating and 'crunching'" (Smart Cities Council 2013). A smart city *collects* information about itself through sensors, other devices, and existing systems. It then *communicates* those data using wired or wireless networks. And finally, it *crunches* (analyzes) data to help explain what is happening now and what is likely to happen next.

Most cities within developed countries have embraced the Smart City approach: Barcelona, Singapore, Nice, Montréal, and thousands of other cities are eager to show the advances developed to improve their citizens' lives through the use of technologies.

According to a literary review by Caragliu et al. (2014), some smart city attributes are as follows:

- A capacity to combine infrastructure and connectivity for social, cultural, and urban development.
- An emphasis on developing an environment that favors business
- Social inclusion as part of an equitable city-growth strategy
- A tendency to promote creative and high-tech industries for generating creative cultures that strengthen urban performance
- Community participation in using technologies, an empowerment that allows them to adapt and innovate
- As a crucial component in their development, smart cities typically aim to achieve social and environmental sustainability.

In their review, Caragliu et al. (op. cit.) conclude by noting that combining the last two dimensions, "community participation" and "sustainability," is actually what best defines the smart city concept.

Considering the above perspectives, one can infer that beyond city infrastructure, it is essential to reinforce the creativity that feeds these intelligent spaces. This idea connects smart cities to education and business.

With the start of the twenty-first century, educational models and competency frameworks have been adjusted and reformed to help insert students in the professional arena and support them throughout life, according to contemporary demands. Among these frameworks are the P21's Framework for 21st Century Learning (P21 2010), enGauge 21st Century Skills (Burkhardt et al. 2003), Nuevo Modelo Educativo (SEP 2017), and the Assessment and Teaching of 21st Century Skills (ATC21S) project (Binkley et al. 2012).

Based on the Québec curriculum program (Gouvernement du Québec 2011) and the P21 Framework (Trilling and Fadel 2009), the #CoCreaTIC framework for the main five competencies for the twenty-first-century includes four competencies – critical thinking, collaboration, creativity, and problem solving – and includes computational thinking (Wing 2006; Romero 2016). Computational thinking is important for developing cognitive and metacognitive strategies for problem-solving based on computer science strategies, knowledge, and techniques (Fig. 1).

Within the #CoCreaTIC #5c21 framework, these five main competencies are considered independent, but the relations among them are an important aspect of the model. In techno-creative activities such as educational robotics team-based projects, learners use all five competencies (Kamga et al. 2016).

The #5c21 proposes the intersection of these five competencies, surrounded by critical thinking, resulting in the following seven combinations:

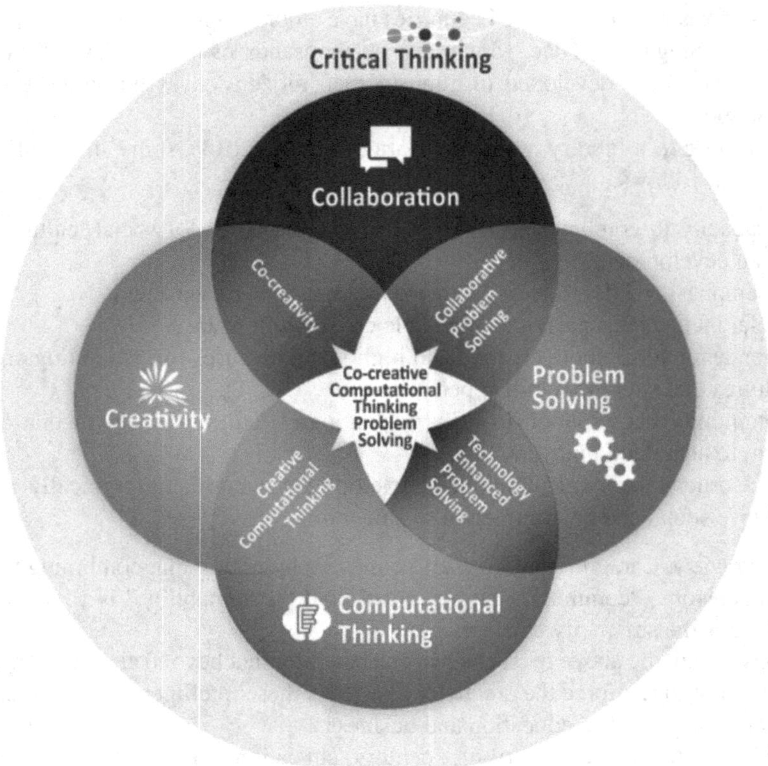

Fig. 1 Five competencies of CoCreaTIC's framework of twenty-first century (Romero 2016). (Reprinted from "Guide d'activités technocréatives pour les enfants du 21e siècle," (1st ed., p.6), Québec: Livres en Lignes du CRIRES, by M. Romero 2016)

- Collaborative problem-solving
- Computing-based problem-solving
- Creative computational thinking
- Co-creativity
- Creative problem-solving (not shown in diagram)
- Collaborative computational thinking (not shown in the diagram)
- The global link of the five competencies as critical, computational, and co-creative problem-solving

Thus, we propose a direct relationship between Caragliu et al.'s (2014) attributes of the smart city and CoCreaTIC's combined twenty-first-century competencies (Romero 2016). We suggest critical thinking is a general competency that influences all combinations; this idea is expressed in Table 1.

Table 1 Relationship between Caragliu et al.'s (2014) attributes of the smart cities CoCreaTIC's combined twenty-first-century competencies (Romero 2016)

Combined 21st century competencies / Attributes of smart cities	1 Collaborative Problem Solving	2 Computer Aid Problem Solving	3 Creative Computer Thinking	4 Co-Creativity	5 Creative Problem Solving	6 Collaborative Computer Thinking
A. Combination of infrastructure and connectivity for social, cultural and urban development.		■	■	■		■
B. Emphasis on the development of an environment that favors business.				■	■	
C. Social inclusion as part of the strategy of equitable growth of the city.	■			■	■	
D. Promotion of the creative and high-tech industries for the generation of creative cultures that strengthen urban performance.			■	■	■	
E. Participation and empowerment of the community in the use of technologies, in order to allow them to adapt and innovate.	■		■	■		■
F. Ownership of a strategy to achieve social and environmental sustainability, and thus the development of the smart city.	■	■		■	■	

Gray-shaded cells indicate cases where the two sets overlap. (Elaborated by J. Sanabria-Z & M. Romero)

The gray-shaded overlaps in Table 1 show the different combinations of competencies that potentially influence smart city attributes. Several attributes of smart cities involve collaborative solution proposals (e.g., C1, E3), while others require technologies used in creative or collaborative ways (e.g., A6, F2). Moreover, co-creativity has a major place in the set of smart city attributes, due to the social characteristics of the attributes which may reflect as organized and productive growth (i.e., column 4).

Twenty-first-century citizens must be able to monitor city performance and devise and implement a general growth strategy toward social and environmental sustainability.

Relationships between smart city attributes and twenty-first-century competencies permit different scenarios for smart city growth. Emerging from public and private spaces, some community-led programs and movements establish powerful learning initiatives. These events are drawing attention to the role of education in modeling the learning process and demonstrably integrate STEAM activities into the curricula, thus further developing student's competencies. We introduce two of these scenarios in the sections below and offer one scenario (relating to chemical bonds) in another chapter in this book.

2 Scenario 1: Competence-Driven Learning Through Makerspaces, Remote Robotics, and Interactive Museums

STEAM curricula can integrate various disciplines, through formal or nonformal education pathways, in spaces with educational services (such as museums) and in settings for co-creation (such as makerspaces). But for any of these routes, a significant challenge is that of helping students understand the importance of cross-disciplinary work. This generally can be accomplished through well-crafted cross-disciplinary projects.

2.1 Makerspaces

Many traditional and nontraditional educational institutions are now incorporating maker laboratories into their facilities. These maker laboratories are emerging because younger generations desire more experiential modes of learning. Several examples follow.

2.1.1 Digital Fabrication Laboratories in Mexico

In 2017, the first network of in-school maker laboratories in Mexico was launched at the University of Guadalajara (UdeG). These laboratories were designed for use by students at the high school level. The network is part of the FabLearn global network organized by the Graduate School of Education of Stanford University. The network started with nine labs in the state of Jalisco and demonstrates the academy's interests in discovering alternatives to traditional formal education. The nine maker labs support informal learning with relevance to the curriculum and social innovation: they evoke solutions from within, to address challenges of the immediate environment.

2.1.2 The EspaceLab of Quebec

This community-managed makerspace, in Québec, Canada, links its activities with those in regional schools. The lab is located in the library Monique Corriveau in Sainte-Foy and is a member of the FabLabs community; it facilitates access for users with diverse technological skills, of all ages. The EspaceLab Junior offers workshops for people between ages 8 and 14, fostering the culture and skills of co-creative problem-solving, technological innovation, scientific approach, and digital manufacturing. Some of the projects make items from recycled or low-cost materials and electrical components, but some students may use Scratch visual programming or educational robots. Teachers from five schools in Quebec City and two schools in Montreal participate in techno-creative activities that can be adapted to each of the classes and different programming technologies, electronics, and educational robotics.

2.1.3   The Project #SmartCityMaker

A Franco-Canadian initiative referred to as #SmartCityMaker brings together researchers, students, and parents. This initiative is organized by the School of the Teaching and Education, the National Institute of Research in Computer Science and Automation, and Laval University (project renamed as #fabville). The project involves co-constructing an ecotech smart city in an interdisciplinary makers lab scenario; citizens help in building a city model that offers techno-creative solutions to sustainable development challenges and risk management. Moreover, through a pedagogically innovative approach, #SmartCityMaker helps develop students' twenty-first-century skills such as collaborative problem-solving, co-creativity, and computational thinking, by relating the students to their territory and to the digital world (Anciaux et al. 2016).

2.1.4   #R2T2 Remote Robotics

Recent international collaborations target lower education levels. Such is the case of the project R2T2, which seeks to activate the intra-country interactions among young engineers (6–15 years old). The R2T2 approach involves a series of space missions that brought together several countries that interact online with tiny Thymio robots to accomplish various adventures. The R2T2 project is sponsored by the Swiss Space Center and manifests through the Federal Polytechnic School of Lausanne and the National Center of Competence in Robotic Research. It is an educational robotics challenge that simulates an international mission where children from several countries control remote Thymio robots located in Switzerland.

"R2T2 Caribbean and America" and "R2T2 Meteor" events were launched in 2017. These events were designed to promote problem resolution, critical thinking, collaboration, and teamwork among elementary-age students in Mexico, France, Canada, Russia, Santa Lucía, and elsewhere.

The four cases above show the relevance of planning with views at developing competencies considering contemporary challenging scenarios. In these cases, competencies of the twenty-first century are developed using artistic components as part of the activities. UdeG labs training, which is based on GIM, requires participants to use digital design tools in 2D, 3D, and RA to construct their idea sketches, which are finalized in 3D physical prototyping. Likewise, students working on the #SmartCityMaker and #fabville projects construct city models with materials such as cardboard and recycled materials. The R2T2 project uses a scale model of power generators in Switzerland and simulates the environment from another planet; participants control their robots remotely to complete a mission. For most of these projects, it would be impossible to imagine the results without the use of art tools and design software, which justifies the importance of considering STEAM to achieve the effective development of co-creation competencies.

3 The Gradual Immersion Method for Enhancing Creativity and Collaboration

We devised the Gradual Immersion Method (GIM) as an "intuitive approach involving learners in creativity-enhancing activities that familiarize them with the use of technologies for digital creation" (Sanabria 2015). The GIM is a pedagogical–cognitive approach based on the creative process and is composed of three cyclical phases of generation and exploration. These phases are applied through three modules: familiarization, digital creation, and exhibition (Fig. 2).

AR stands for augmented reality. Adapted from "The Gradual Immersion Method (GIM): Pedagogical Transformation into Mixed Reality" by J. Sanabria, 2015, *Procedia Computer Science*, 75, p. 371

The three modules work by aiding participants' navigation into the appropriation of a certain topic through various activities. For example, during an introduction to the work of iconic architect Luis Barragan (see Fig. 3), various scenarios may be proposed to help student understand his building-design technique.

Module I (familiarization) identifies a Learning Object (LO) comprised of structured activities. The LO includes instructions, goals, tasks, and evaluations (Wiley 2000). Each set of interactive activities allows students to familiarize themselves on a topic by solving short challenges and reflecting on their learning experience and/or perception of the activities. A LO about Luis Barragan's buildings, for example, uses an interactive screen and may first instruct the participant to observe a number of his works to identify their common features. The students then explore how Barragan's color palette changes the image of a gray building by adding colors digitally into pictures. Finally, the students associate 3D geometric figures with a Barragan masterpiece by dragging them to it. At any time, the participant may be instructed to reflect on the features of Barragan's style and input those conclusions into the system, what in the GIM are called "criteria," a sort of self-description of

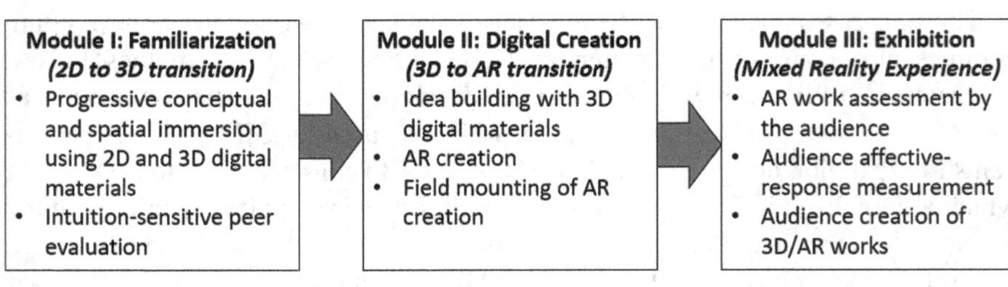

Fig. 2 The three GIM modules: familiarization, digital creation, and exhibition

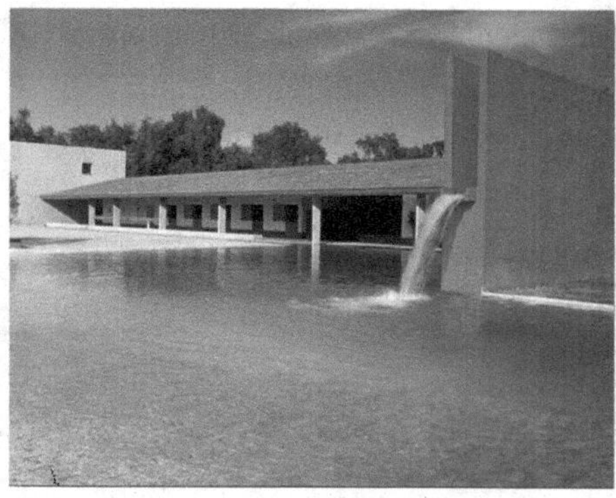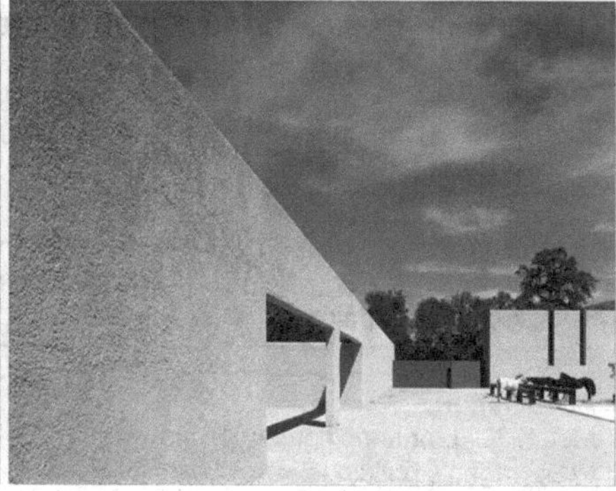

Fig. 3 Architect Luis Barragan's "Cuadra San Cristobal," Los Clubes, Mexico City, 1966–1968. Photos: Armando Salas Portugal (Barragan Foundation, 2019). Reprinted with permission; D.R. © Barragan Foundation / SOMAAP / 2019

their understanding of how activities are perceived or what the activities taught them (e.g., "In this activity I realized that Barragan tended to design wide spaces harmonized with giant geometric structures on bright colors"). At the end of the module, students evaluate their level of learning by identifying Barragan's buildings among a provided variety of famous artists' creations. The different activities allow students to interact with didactic materials, gradually advancing from 2D to 3D models. The module also allows peer evaluation as an activity is completed, depending on the interests of the instructor or facilitator.

Module II (digital creation) starts when participants understand the theme from different perspectives. By this point, it is expected that participants have characterized their own criteria about the LO in Module I. Module II provides a digital-creation LO, instructing students to select tools for constructing 3D concepts/ models or even audios, based on their previous criteria. Here, an example of the instruction is "Design your own Barragan's style studio using the 3D digital tools on the screen." Module II culminates by transferring their 3D digital creation into an augmented reality (AR) artistic expression, by anchoring it into the physical context (e.g., a 3D digital window frame based on Barragan's style, positioned on the façade of an ancient building using a geolocation function on a mobile device).

Module III (exhibition) involves exhibiting the participants' proposals as a mix between the physical and digital work, known as a "mixed-reality experience." For example, by pointing a mobile device to an apartment building, different shapes and sizes of Barragan architecture 3D digital elements, such as windows or fountains, may be seen floating or anchored on the facades. Students seek ways to obtain feedback on the project from a target audience. The audience, while involved in evaluating the project, also comes to understand the exercise of overlapping 3D digital elements on buildings to simulate Barragan's work. In this way, the audience is encouraged to co-create other versions of the project components, their own digital windows using Barragan's style, for instance, thus further disseminating learning.

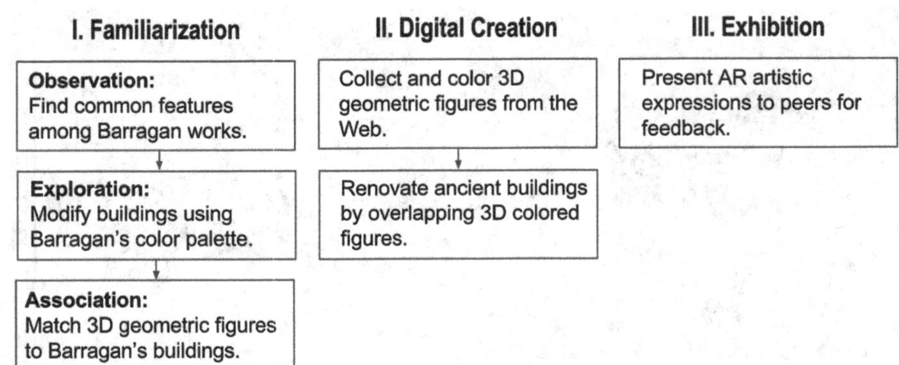

Fig. 4 Example representation of the three GIM modules for learning about Luis Barragan's architectural style. (Elaborated by J. Sanabria-Z and M. Romero)

The diagram below shows how the activities described above can be structured through the GIM three modules so students can learn about Luis Barragan's architectural style (see Fig. 4).

We offer two examples, below, showing how the GIM was implemented. These examples emphasize creative aspects of learning through GIM, with discipline focused on artistic creation through Surrealism and on smart city habitats, respectively.

3.1 Applying GIM to "Creating Surreal Expressions"

This example centered on learning features about the artistic movement of Surrealism. It was a collaborative activity accomplished by high school students in a museum environment. The project focused on the concept of bisociation (Koestler 1964). Bisociation is understood as interactions between two elements (e.g., photographs of people, animals, or things) that appear to be unrelated, but which, when combined, establish a third element with a meaning not evident in the isolated original elements. We track the bisociation project successively through the three GIM modules in the section below (Sanabria et al. 2015).

3.2 Module I: Familiarization with Surrealist Works

Students from technology, art, and competency-centered high schools were organized into mixed teams (five to six students per team). The team members then were allowed to become familiar with the idea of Surrealism by solving a set of activities (the Learning Object) based on bisociation. In these activities, students were challenged to solve topics by observation, combination, association, grouping, and discernment before evaluating their learning. Figure 5, for example, shows performance during the activity "Grouping," in which the students drag images of surrealist

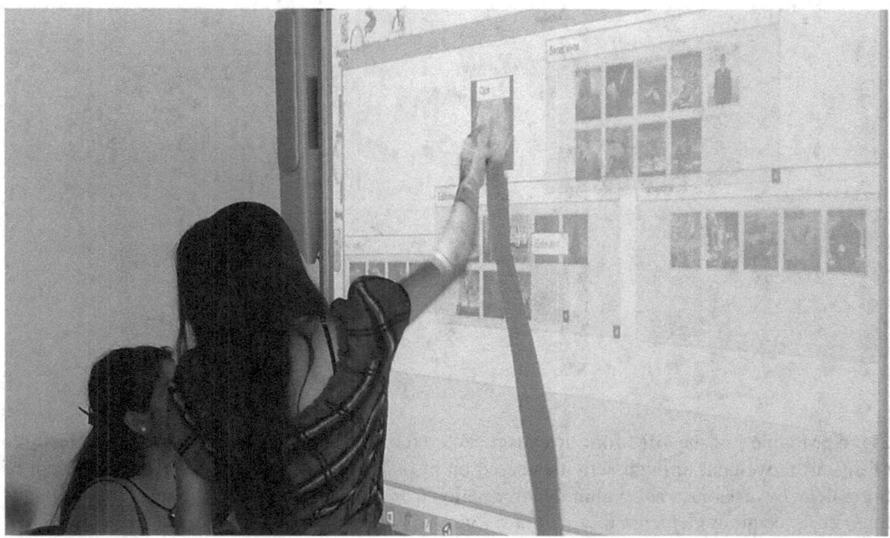

Fig. 5 A team of students at a museum, solving the activity "Grouping" by dragging pictures of surrealist paintings into groups and labeling them. Sanabria-Z, J. (photographer) (2017). *Grouping on AMLA* [photograph]. Guadalajara, Jal

paintings into groups, which are labeled by the team members regarding their own perceptions (paradise, nature, joy, etc.).

The learning objectives included five activities and an evaluation task and were displayed on interactive whiteboards. After accomplishing the activities, students assimilated the topics by evaluating other teams' work. In the figure below, the interface for each activity is shown: *Observation* aimed at building criteria about the common characteristics of various bisociated images. *Combination* aimed at building their own bisociations using animals, objects, and persons. *Association* focused on realizing what elements correspond to the painting by dragging them to it. *Grouping* was a perception task to understand what features are common among surrealist paintings. *Discerning* showed two similar paintings from different art movements, so that students could choose the one portraying surrealist features. The *Evaluation* task showed the bisociation proposals made on the *Combination* activity, asking other teams to score them on a creativity scale showing levels of originality, flexibility, and fluency (see Fig. 6).

3.3 Module II: Creating Surreal Expressions with Augmented Reality

Once the surrealist concept was appropriated, the student teams propose their own version of bisociated elements. This was done using an augmented reality (AR) application in a mobile device to anchor provided 3D digital objects in the physical environment (Arámburo-Lizárraga and Sanabria 2015) (Fig. 7).

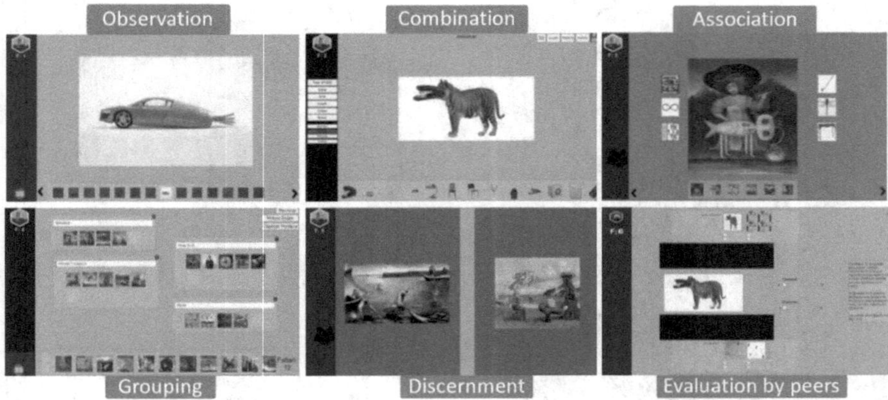

Fig. 6 Module I of the GIM for Surrealism centered on students' appropriation of some features of this art movement through activities based on bisociation. Students assimilated the concept of Surrealism by tasks such as combining information from pairs of images, grouping surrealist paintings, or associating elements to a painting (Sanabria and Arámburo-Lizárraga 2017). Reprinted from "Enhancing 21st Century Skills with AR: Using the Gradual Immersion Method to develop Collaborative Creativity" by J. Sanabria and J. Arámburo-Lizárraga, 2017, *Eurasia Journal of Mathematics, Science and Technology Education*, 13(2), p. 497

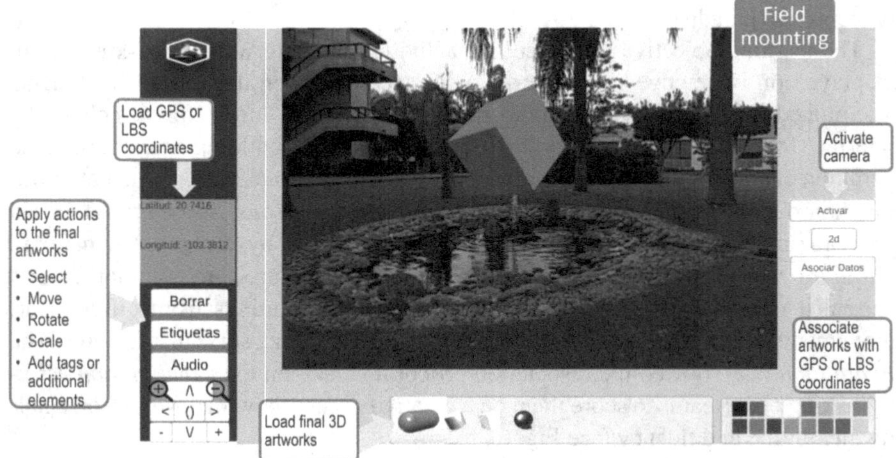

Fig. 7 Module II of the GIM applied to surrealistic expression. Creation was accomplished with the help of AR operated on a mobile (Sanabria and Arámburo-Lizárraga 2017). Reprinted from "Enhancing 21st Century Skills with AR: Using the Gradual Immersion Method to develop Collaborative Creativity" by J. Sanabria and J. Arámburo-Lizárraga, 2017, *Eurasia Journal of Mathematics, Science and Technology Education*, 13(2), p. 498

3.4   Module III: Exhibition of Surrealist Expressions

In Module III, the students displayed their surrealist expressions to an audience, a researcher, or to their peers (Fig. 8). The audience may provide feedback using instruments or scales provided by the researcher or instructor. For instance, evaluators could visit the physical location where the digital elements are anchored and used the scale in the application to record their impressions. For this project, the AR surrealist expressions created by the students were rated by the researchers using the same scale as in the evaluation or Module I to measure talent through originality, fluency, and flexibility of the proposed works.

3.5   Learnings from the Co-creative Experience

In the research of surrealist expressions created by participants with technological, artistic, and competency-centered profiles, we compared GIM results with the Multifactorial Evaluation of Creativity (MUEC) (Sánchez Escobedo et al. 2009), an instrument that measures creative talent. This comparison was based on the same parameters (originality, flexibility, and fluency) and showed that the GIM reduces differences between student profiles, suggesting that its modules facilitated the creative talent of the participants. In general, the significance or the method is how it stimulated students and allowed us to evaluate the creative competency from

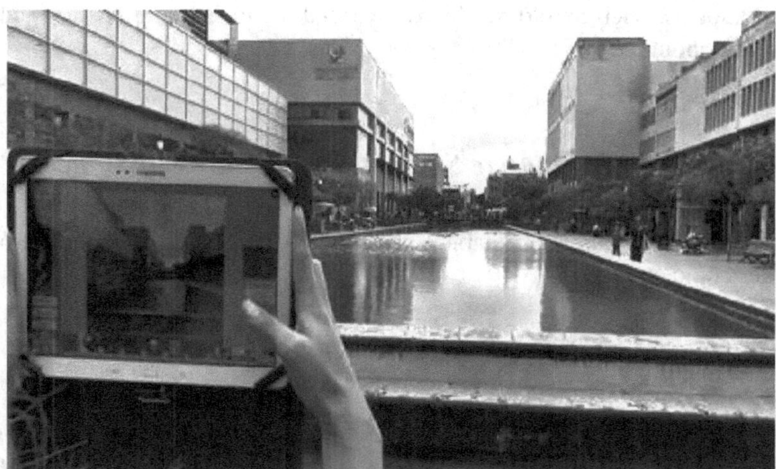

Fig. 8 Module III of the GIM on surrealistic expressions, using AR supplied through a mobile device (Sanabria and Arámburo-Lizárraga 2017). Reprinted from "Enhancing 21st Century Skills with AR: Using the Gradual Immersion Method to develop Collaborative Creativity" by J. Sanabria and J. Arámburo-Lizárraga, 2017, *Eurasia Journal of Mathematics, Science and Technology Education*, 13(2), p. 498

students of a wide range of profiles, using digital technologies in an intuitive fashion (Sanabria et al. 2015).

4 Scenario 2: GIM-Based Co-creative Proposal for Smart City Habitats

A GIM approach can increase student creativity by supporting "community participation" and "sustainability" objectives under smart city attributes. This GIM-based project, dealing with habitats, involves junior high school students, 11 to 15 years old. The students identify smart city characteristics through team-constructed activities. Accomplishing these objectives allows the students to generate proposals and receive feedback on their implementation plans. The instructional design for the selection of the stimuli to be provided to students – that is, the images and models – should be carefully planned by the researchers, considering Caragliu et al.'s (2014) summary of attributes of smart cities introduced at the beginning of this chapter ("Combination of infrastructure and connectivity for social, cultural, and urban development," "Emphasis on the development of an environment that favors business," and so forth).

Familiarization activities for the Learning Object in Module I for smart city habitats, the steps for digital creation in Module II, and guidelines for conducting the exhibition are shown in Fig. 9. These items may inspire guidelines for designing a more integral set of activities, in which the topics should relate to smart city attributes and the Learning Object might include different inputs besides pictures and 3D models, such as blogs, videos, and audios, to help enrich the students' knowledge about the topic.

4.1 Module I: Familiarize Students with Smart City Concepts

To familiarize students about habitat co-creation for smart cities, the following activities could apply:

4.1.1 Reflecting

Futuristic habitat solution videos are shown representing various living challenges, such as transportation, access to food, waste disposal, green space, affordable housing, etc. In this activity, the participants should reflect on needs and resources and generate criteria for these things (see Fig. 10).

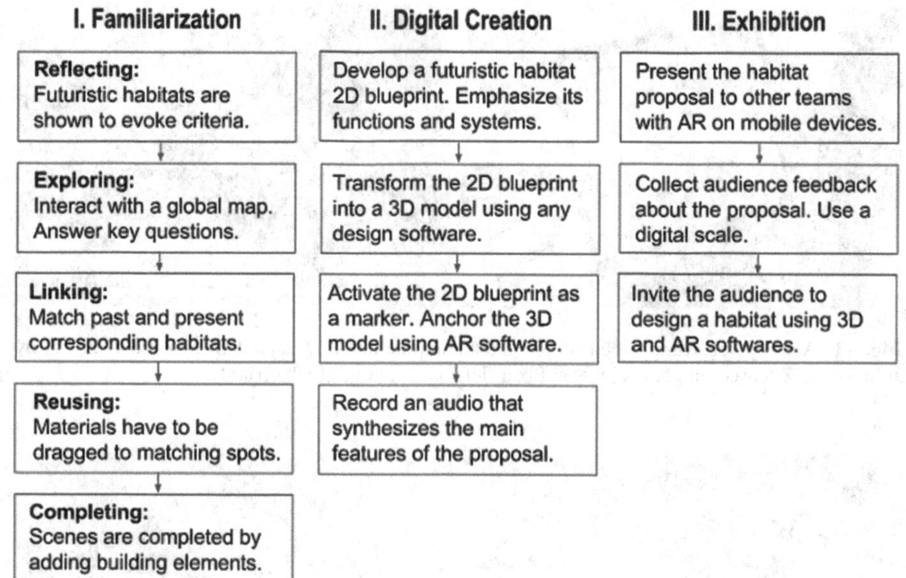

Fig. 9 GIM-based activities for the Smart City project focusing on habitats. (Elaborated by J. Sanabria-Z and M. Romero)

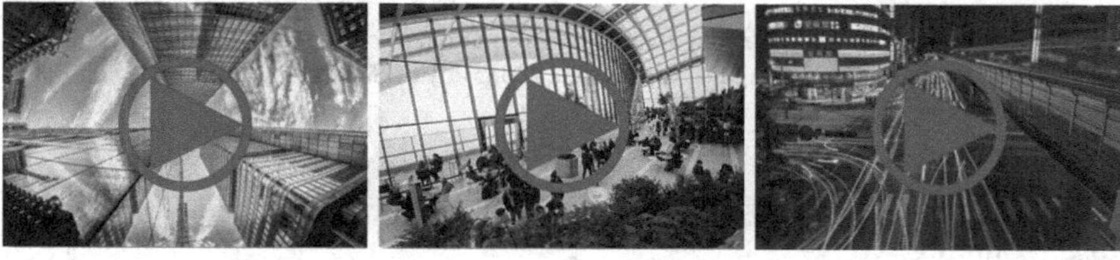

Fig. 10 Opening activity with videos of futuristic habitat solutions to evoke students' reflection, such as the "Dynamic Tower" (Fisher, 2017), the "House on ile renelavasseur" (Mark Foster Gage Architects, LLC 2015), or "Sanbock Doro" (Gonzalez, 2017). Adapted from Pexels (2015) by J. Sanabria-Z, based on photos by Philipp Birmes (left), Riccardo Bresciani (center), and Pixabay (right)

4.1.2   Exploring

For this activity, we suggest a third-party online resource, but instructors may decide to modify it, translate it, or fully design their own blog or site. Students are instructed to visit the website "Global cities of the future: An interactive map" (McKinsey and Company 2017), which displays a short introduction to future cities and emerging urban clusters. They are prompted to answer questions about the future of cities regarding largest number of children and elderly to the world's population, top 25 cities by per capita GDP, and about how their regional patterns of growth will differ (see Fig. 11).

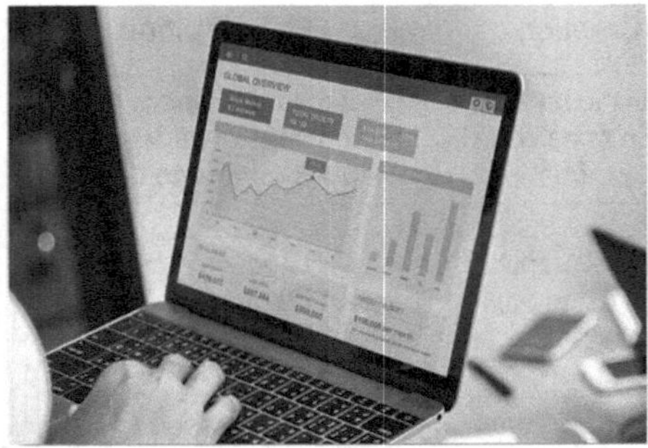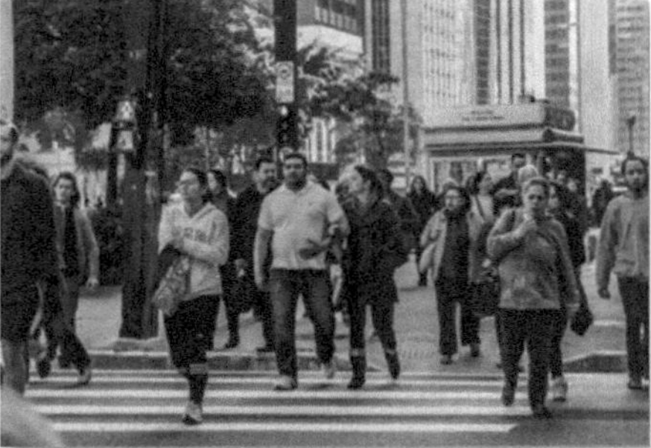

Fig. 11 Activity regarding global cities and their population. Adapted from Pexels (2015) by J. Sanabria-Z, based on photos by RawPixel (left) and Kaique Rocha (right)

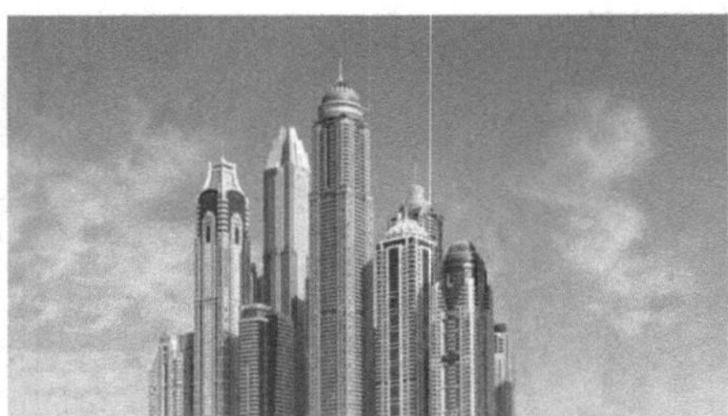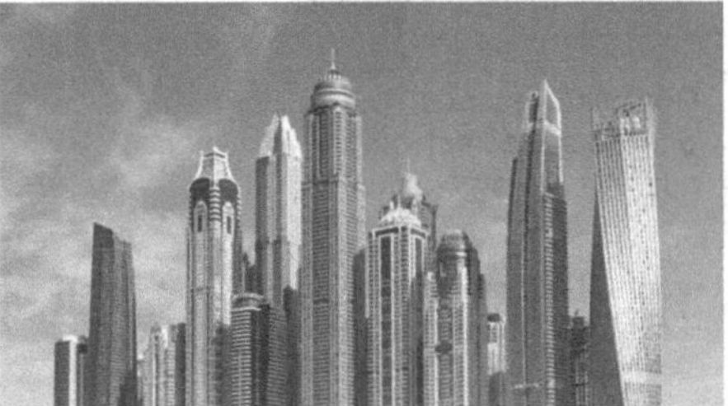

Fig. 12 Pictures representing the longitudinal transformation of cities are shown in random order for students to match them. Adapted from Pexels (2015) by J. Sanabria-Z, based on a photo by Aleksandar Pasaric

4.1.3 Linking

Pictures of different habitats that have drastically changed are presented randomly, before and after current conditions. Participants are instructed to link the pictures that correspond to the same location, expected to grasp the idea of transformation (see Fig. 12).

4.1.4 Reusing

Pictures of buildings made of atypical materials are presented, surrounded by images of these materials; participants have to drag images of the materials to their rightful place. Through this activity, students realize the possibilities for building using unexpected materials (see Fig. 13).

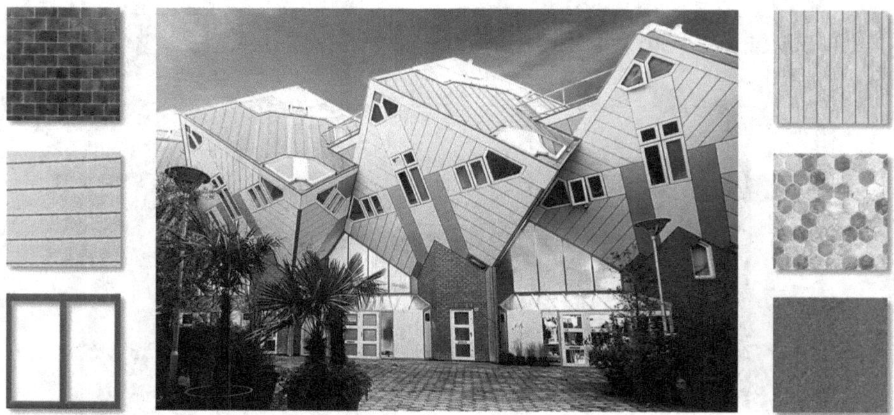

Fig. 13 Images of buildings with atypical materials are shown for students to find the rightful choice by dragging it to the main object. Adapted from Pexels (2015) by J. Sanabria-Z, based on a photo by Karin S.

4.1.5   Completing

Three-dimensional (3D) modeling elements are shown on a design interface such as SketchUp,[1] so students can complete a semi-built structure or create one from scratch. By accomplishing this task, students become familiar with 3D interfaces (see Fig. 14).

4.2   Module II: Digital Creation of Habitat in the Smart City

According to the criteria produced and the learnings from the familiarization module, students are instructed to create their own 3D elements using the SketchUp software. They are also prompted to build their own proposal of a futuristic habitat based on the activities of Module I (see Fig. 15). To reach this goal, they are advised to follow these steps:

- Develop a 2D blueprint of the proposal, emphasizing with arrows, illustrations, and short texts, the functions and systems it will provide.
- Transfer the 2D project to a 3D model using the SketchUp tools.
- Generate an AR marker (level 2) using the 2D blueprint on paper, and anchor the 3D model to it using an AR tool, such as AumentatyCreator.[2]
- Additionally, an audio that synthetizes the proposal main features may be recorded.

---

[1] SketchUp is a 3D modeling computer program available as a free or a paid version, owned by Trimble, Inc. [https://www.sketchup.com].

[2] Aumentaty Creator is an augmented reality computer program available as free or paid version, owned by AUMENTATY, S.L. [http://www.aumentaty.com].

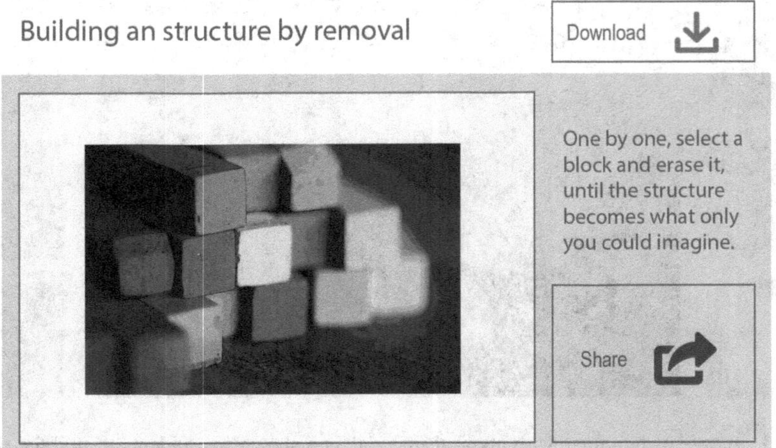

Fig. 14 Example interface for designing by removing blocks. Created by J. Sanabria-Z based on Bartels (2017), using a photo from Pexels (2015) by Sharon McCutcheon

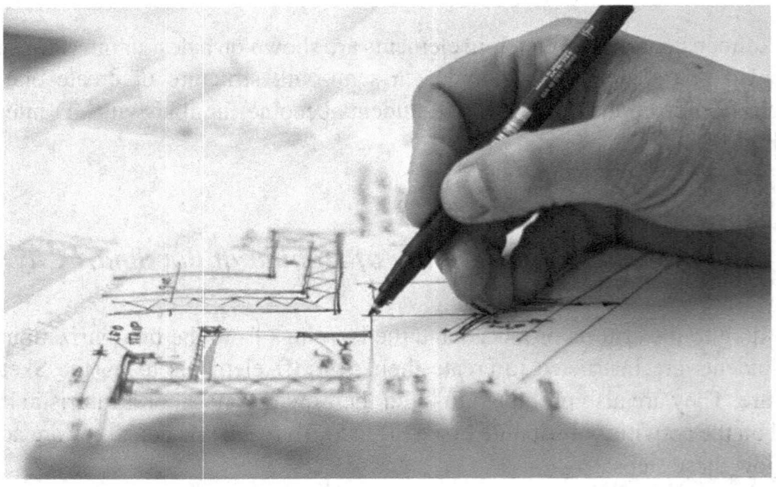

Fig. 15 Expectation of 2D proposals of habitat for the smart city. Adapted from Pexels (2015) by J. Sanabria-Z, based on a photo by Lex Photography

4.3 Module III: Exhibition of Augmented Reality Creations for the Smart City

When the markers are generated, participants can exhibit their habitat proposals for the smart city using the augmented reality tools described earlier. The setting should be large enough to accommodate paper blueprints, anchored with 3D habitat designs, which will be distributed around the space for viewing by the audience.

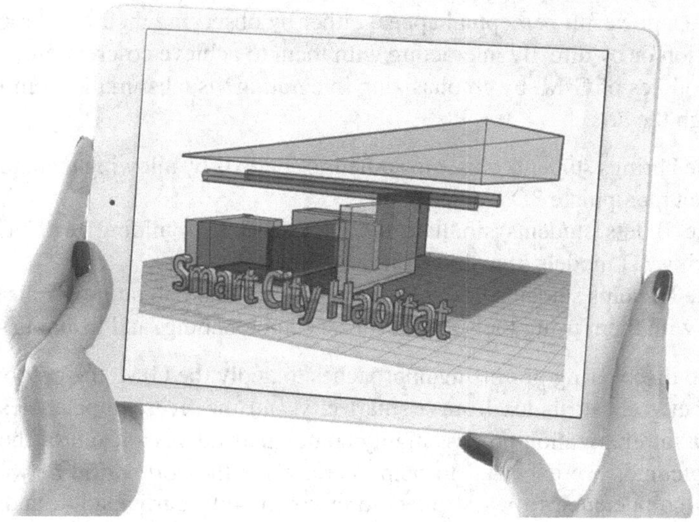

Fig. 16 Expectation of a 3D smart city habitat model to be subsequently shown using AR. Created by J. Sanabria-Z with Tinkercad (Autodesk, 2019), based on a photo from Pexels (2015) by Pixabay

The audience will be provided with tablets already activated with the software. When pointing the camera to a marker, the anchored 3D model will appear on the screens and the audio will start describing its features (see Fig. 16).

The evaluation of the experience can be collected from both the audience and peers that designed the proposals. To accomplish this step, tablets may be provided with an interactive scale that, after visiting a team's exhibition, measures some parameters related to the attributes of smart cities according to Caragliu et al. (2014) study. The goal of the evaluation is to identify how well the proposal integrates some of the attributes, toward a sustainable city that acknowledges community participation.

4.4 Expectations About Co-creation for Smart Cities Based on GIM

Through the GIM, we anticipate helping students develop multiple competencies urgently needed to thrive in twenty-first-century society and engage in the environmental challenges the world is facing. Based on the experiences and examples in art and sciences to illustrate the GIM application, the proposal for co-creating a futuristic habitat for the smart city aims at getting across to students the importance of environmental care by integrating art-based dynamic activities. During the deployment of activities, an active creative process is expected, where participants collaborate in the construction of projects, aware of their social and environmental influence.

Dynamic websites, rendered videos, 3D models, and more, all resources are thought to provoke a powerful perceptual appeal either by observing their aesthetic designs for reflection or by directly interacting with them to achieve co-creation.

The modules of GIM, by emphasizing integrating visual stimuli, promote learning through the arts:

- Module I brings students closer to authentic contexts by allowing them to observe and then manipulate 3D elements.
- Module II lets students simulate the experience of building real habitats by anchoring 3D models in a physical environment.
- Module III helps students to select appropriate virtual habitat proposals by interacting with them using their senses (observing, touching, and listening).

Beyond discovering promising approaches to apply the GIM, the proposed relationship between attributes from a smart city and the five competencies will be tested as a familiarization process highly contextualized. As a result of the project, certain outcomes are expected to help characterize the correlation between attributes and competencies toward integrating community participation and sustainability in twenty-first-century smart cities:

- Meaning of roles taken by the participants as well as their degree of engagement
- Threshold/limits of project development based on the available design tools
- Deeper understanding of the significance of their projects to improve their cities
- Awareness of their contribution to the team habitat proposal during co-creation
- Self-transformation throughout the learning experience of building for the future
- Positive changes on self-confidence due to empowerment during collaborative tasks
- Comprehensive assimilation of technology potential for high-definition prototyping

The two projects offered here as examples should reach a wide committed audience to increase the possibilities of implementing them and renovating the habitat of future cities and their surroundings.

References

Anciaux, N., Grumbach, S., Issarny, V., Mitton, N., Morin, C., Pathak, A., & Rivano, H. (2016). *Villes intelligentes: défis technologiques et sociétaux*. MOOC, FUN.

Arámburo-Lizárraga, J., & Sanabria, J. C. (2015). An application for the study of art movements. *Procedia Computer Science, 75*, 34–42. https://doi.org/10.1016/j.procs.2015.12.196.

Autodesk. (2019). *Tinkercad*. [Website]. Retrieved from https://www.tinkercad.com/

Barragan Foundation. (2019). *Barragan Foundation*. Retrieved from http://www.barragan-foundation.org/

Bartels, F. (2017). *Building by removal*. Retrieved from https://3dwarehouse.sketchup.com/model/fbbbcfc92823200a98b0cf5f411dc520/Building-by-removal

Binkley, M., Erstad, O., Hermna, J., Raizen, S., Ripley, M., Miller-Ricci, M., & Rumble, M. (2012). Defining twenty-first century skills. In P. Griffin, E. Care, & B. McGaw (Eds.), *Assessment and teaching of 21st century skills*. Dordrecht: Springer.

Burkhardt, G., Monsour, M., Valdez, G., Gunn, C., Dawson, M., & Lemke, C. (2003). *enGauge 21st century skills: Literacy in the digital age* (p. 88). Retrieved from North Central Regional Education Laboratory and the Metiri group website: http://www4.unescobkk.org/ict/elearning/pdf/enGauge21st_Century_Skills_Literacy_in_the_Digital_Age.pdf

Caragliu, A., del Bo, C., & Nijkamp, P. (2014). Smart cities in Europe. In M. Deakin (Ed.), *Smart cities: Governing, modelling and analysing the transition* (pp. 173–195). Oxon: Routledge.

FabLearn Organization. (2017). *Fablearn.org*. Graduate School of Education, University of Stanford. Retrieved from http://fablearn.org/

Fisher, D. [Green]. (2017, December 5). *The dynamic tower*. [Video file]. Retrieved from http://www.dynamicarchitecture.net/revolution/

Florida, R. (2002). Bohemia and economic geography. *Journal of Economic Geography, 2*(1), 55–71.

Florida, R. (2014). *The rise of the creative class – Revisited: Revised and expanded*. Basic Books (AZ).

Gonzalez, D. (2017). *Dionisio Gonzalez: Sandock Doro*. [Website]. Retrieved from http://www.dionisiogonzalez.es/SanbockDoro.html

Gouvernement du Québec. (2011). *Programme de formation de l'école québécoise*. Retrieved from http://www.education.gouv.qc.ca/contenus-communs/enseignants/programme-de-formation-de-lecole-quebecoise/

Hutton, T. (2017). *The creative city* (A Research Agenda for Cities, 137). Edward Elgar Publishing Limited.

Kamga, R., Romero, M., Komis, V., & Mirsili, A. (2016, November). Design requirements for educational robotics activities for sustaining collaborative problem solving. In *International conference EduRobotics 2016* (pp. 225–228). Cham: Springer.

Koestler, A. (1964). *The act of creation*. London: Hutchinson & Co.

Mark Foster Gage Architects, LLC. (2015). *House on Ile Rene-Levasseur*. Retrieved August 9, 2019, from MFGA 080219 website: https://www.mfga.com/copy-of-penkridge-hall-annex

McKinsey and Company. (2017). *Global cities of the future: An interactive map*. Retrieved on December 5, 2017, from https://www.mckinsey.com/global-themes/urbanization/global-cities-of-the-future-an-interactive-map

P21. (2010, June 18). *Framework for 21st century learning*. Retrieved December 3, 2017, from http://www.p21.org/index.php?option=com_content&task=view&id=254&Itemid=119

Pexels. (2015). *Pexels: Empowering creators*. [Website]. Retrieved from https://www.pexels.com

Robinson, K., & Aronica, L. (2016). *Creative schools: The grassroots revolution that's transforming education*. Penguin Books.

Romero, M. (2016). *Guide d'activités technocréatives pour les enfants du 21e siècle* (1st ed., p. 6). Québec: Livres en Lignes du CRIRES. Retrieved from https://lel.crires.ulaval.ca/oeuvre/guide-dactivites-technocreatives-pour-les-enfants-du-21e-siecle.

Romero, M., Lille, B., & Patiño, A. (2017). *Usages créatifs du numérique pour l'apprentissage au XXIe siècle*. Canada: PUQ.

Sanabria, J. (2015). The Gradual Immersion Method (GIM): Pedagogical transformation into mixed reality. *Procedia Computer Science, 75*, 369–374. (Elsevier).

Sanabria, J., & Arámburo-Lizárraga, J. (2017). Enhancing 21st century skills with AR: Using the gradual immersion method to develop collaborative creativity. *Eurasia Journal of Mathematics, Science and Technology Education, 13*(2), 487–501.

Sanabria, J. C., Chan, M. E., Mateos, L. R., Mariscal, J. L., Arámburo-Lizárraga, J., & De la Cruz, G. (2015). Applying the gradual immersion method to surrealist art creation using

Interactive Whiteboards (IWBs). *International Journal of Arts & Sciences, 08*(06), 627–638. ISSN:1944-6934.

Sánchez Escobedo, P. A., García Mendoza, A., & Valdés Cuervo, A. (2009). Validity and reliability of an instrument to measure creativity in adolescents. *Revista Iberoamericana de Educación*, 3–16. (In Spanish).

SEP. (2017). *Nuevo Modelo Educativo*. Retrieved from https://www.gob.mx/sep/documentos/nuevo-modelo-educativo-99339

Smart Cities Council. (2013). *Smart cities readiness guide*. Redmond: Smart Cities Council.

Trilling, B., & Fadel, C. (2009). *21st century skills: Learning for life in our times*. Wiley.

Wiley, D. A. (2000). Connecting learning objects to instructional design theory: A definition, a metaphor, and a taxonomy. In D. A. Wiley (Ed.), *The instructional use of learning objects* [online]. Retrieved from http://reusability.org/read/

Wing, J. M. (2006). Computational thinking. *Communications of the ACM, 49*(3), 33–35.

World Economic Forum. (2016). *The future of jobs: Employment, skills and workforce strategy for the fourth industrial revolution*. Geneva: World Economic Forum.

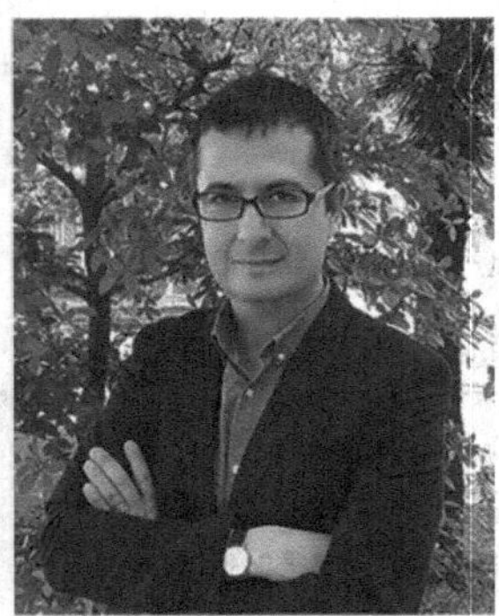

Jorge Sanabria-Z holds a doctorate degree in Kansei Sciences from the Advanced Research Centre in Neuroscience, Behaviour and Qualia at the University of Tsukuba, Japan. He completed a postdoctoral course at the University of Guadalajara, Mexico, from which he has developed the Gradual Immersion Method, a tech-based approach to promote creative cognition through collaboration. He is currently assistant research professor at the University of Guadalajara, where he focuses on the development of training methods and assessment of twenty-first-century skills in digital fabrication and educational robotics environments.

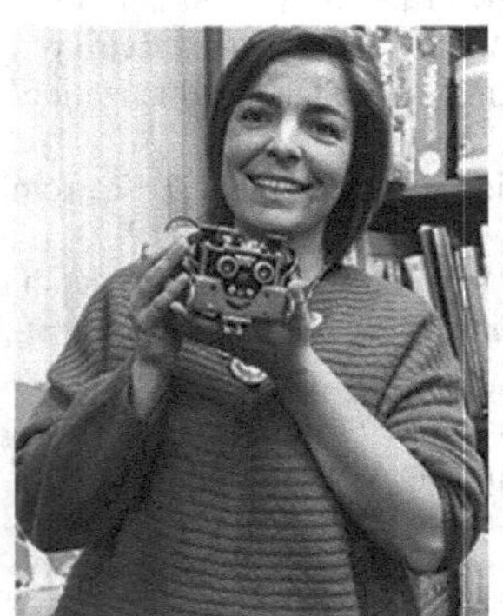

Margarida Romero is research director of the Laboratoire d'Innovation et Numérique pour l'Éducation (#fabLINE), a research lab in the field of Technology-Enhanced Learning (TEL). She is also a professor at Université de Nice Sophia Antipolis (FR) and assistant professor at Université Laval (CA). Her research is oriented toward the inclusive, humanistic, and creative uses of technologies (co-design, game design, and robotics) for the development of creativity, problem-solving, collaboration, and computational thinking.

Applying Gradual Immersion Method to Chemistry: Identification of Chemical Bonds

Guillermo Pech, Jorge Sanabria-Z, and Margarida Romero

1 Typical Issues on Teaching Chemistry

As with other Science subjects, Chemistry deals with some concepts that are impossible to see by the naked eye, or even difficult to imagine. Principles behind atoms and molecules, which are constituents of the products, materials, and organisms around us, belong to a plane called "microscopic" (and also nanoscopic). For their understanding, teachers often must resort to detailed explanation involving abstract ideas represented by models, symbols, and mathematical formulas that run counter to the daily sensory experience (Kind 2004). This traditional approach can generate erroneous ideas in the learner, bearing little relation to reality and stepping away from the scientific way or learning. By not generating meaning or capturing student interest, this approach can imbue students with the impression that Science is a very difficult disciplinary area, with very little application (Pech et al. 2015; Galagovsky 2007; Acevedo 2004; Gómez and Kent 2004; Furio et al. 2001).

In Mexico, Chamizo et al. (2004) found that by the end of K–12, most of the students did not understand basic concepts of Chemistry. Rather, they accomplished this only after concluding an undergraduate major in Chemistry. This example

G. Pech (✉)
Technical Middle High School 157, Secretariat of Public Education, Guadalajara, Mexico
e-mail: guillermo-pech@docentes.sej.gob.mx

J. Sanabria-Z
Innovation Generation and Management Program, University of Guadalajara,
Guadalajara, Mexico
e-mail: sanabria@suv.udg.mx

M. Romero
Laboratoire d'Innovation et Numérique pour l'Éducation, Université de Nice Sophia
Antipolis, Nice, France
e-mail: Margarida.Romero@unice.fr

© Springer Nature Switzerland AG 2019
A. J. Stewart et al. (eds.), *Converting STEM into STEAM Programs*,
Environmental Discourses in Science Education 5,
https://doi.org/10.1007/978-3-030-25101-7_15

offers a reason for proposing alternatives for better ways to illustrate microscopic principles proposed in the classroom, which could allow students to better connect their learning to the everyday environment, making learning more durable. Some of the learning objectives in these abstract areas center on the interaction of particles: states of aggregation of matter, the corpuscular nature of matter, differences between elements, compounds and mixtures, chemical bonds, processes and chemical reactions, acids and bases, stoichiometry and mole concept, and thermodynamics (Kind 2004).

In basic education (i.e., elementary and middle high school), the first formal experience with chemical language in Mexico occurs during the last year of middle high school education. At this level, competency-based training is proposed, aimed at basic scientific training. In the discipline of Chemistry, the objective, in this case, is to have students understand natural phenomena and processes from the scientific perspective. The concept of chemical bonds helps transition student understanding from the world of atoms to the formation of the chemical compounds represented in chemical reactions that occur in the environment. Chemistry teachers are generally in accord with the idea that students have difficulty in understanding these concepts.

2 Augmented Reality and Educational Possibilities

Augmented reality (AR) is a recent technology that is considered useful for teaching chemistry. AR is a system (or technique) that allows teachers to integrate digital elements, such as audio or 3D models, into the physical environment by superimposing them, and thus "enrich reality," assisted by several interactive technologies (Sanabria 2017; Prendes 2015; Azuma 1997). The combination of digital information and information in real time can serve different purposes, including education (Cabero and Barroso 2016). In particular, AR can highlight the following aspects of teaching:

(a) Enabling learning content in 3D perspectives
(b) Ubiquitous, collaborative, and situated learning
(c) Presence, immediacy, and immersion by students
(d) Visualizing concepts not visible by the senses
(e) Reducing formal and informal learning

We highlight the use of AR for understanding complex phenomena and concepts, for it allows a phenomenon or object to be decomposed into its different phases or parts and observed from different points of view (García et al. 2010, in Cabero and Barroso 2016).

As applied in the area of chemistry, AR can allow students to bring to the real world those concepts of the microscopic world that are not observable by the human eye. It does this by linking concept to the authentic environment so as to illustrate the principles that govern the everyday environment. AR also promotes the development of spaces in which the student can interact with the unobservable universe, thus generating significant learning experiences.

However, to adequately implement AR in educational environments, teachers must introduce users to the system based on their intuitive abilities for digital environments. A means for such incorporation is proposed by Sanabria (2015). With the Gradual Immersion Method (GIM), students use interactive devices for collaborative learning, supported by AR. The GIM approach promotes creativity and learning through intuitive interfaces.

3   The Gradual Immersion Method

The GIM considers two leaning aspects: It (1) aims to deepen the appropriation of knowledge through interactive tasks related to a particular topic; and (2) allows the spatial transition from 2D to 3D, and then to AR. The GIM has been designed for interactive activities involving teams of four to five students. This number promotes a "progressive familiarization" (Sanabria 2017) for constructing and exhibiting in AR, achieved through a sequence of three modules (Fig. 1).

In one application of the GIM, high school students created surrealist expressions in the context of a museum, resulting in intuitive learning and the development of creativity oriented to works of art in a nonformal context (Sanabria et al. 2015). Experience of this sort leaves a window open for other disciplines of a formal context, such as school sciences, so that the participants, in addition to learning in scientific areas, may develop competences such as collaboration and creativity, which are essential for resolving the new challenges of the twenty-first century (SEP 2016).

In another application, the GIM was used in a lower-level high school module. This project was designed according to the science curriculum of Chemistry in lower-level high school (SEP 2013). Twenty-eight students between 14 and 16 years of age participated in the project. The students completed a diagnostic questionnaire

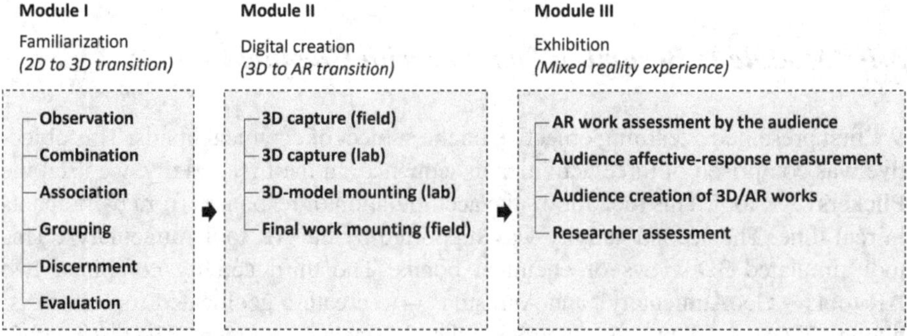

Fig. 1 The three GIM modules. (Reprinted from Sanabria, J. (2015). The Gradual Immersion Method (GIM): Pedagogical Transformation into Mixed Reality. *Procedia Computer Science*)

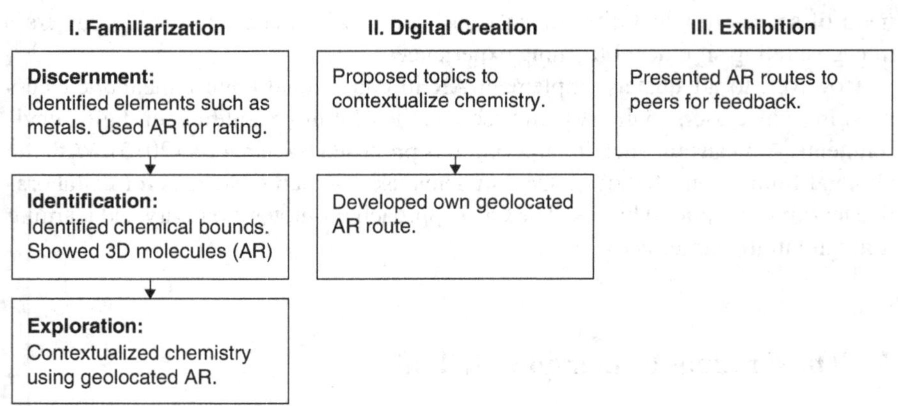

Fig. 2 Modules of the GIM applied in a chemical bond project. (Elaborated by Jorge Sanabria-Z and Guillermo Pech)

on their knowledge of the concept of molecules, bonds, atoms, models, and chemical bonds. Results showed that 85.7% of the students did not have much in-depth understanding about chemical bonds, their types, or other related ideas such as molecules, or where chemical bonds occurred. These deficiencies justified applying the GIM, to determine if the deficiencies could be overcome.

The instructor adapted the three GIM modules before starting the experiment, as follows: For Module I, a Learning Object with 3 activities aimed at introducing students to the world of chemical bonds; for Module II, the students were asked to co-create their own versions of atoms, bonds, and molecules, with AR; and for Module III, the students shared their experiences to obtain feedback. Figure 2 shows the steps taken to implement this GIM, in which we sought to transition from the 2D representations to the 3D of the molecular world while seeking for a transition through different types of AR.

3.1 Module I: Becoming Familiar with Chemical Bonds

We first presented a learning objective on the subject of chemical bonds. This objective was comprised of three activities in sequence. In the first activity, we used the Plickers® AR tool. This tool allows the accumulation of responses from participants in real time. The second activity was supported by the AR tool Aumentaty®. This tool simulated 3D views of chemical bonds. The third activity combined two AR tools – GeoAumentaty® and Aurasma® – to create a geolocated route for displaying 3D digital elements anchored in the physical environment. The three activities were discernment, identification, and exploration. The processes used to support these objectives are described below:

- *Discernment*: Students were asked to discriminate among metallic and nonmetallic elements, based on information in the periodic table. Supported by the technology of AR type 0 (quick response "QR"-type markers), students completed an activity in which questions were projected that gradually showed 2, 3, and 4 multiple-choice answers – these prospective answers were accumulated by the teacher through the camera on a mobile device.

The experiment used Plickers® online software; the teams registered on this platform and entered the questions and their answers there. Next, we distributed paper markers to the participants for anonymous responses. Upon reviewing each project question, the participants showed to the mobile camera the side of the marker with the answer option they considered correct, which was registered by the software (see Fig. 3).

- *Identification*: Students needed to identify the type of chemical structure in a 3D chemical bond. The challenge-type activity used the Aumentaty® program, an augmented reality level 1 tool. This tool allowed the 3D digital elements to be embedded in markers. The teams contained three to five students, to whom we distributed paper cards with markers and a series of questions, such as the following: What is this element? What is its 3D structure according to Lewis? What must happen to make it stable? When the questions had been answered, a representative for each team took turns to show one of the cards to the camera, revealing the 3D bond of the electronic configuration of an element and its Lewis representation. This method helped elicit group discussion about the teams' adequacy of response to the question (Fig. 4).

- The Aumentaty platform allowed us to upload comments in a "digital chat" mode. The activities allowed the students to reflect on, and discover in some

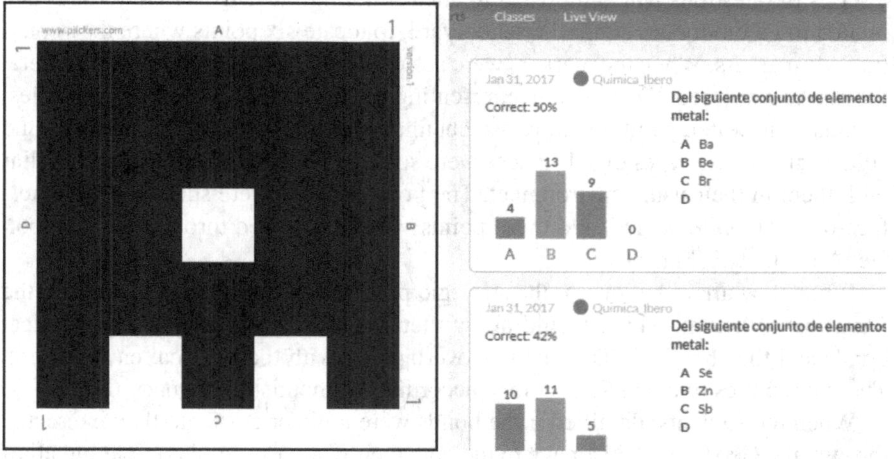

Fig. 3 AR marker (left) and projected screen, showing the percentages of responses to questions (right) on the Plickers® platform. (Elaborated by Guillermo Pech)

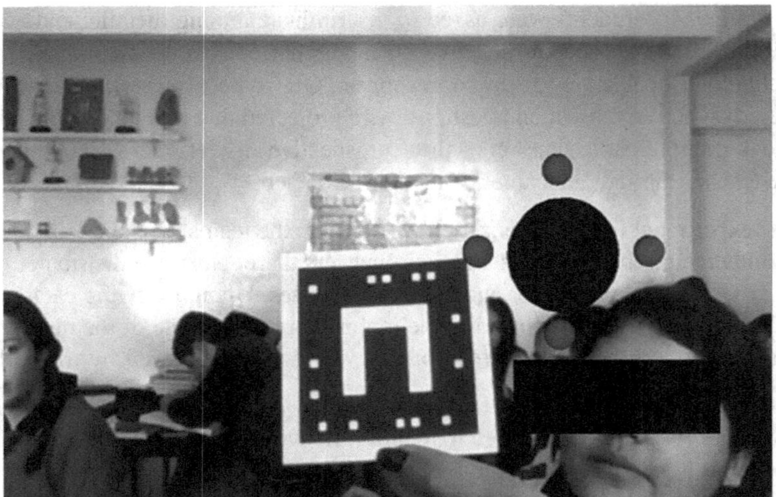

Fig. 4 A 3D representation of a chemical bond anchored to an augmented reality marker in Aumentaty®

detail, how to identify structure of ionic and covalent bonds. At the end of the reflection and general conclusion session, the students prepared and presented a portfolio of their activities.

- *Exploration*: To contextualize learning, we introduced the students to an exploration route inside the school facilities, where they could find 3D chemical bonds with using mobile devices equipped with AR.

To do this, the students first observed a video to learn the procedure for the desired activity. They then were assigned a mobile device (one per team) and given a series of questions which they had to answer during the activity. They were then invited to go outdoors, within the schoolyard, to locate six points where they could find AR markers. When the markers were activated on their mobile device's camera, the markers showed 3D elements representing different relationships of molecules: chemistry in water, paint, metal items, chemistry in plants, plastic in products, and cola drinks. These types of substances were selected because students were familiar with them in their usual environment. The points to visit were shown on the interface of the mobile device, and these points were geolocated through the AR tool, GeoAumentaty® (Fig. 5).

When they arrived at one of the strategic points, the students had to activate the Aurasma® AR tool, which would allow them to observe the simulated chemical bonds and the chemical 3D structures overlapping with the physical environment. These structures retained CPK colors, according to chemistry language (Fig. 6).

When the students identified if the bonds were ionic or covalent, they interacted through the GeoAumentaty® page to indicate their responses until completing all of the points comprising the route.

Fig. 5 Interface of the path of chemical bonds in GeoAumentaty® AR tool. (Elaborated by Guillermo Pech)

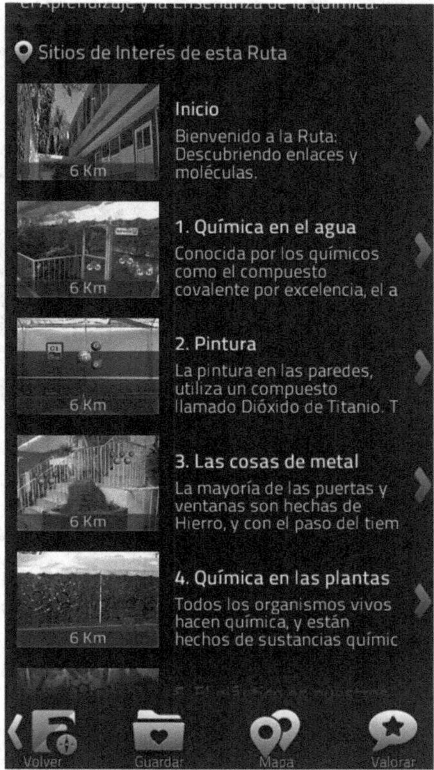

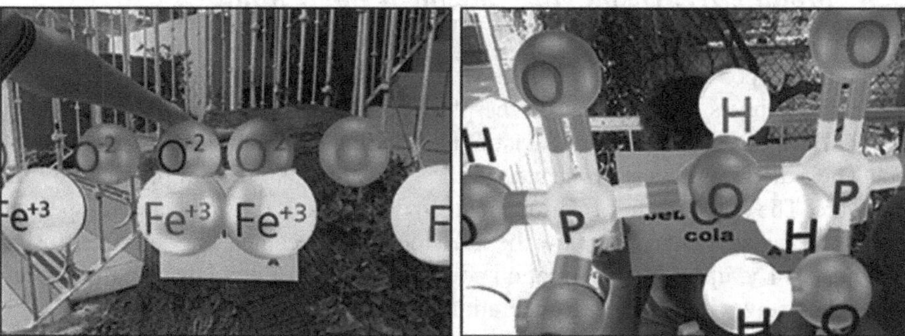

Fig. 6 3D chemical bonds in a metal handrail (left) and in cola beverages (right) using the Aurasma® AR tool in a mobile device. (Elaborated by Guillermo Pech)

3.2 Module II: Creating Contextualized Chemical Bonds

After becoming familiar with different contextualized chemical bonds in Module I, and to relate other chemical compounds around, the students co-created their own geolocated route using GeoAumentaty®. Similar to the activity that was done with

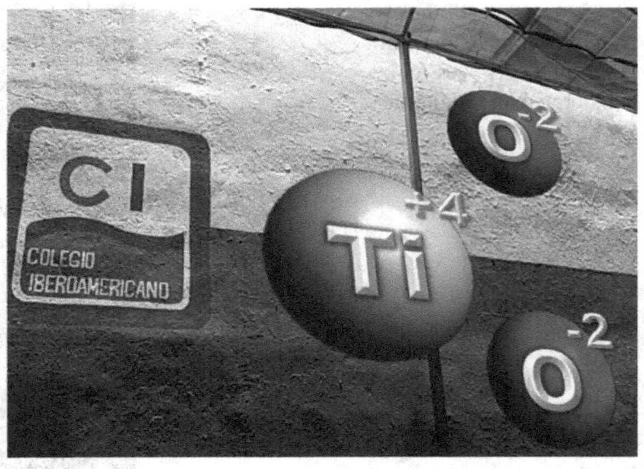

Fig. 7 3D digital chemical bond of a wall's surface as seen through a mobile's device camera in Aurasma® AR tool. (Elaborated by Guillermo Pech)

Aurasma®, the different points of the route should show the chemical bonds that belong to the different components of the landscape and space. The theme was free, and teams (comprised of up to four students) researched and prepared original route proposals such as "The path of life," which included a water molecule, components of the atmosphere, the structure of a tree, and the compounds for life in animals, or "miscellaneous," which accumulated objects made of materials such as cellulose derived from cardboard and cotton, compounds in the stars and the components of plastic (Fig. 7).

3.3 Module III: Displaying Chemical Bond Routes

When the proposals had been co-created, the students met in the classroom to present, to their peers, information on the places they visited and the molecules they found. Each team obtained feedback on their findings during this phase of the project. Their discoveries were uploaded to the GeoAumentaty forum provided by the software. The uploaded information included descriptions of each team's respective journey and statements about how the team members became aware of the diffusion of science. The uploaded information could then be consulted later, by the participants themselves and by future generations of students at the school (Fig. 8).

4 Learnings from the Co-creative Experience

Some twenty-first-century competencies were evaluated in this project using the CoCreaTIC tool (Lepage and Romero 2017). Components considered as strongly present by the students included those related to computational thinking, creativity, and collaboration. Below, we provide a general analysis of that process.

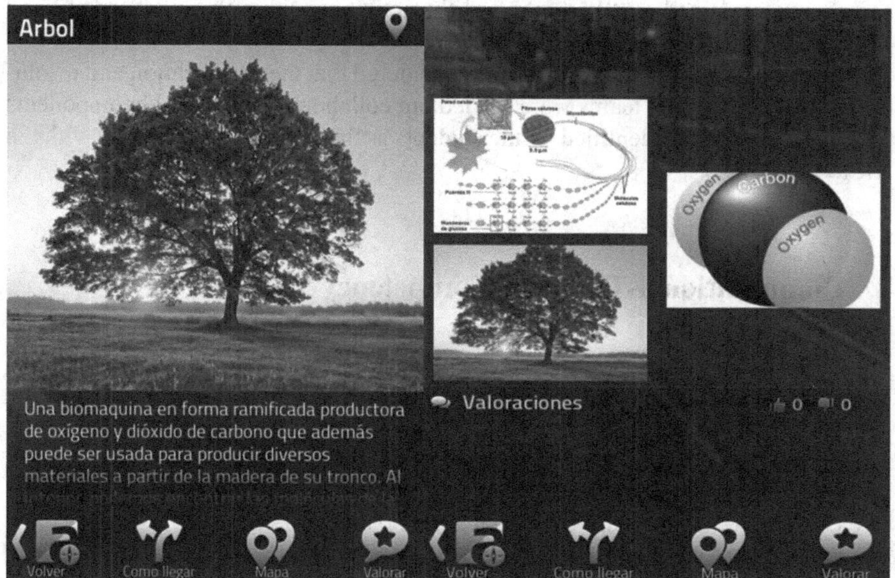

Fig. 8 Information presented about one of the points of interest in "The path of life" and associated compounds in GeoAumentaty®. (Elaborated by Guillermo Pech)

5 Creativity Competency

In particular, we measured creativity by using the "generation/creation" and "attitudes associated with creativity" components, which, in turn, were divided into subcomponents named "generation of ideas," "brainstorming," "divergent exploration," and "creative orientation toward the world ("creattitude")." In the latter category, the observables proposed were "We generate ideas together (as a team)" and "We think of ideas to apply to our surroundings."

We found that 60% of the participating students became aware of the collaborative generation of ideas within the GIM activities and 55% of the participating students thought about possible alternative applications after learning to use the technologies. Survey results showed that the provided challenges incited the students to co-creativity and promoted reflection on the possible application of the AR tools in other subjects of STEAM. Further, targeted learning about chemical bonds appeared effective through the GIM process.

6 Collaboration Competency

For the collaboration competency of the CoCreaTIC instrument, the components "Establish and maintain a knowledge and a shared organization" and "(Co)-construction of knowledge and/or artifacts" were chosen. The subcomponents of the same name remained as "We organized the activities within the work teams" and "We assembled the solution as a team."

Sixty-five percent of the participants considered that the organization and resolution of the activities by teams were carried out collaboratively. The subcomponents were most frequently identified by the students as they developed activities with the AR tools.

7 Computational-Thinking Competency

The components of "Modeling" and the "Attitudes associated with computational thinking" were considered, since they are the most evident due to the nature of the activities carried out, the manipulation of the tools of AR, and the content. The subcomponents chosen were "Became aware of the importance of planning a solution before creating it" and "Orientated to quality," respectively. The sentences were "we considered it important to have an action plan before creating a solution" and "we wanted to make the best product or solution possible."

Fifty percent of the participants considered the action plan to be very important, and 40% scored it as present. Additionally, 75% of the participants considered very present the idea of making the best product or solution possible during the activities with AR. This subcomponent of the twenty-first-century skills was chosen most often by the students.

8 Learning of Chemical Bonds

During the three GIM modules, conventional and nonconventional assessment instruments were applied, divided into diagnostic, formative, and summative assessments. Three criterion reference tests of conventional type (CRT) were applied (Dick and Carey 2001), named as diagnostic, intermediate, and final tests (see Fig. 9).

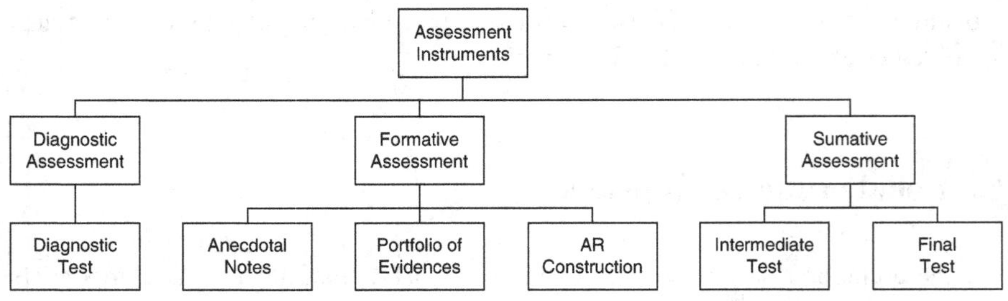

Fig. 9 Types of assessment instruments applied to the chemical bond experience

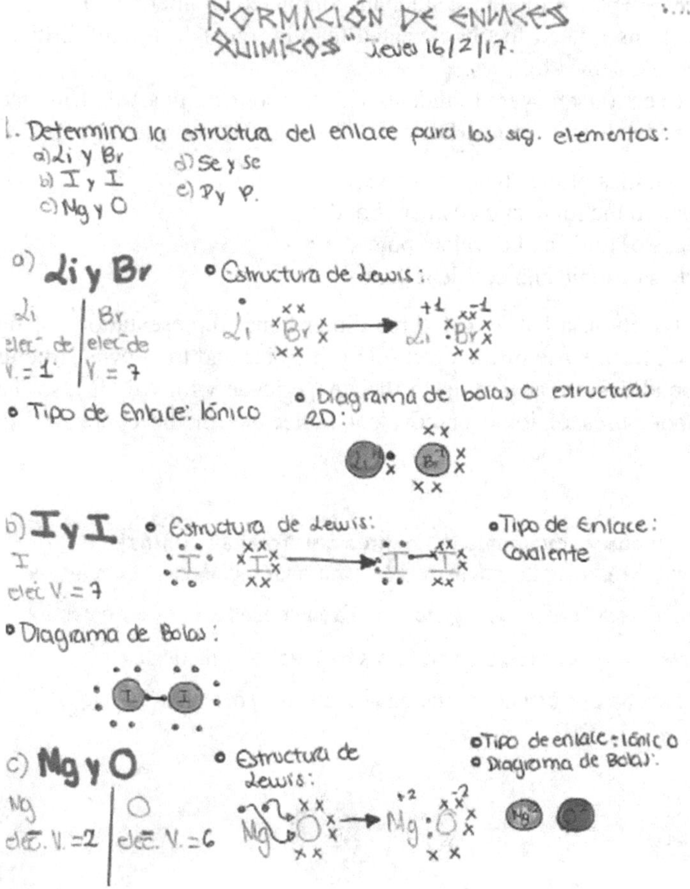

Fig. 10 Activities included in the portfolio of evidences of GIM's Module I, prepared by the students. (Elaborated by Guillermo Pech)

Before Module I, "Becoming Familiar with Chemical Bonds," students completed a diagnostic test on their knowledge of the concept of molecules, bonds, atoms, models, and chemical bonds. Results showed that 85.7% of the students did not have much in-depth understanding about chemical bonds, their types, or other related ideas such as molecules, or where chemical bonds occurred.

During the development of Module I, when AR activities were evaluated, students registered their constructions by writing a portfolio of evidences (see Fig. 10). They were requested to determine the bond structure for some chemical elements, including the type of bond. The portfolio showed the structures forming the links for various molecules. In the same way, the students registered their learning progress by writing anecdotal phrases such as "Today I learn that" or "To improve I need to."

After the three GIM modules, students completed an intermediate test. This test had two sections related to the expected learning goals proposed in the module, which are explained as follows.

The first section recovered students' ideas about particles and interactions in chemical bonds, through items (Dick and Carey 2001) considering:

(a) Characteristics of the chemical bond
(b) Elements in the ionic and covalent bond
(c) Examples of ionic and covalent bonds
(d) Properties of ionic and covalent bonds

In a second section, a list with 6 three-dimensional representations of molecules was provided for the use of the octet rule principles and the Lewis structure in the chemical bond design, according to their experience with AR. It also included a question about uses of ionic or covalent molecules in the context of daily life (Fig. 11).

II. Enlaces iónicos y covalentes [10 puntos c/u. Total 60 puntos].
   1.   Revisa la representación de bola (2D) para los siguientes compuestos.

   2.   De acuerdo a la regla del octeto, ¿qué debe suceder para ser estable?

   3.   ¿Cuáles son las estructuras de Lewis para estos compuestos?

   4.   ¿Se trata de un compuesto con enlace iónico o covalente?

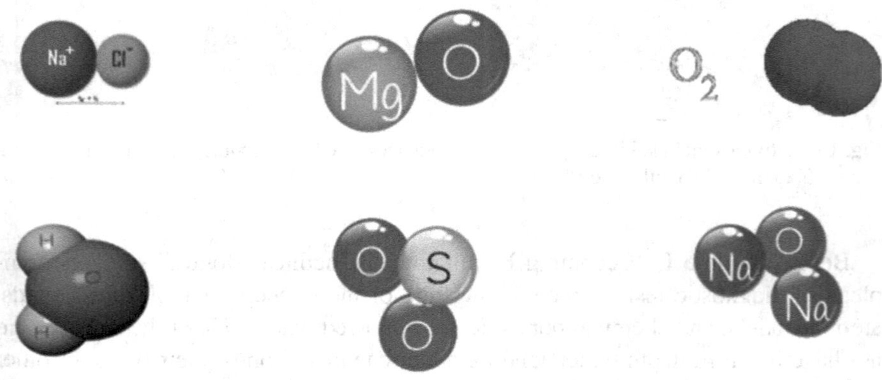

III. Aplicaciones de los compuestos iónicos y covalentes [5 puntos extra].
Menciona un compuesto iónico o covalente de uso en la vida cotidiana:

Fig. 11 Section 2 of the CRT for chemical bond assessment. (Elaborated by Guillermo Pech)

As a result of the intermediate test, 96.6% of the students demonstrated mastery related to chemical bonds when describing their characteristics and identifying the ionic and covalent compounds in their evaluation of learning.

To verify the permanence of the learning, students completed a final test at the end of the course, 4 months after the intermediate test. The final test included items similar to those in the first test, with three-dimensional representations (spherical structures) of compounds in which the student had to identify the type of bond (ionic or covalent) and represent it using Lewis structures for the compounds. Results from the final test showed that 92.3% of the course participants retained the information, demonstrating durability in learning about the chemical bonds by identifying ionic and covalent compounds and their corresponding Lewis structures.

9 User Experience and Recommendations

In reference to the perception of the subject as an abstract science, 72.7% of the participants changed their way of viewing the subject: They considered it as more interesting. These findings registered in experience and user evaluations.

Students taking the course commented that the tools allowed a clearer visualization of the molecules and how things are from the inside (i.e., of what they were composed). Now they consider the use of the mobile device to learn and the use of new augmented reality tools for use in their other science classes. They also mention that it was a different way of working and that they would like to have other activities of this type, such as chemical reactions and mixtures, and also apply them to other subjects such as history, biology, and physics.

As key ideas, the activities were different from those considered conventional. They also were illustrative and converted the students from a passive role to an active one in the learning process.

10 Conclusions

The Gradual Immersion Model may serve as an example for developing STEAM approaches, mainly for science teachers but for other areas as well. In this initial approach for using augmented reality tools to teach chemistry at lower-level high school with GIM, we found that the instructional design and tools greatly benefited student learning, compared to students who did not use these tools. We also found that the tools augmented computational thinking, creativity, and collaboration, even as they helped the students learn about chemical bonds. The strategy we propose, of applying GIM with AR tools, fosters an innovative learning experience which arouses student interest and facilitates their learning process, according to the students' own opinions.

11 Future Plans

Among future projects, in addition to improving the tools based on the experience and the participants' opinions, we anticipate validating the registration and evaluation instruments, such that they can be shared with other educational centers and investigators. We also expect to translate these instruments, making them more available to others internationally.

We hope that our proposal for using GIM will encourage its development for use by science teachers and teachers in other discipline areas, such as History, Mathematics, Computing, Spanish, Physics, and Biology.

References

Acevedo, J. (2004). Reflexiones sobre las finalidades de la enseñanza de las ciencias: educación científica para la ciudadanía. *Revista Eureka sobre enseñanza y divulgación de las ciencias, 1*(1).

Azuma, R. (1997). A survey of augmented reality. *Presence: Teleoperators and Virtual Environments, 6*(4), 355–385. Recuperado http://www.cs.unc.edu/~azuma/ARpresence.pdf.

Cabero, J., & Barroso, J. (2016). The educational possibilities of augmented reality. *Journal of New Approaches in Educational Research, 5*(1), 44.

Chamizo, J. A., Nieto, E., & Sosa, P. (2004). La enseñanza de la química. Tercera parte. Evaluación de los conocimientos de química desde secundaria hasta licenciatura. *Educación química., 15*(2), 108–112.

Dick, W., & Carey, L. (2001). The systematic design of instruction: Origins of systematically designed instruction. *Classic Writings on Instructional Technology, 2*, 71–80.

Furio, C., Vilches, A., Guisasola, J., & Romo, V. (2001). Finalidades de la enseñanza de las ciencias en la secundaria obligatoria. ¿alfabetización científica o preparación propedéutica? *Revista Enseñanza de las ciencias, 19*, 3.

Galagovsky, L. R. (2007). *Enseñar química vs. Aprender química: una ecuación que no está balanceada* (Vol. 6). *Química Viva*.

Gómez, C. M., & Kent, V. (2004). El aprendizaje de la química. In J. A. Chamizo (Ed.), *Antología de la enseñanza experimental* (pp. 109–124). México: Universidad Nacional Autónoma de México, Facultad de Química.

Kind, V. (2004). *Más allá de las apariencias: ideas previas de los estudiantes sobre conceptos básicos de química*. México: Universidad Nacional Autónoma de México, Facultad de Química. Retrieved from http://depa.fquim.unam.mx/sieq/Documentos/masallaapariencias.pdf.

Lepage, A., & Romero, M. (2017). Évaluation par compétences d'activités de programmation créative avec l'outil #5c21. In *Actes du colloque CIRTA 2017* (Vol. 1). UQAM, Québec: CRIRES.

Pech, G. E., Brito, W., & Rubio, N. (2015). Desarrollo de un módulo instruccional para la enseñanza de la tabla periódica. *Educación y Ciencia, 7*(42), 44–55. (ISSN:2448-525X).

Prendes, C. (2015). Realidad aumentada y educación: análisis de experiencias prácticas. *Pixel-Bit, Revista De Medios y Educación, 46*. https://doi.org/10.12795/pixelbit.2015.i46.12

Sanabria, J. C. (2015). The Gradual Immersion Method (GIM): Pedagogical transformation into mixed reality. *Procedia Computer Science, 75*, 369–374. https://doi.org/10.1016/j.procs.2015.12.259.

Sanabria, J. C. (2017). Chapter 14: Alcances del Método de Inmersión Gradual (MIG): Percepción y creación colaborativa con realidad aumentada. In *Desafíos de la Cultura Digital para la Educación* (1st Ed.). Universidad de Guadalajara. Editors: Coordinación de Recursos Informativos de UDGVirtual.

Sanabria, J. C., Chan, M. E., Mateos, L. R., Arámburo-Lizárraga, J., & De la Cruz, G. (2015). Applying the gradual immersion method to surrealist art creation using Interactive Whiteboards (IWBs). *International Journal of Arts & Sciences, 8*(6), 627–638.

SEP. (2013). *Secretaría de Educación Pública. Educación Básica Secundaria. Ciencias. Programa de estudio 2011 Guía para el Maestro.* México: Secretaría de Educación Pública.

SEP. (2016). *Secretaría de Educación Pública.* Nuevo Modelo Educativo. https://www.gob.mx/sep/documentos/nuevo-modelo-educativo-99339

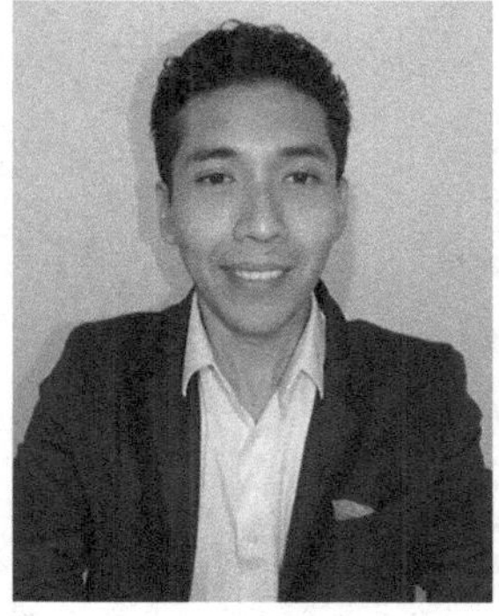

Guillermo Pech has a Master's degree in Learning Technologies from the University of Guadalajara and chemist for Autonomous University of Yucatán and Science and Technologies teacher in middle high school at Jalisco. He works in instructional design for STEAM activities that involve augmented reality and robotic and digital fabrication strategies for creating scientific vocations and twenty-first-century skills development. He collaborates with Drs. Sanabria and Romero on various teacher and student training programs.

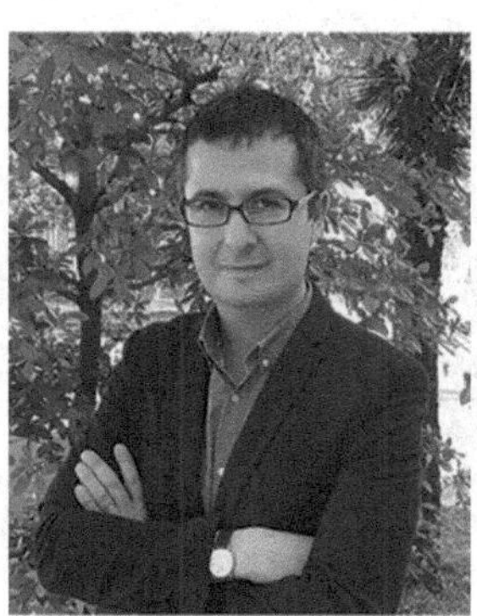

Jorge Sanabria-Z holds a doctorate degree in Kansei Sciences from the Advanced Research Centre in Neuroscience, Behaviour and Qualia at the University of Tsukuba, Japan. He completed a postdoctoral course at the University of Guadalajara, Mexico, from which he has developed the Gradual Immersion Method, a tech-based approach to promote creative cognition through collaboration. He is currently assistant research professor at the University of Guadalajara, where he focuses on the development of training methods and assessment of twenty-first-century skills in digital fabrication and educational robotics environments.

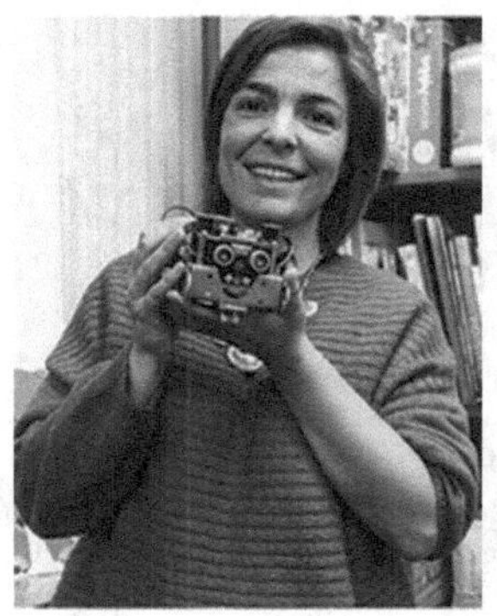

Margarida Romero is research director of the Laboratoire d'Innovation et Numérique pour l'Éducation (#fabLINE), a research lab in the field of Technology-Enhanced Learning (TEL). She is a professor at Université de Nice Sophia Antipolis (FR) and assistant professor at Université Laval (CA). Her research is oriented toward the inclusive, humanistic, and creative uses of technologies (co-design, game design, and robotics) for the development of creativity, problem-solving, collaboration, and computational thinking.

From Conceptualization to Implementation: STEAM Education in Korea

Sophia Jeong, Hyoungbum Kim, and Deborah J. Tippins

1 Introduction

In many developed countries, a presumed goal of STEM (Science, Technology, Engineering, and Mathematics) education is to develop excellent science and technology talents and secure economic competitiveness in an increasingly market-driven global economy. Building upon the past two decades of STEM education reform efforts, Sanders (2009) argues that integrative STEM education focuses on integrating the four disciplines in science, technology, engineering, and mathematics and on new integrative approaches that explore teaching and learning between and among any two or more of the STEM subject areas. Engineering design-based learning can be an effective pedagogical approach for effectively teaching crosscutting concepts across these four content areas, and some investigators argue that "the flavor of integrative STEM education resonates in several of the engineering accreditation standards that grew out of the engineering education reform efforts" (Sanders 2009, p. 23). Further, STEM education emphasizes that it is critically important to foster opportunities to increase problem-solving skills, especially in light of developing economic competitiveness (Maes 2010).

The current framework of integrative STEM was extended to include the arts. MEST (2011) conceptualized STEAM education in which the arts were grafted onto the integrative STEM and implemented STEAM at the elementary and second-

The original version of this chapter was revised: The chapter author's, "Sophia (Sun Kyung) Jeong", first and last names has been corrected now. The correction to this chapter is available at https://doi.org/10.1007/978-3-030-25101-7_19

S. Jeong · H. Kim (⊠) · D. J. Tippins
Department of Earth Science Education, Chungbuk National University,
Cheongju, Chungbuk, South Korea
e-mail: sj33678@uga.edu; hyoungbum21@chungbuk.ac.kr; dtippins@uga.edu

A. J. Stewart et al. (eds.), *Converting STEM into STEAM Programs*,
Environmental Discourses in Science Education 5,
https://doi.org/10.1007/978-3-030-25101-7_16

ary level in Korea (Kim and Chae 2016) to meet the needs of educational reforms in science and mathematics education. These needs were identified in two international reports: (1) Trends in International Mathematics and Science Study (TIMSS) and (2) the Programme for International Student Assessment (PISA). TIMSS measures the mathematics and science achievements of students in grades 4 and 8. In 2011, 42 countries participated in TIMSS, and students of Korea ranked among the highest in the world. For example, the achievements in science and mathematics for the fourth grade and eighth grade students in Korea ranked among the highest, and academic achievements of the seventh grade students in Korea ranked third overall for science and first overall for mathematics. However, the 2011 TIMSS result reported that Korean students' interests in science were very low: 11%, compared with the international mean of 35% (Martin et al. 2012; Mullis et al. 2012). The magnitude of this discrepancy highlighted the need for educational reforms to promote Korean students' interests and self-confidence in science and mathematics. Supporting the findings from TIMSS, Korean students tend to perceive science and mathematics as difficult subjects, with students having little to no interests due to the less-than-positive association with the college entrance examination (MEST 2011). Further, a survey found that, coupled with negative attitude or perception toward careers in engineering, the level of awareness about possible future career choices in science and engineering was lacking among the elementary school students in Korea (Lee and Park 2010; MEST 2012a, b).

Many education experts in Korea attribute a widely pervasive teaching method, which focuses on rote memorization and content delivery by lecture, as a contributing factor for the observed trends in the lack of interests in STEM. In general, even though students exhibit high levels of academic achievements in science and mathematics, they may not concurrently have high interests and self-efficacy in science and mathematics (KOFAC 2012). STEAM education began being promoted in Korea as a way to foster students' interests and self-efficacy in science and mathematics and to improve and enhance their academic achievements, creativity, and problem-solving skills.

2 Overview and Specific Directions of STEAM in Korea

Georgette Yakman is attributed to have first proposed the term STEAM, as her work on the STEAM pyramid. Here, Yakman expanded the concept of STEM to promote interdisciplinarity by emphasizing that science and technology can be interpreted through engineering and the arts, which are based in the elements of mathematics (Yakman 2008). The STEAM pyramid aims to propose a framework for teaching across the disciplines of science, technology, engineering, and the arts where by science and technology are interpreted through engineering and the arts; all four disciplines are based in mathematical elements. The STEAM pyramid is divided into five levels. The first and the lowest level is content specific to each of the four disciplines. For example, science content includes the history and nature of science, concepts and processes in science, inquiry, physics, biology, chemistry subjects as well as geosciences and biochemistry, etc. (see www.STEAMedu.com). The first level elaborates similar details for the other disciplines in technology, engineering, mathematics, and the arts. The second level above the first level are silos divided based on the four disciplines. The third level is where STEM and the arts are distinguished into two areas. On the fourth level is where STEM and the arts become STEAM. The fifth and the top level indicates lifelong, holistic learning. In the STEAM pyramid, as one moves vertically from a lower level to a higher level, toward lifelong, holistic education, each of the specialized subject areas gradually become integrated and achieve a holistic education (Yakman 2008).

In the current rhetoric of facing the new changes and challenges of the twenty-first century and developing economic competitiveness in the global economy, there is an increasing demand for nurturing creativity and affect, while pushing the boundaries of academic disciplines toward more integrative and holistic learning (Cho and Kim 2017). Consistent with these new educational goals, STEAM education can be conceptualized as the new educational paradigm that emerged from the need to respond to these changes (Yakman 2008). That is, some educators propose that STEAM education effectively fosters critical thinking abilities and nurtures affective domains such as emotions (KOFAC 2012). As such, one goal of STEAM education in Korea is to foster students' genuine curiosity toward the sciences, such that they not only KOFAC 2015learn to define and solve new problems, but also develop superior knowledge in science and technology (KOFAC 2012).

Similarly, MEST (2011) defines an aim of STEAM education to elevate students' interest in and understanding of science and technology, as well as to cultivate their STEAM literacy, creative thinking, and problem-solving skills. To this end, STEAM curriculum involves restructuring how science, technology, engineering, and mathematics are taught, at the elementary and secondary levels, so as to highlight ideas that crosscut the four subject areas while grafting humanities and the arts onto the integrative STEM education framework (MEST 2011). For example, Kim and Chae (2016) developed a STEAM program which incorporated different aspects of traditional Korean culture into the school curriculum. These aspects included *hanji* (Korean traditional paper) in the third and fourth grades, Korean foods in the fifth and sixth grades, *hanok* (Korean traditional house) in middle school, and *danso* (Korean traditional musical instrument) in high school. This STEAM program was based on the theme of Korean traditions and helped students learn about, and explore, the principles of Korean traditional cultures. In this program, students developed an understanding of scientific, engineering, and technology principles as they learned about their ancestors' cultural heritage, and they advanced their creativity and tapped into their emotions by exploring the artistic beauty of Korean cultures.

With the aforementioned goals of STEAM, the Korean government announced that STEAM education would be enforced in all elementary and secondary schools as part of the 2009 national revised curriculum (MEST 2011, 2012a, b). STEAM education is now widely promoted in Korea and various aspects of STEAM are being investigated.

3   STEAM, Creative Design, and Problem-Based Learning

Through contextualizing problems relevant to real life, students can develop genuine interests in and understanding of the sciences, as they cultivate problem-solving skills (KOFAC 2015). STEAM education builds upon the principles of science and mathematics in order to develop students' problem-solving skills as they learn to apply the concepts of engineering and technology, which are grounded in creative design (KOFAC 2015; Yakman 2008).

While STEM disciplines have generally been dealt with separately, integrative STEM/STEAM education developed from the need to face new and complex problems that are becoming harder to define and which cannot be easily solved by applying one specific discipline (MEST 2012a, b). Further, students will likely to encounter new challenges whose nature of the problem may look drastically different from the challenges we know today (MEST 2012a, b). Thus, integrative knowledge of the different disciplines is becoming increasingly important.

Creative design, based in elements of engineering and technology, is critically important in STEAM. A STEAM lesson during which students have opportunities to engage in creative design-based learning can enhance students' problem-solving abilities (KOFAC 2015). Generally speaking, the discipline in the sciences asks the question why. For example, some "why" questions could be "Why does the wind blow?" or "Why do rainbows form?" Similarly, science largely explores causal relationships. As such, the types of questions that deal with natural phenomena are likely to belong to the category of the sciences (KOFAC 2015). On the other hand, engineering represents a process through which "how-type" questions are answered (i.e., questions that require problem-solving skills). For example, a "how" question could be "How can I make an environmentally friendly automobile?" or "How can I build a house with high thermal efficiency?" (KOFAC 2015). Given that creative design involves a comprehensive process by which engineering concepts are applied to effectively solve a problem, engineering could be considered to be at the forefront of STEAM education (KOFAC 2015). Similarly, Sanders (2009) stated that there could be no STEM or STEAM without the T and E, emphasizing the critical importance of technology and engineering in integrative STEM or STEAM education. Therefore, an effective STEAM lesson grounded in creative design would help students not only learn scientific principles, but also how to apply engineering and technology principles with greater self-efficacy, self-motivation, and creativity (KOFAC 2015).

4   The Importance of the Arts in STEAM: "Emotional Touch" in Problem Solving

In general, the arts refer to the disciplines of literature and visual arts, which can facilitate an expression of creativity. In the context of STEAM in Korea, the definition of the arts extends to include communications and humanities, in order to foster "emotional touch" as part of students' learning experience (MEST 2012a, b). Science education experts in Korea propose that emotional touch is an essential element of a successful problem-solving skill (KOFAC 2015). In some ways, the concept of emotional touch is similar to the concept of affective domain. However, the difference lies in that emotional touch is student centered, rather than teacher centered (KOFAC 2015). Park et al. (2016) refer to emotional touch as "experiences that enable a positive cycle of self-directed learning where students feel interest, confidence, intellectual satisfaction and a sense of achievement, as they find motivation, passion, flow and personal meaning in learning" (p. 1742). In this definition, positive attitude toward learning is fostered, including self-interest, motivation, and desire, to further promote self-directed learning (KOFAC 2015).

> Emotional touch arises from the presentation of a real world problem related to the learning content, and the students who have undergone emotional touch come to examine whether the problem leads to another situation in the real world, not just by recognizing the necessity of learning. (KOFAC 2015)

Baek et al. (2011) define this process as a virtuous cycle during which a sense of achievement increases and reinforces one's interest and confidence toward learning and vice versa. Moreover, an aspect of emotional touch includes not only the success that a student may experience but also positive feedback of a teacher, which can help encourage a student's curiosity and desire to learn (KOFAC 2012). For example, when students attempt to solve a problem that is contextualized in everyday life, they can hone their problem-solving skills while they immerse themselves in the process of defining the problem. They can uncover clues about the nature of the problem, and then solve the problem through creative design (KOFAC 2015). Consequently, Baek et al. (2011) posit that this virtuous cycle should be repeated so that, upon experiencing success at solving a challenging problem, students' interests in mathematics and science are enhanced, motivation is rewarded, and the student develops a sense of achievement that reinforces positive feelings about their success.

5   Theoretical Framework of STEAM in Korea

The fundamental components of the STEAM education framework in Korea include contextualizing problems, creative design, and emotional touch. These three things require integrative STEAM literacy and promote a self-directed learning experience (Baek et al. 2011). Here, we adapted three design principles from the theoretical framework of STEAM from the Korea Foundation for the Advancement of Science and Creativity (KOFAC). This framework can provide a guideline for practitioners to design STEAM lessons or programs (Table 1).

Contextualizing a problem in a real-life situation is important because it provides a rationale for problem-solving that may be meaningful and personal to students (Lee 2013). Creative design (step 2, Table 1) begins when a student identifies a need

Table 1 Adapted from the theoretical framework of STEAM in Korea adapted from KOFAC (2012)

Steps	Design principles	Aims
1	Contextualization of a problem in a real-life situation	To provide a rationale for problem solving
2	Creative design	To facilitate self-directed learning to define a problem and find a solution
3	Emotional touch	To foster students' interest, motivation, and sense of achievement through a successful problem-solving experience

to solve a particular problem, and thus initiates the self-directed design work, which seeks adequate solution to the problem (Baek et al. 2011; Yakman and Lee 2012). Thus, creative design encourages an open-ended approach and provides the opportunity to reflect on their problem-solving strategy (KOFAC 2012). Particularly grounded in the notion of "hands-on" or "hands-in" activities (i.e., active engagement), design-based learning promotes communications, collaborations, and creativity (Baek et al. 2011). Finally, emotional touch extends the existing affective domain and engages students in hearts-on activities based on interests (KOFAC 2015). Therefore, emotional touch fosters a student's interests in science learning through immersion, heightened sense of achievement, and motivation (KOFAC 2015).

The theoretical framework of STEAM that includes contextualization, creative design, and emotional touch can be implemented and adapted in various curriculum models used in Korea such as (1) in-subject curriculum model, (2) across-subjects curriculum model, or (3) creative experience activities model which requires the restructuring of an existing model with a specialization area. The in-subject curriculum model is similar to disciplinary core ideas in the Next Generation Science Standards (NGSS 2013). Within a specific content area, the in-subject curriculum model emphasizes core ideas that have broad importance within or across multiple disciplines such as technology, engineering, the arts, and mathematics. Across-subjects curriculum model is similar to crosscutting concepts in the NGSS in that, given a concept, students explore the connections across different discipline domains. Lastly, the creative experience activities model involves restructuring an existing curriculum model with a specialization area or interest that can include but is not limited to (1) an after-school program (e.g., a club activity), (2) an extracurricular activity program, (3) a volunteer activity), or (4) career planning activity, which are all associated with a specific educational goal. An existing curriculum model can be restructured to integrate any of the four aforementioned specific interests (club, extracurricular, volunteering, and career-planning activities) and be transformed into a new curriculum model (KOFAC 2015). Within the program of interest, the creative experience activities model can be adapted to highlight its programs mission that may include creative thinking, experiential learning, a specific career path planning, etc. (KOFAC 2015).

6 Previous Literature on STEAM in Korea

The number of research studies reporting on STEM or STEAM education in areas such as its development, implementation, and content knowledge has increased substantially since 2010 (KOFAC 2015). For example, Baek et al. (2011) reported on the development and directions of STEAM in light of the current trends and issues of science education in Korea. Kim et al. (2012) defined core STEAM knowledge and competency and proposed a theoretical framework that detailed the potential development and implementation of STEAM education. Jeong and Kim (2015) developed a STEAM program that used six structured inventive thinking strategies to increase students' interest in STEM, problem-solving skills, and STEM content knowledge. More recently still, Jeong and Kim (2015) developed a STEAM program and implemented it in the context of problem-based learning: This program required students to participate in global climate change monitoring activities. The effectiveness of the STEAM program developed in Jeong and Kim (2015)'s study, especially its success with girls, is described in a later section on promoting girls in STEAM.

Other studies have investigated Korean teachers' perceptions of STEAM. For example, Lee et al. (2012) analyzed the perception of STEAM education of elementary and secondary school teachers, based on self-reported survey responses from 251 school teachers who taught science, mathematics, and/or technology. The Lee et al. (op. cit.) study reported that teachers perceived integrated STEAM education to be highly efficient, not only in developing integrated problem-solving skills and STEM-related literacy but also in positively impacting workforce by identifying talents in the fields of science and technology.

Similarly, Shin and Han (2011) surveyed 93 elementary school teachers who have taught mathematics and science in gifted classes. The survey involved a 15-item questionnaire designed to elicit teachers' awareness of STEAM education. The Shin and Han (op. cit.) qualitative study reported that after having participated in STEAM educational activities, teachers' awareness about the need of STEAM education increased and that the teachers' perceptions about STEAM education and its impact on elementary education were positive.

In summary, these studies show a general trend that acknowledges the need for integrative STEAM education in Korea. To enhance teachers' understanding of STEAM and its implementation in the classroom, science education experts are now calling for educational reforms that promote professional development programs designed to increase teacher's expertise about STEAM teaching.

Many studies report the effectiveness of STEAM program on increasing students' interest and self-efficacy toward STEAM disciplines. For example, Park and Shin (2012) investigated and compared an intervention class of 30 students whose science curricular unit on "Our Body" was implemented as a STEAM curriculum. This unit was introduced during the second semester of the 5th grade, while the comparison (control) class of 29 students was taught by implementing lecture-style teaching. All students responded to questionnaires on several affective domains in science learning, such as self-efficacy and interest in and attitude toward science. The Park and Shin (op. cit.) study reported that the science lesson on "Our Body" as a STEAM curriculum positively influenced the elementary students' self-efficacy toward the science subject. Mean scores for the control group and the STEM-intervention group were similar at the start of the study (3.32 vs. 3.40, respectively; $p = 0.65$), but the intervention effect was significant at the end of the study (3.30 for the control group vs. 3.79 for the intervention group; $p = 0.003$). The students' interest in science, science learning, scientific activities, and pursuing science-related jobs also increased in response to the STEAM intervention (by 19.3%, 20.7%, 12.0%, and 20.3%, respectively). All of these differences were statistically significant ($p < 0.05$).

Lee and Kang (2014) developed a STEAM program that they implemented as an after-school program with middle school students. In this study, they investigated the program's effects on improving students' creative thinking and critical thinking skills through inquiry. Their study reported that the intervention group demonstrated substantial increases in the following areas: creative thinking skills, fluency, and originality. The intervention group's overall score was ~34.8, which was 34% higher than the comparison group's overall score of ~25.9 on critical thinking skills. The difference between the comparison and intervention groups in this case was statistically significant.

Kim et al. (2013) developed a STEAM program based on four topic areas in high school life sciences – circulation, plants, genetics, and stimulus and response. These topic areas were integrated with technology, engineering, the arts, and mathematics. The Kim et al. (op. cit.) study reported that students' confidence and interest in science and values for science increased. The STEAM program also increased students' creativity and subcategory elements (fluency and originality) and elaboration.

Kim and Chae (2016) developed a STEAM program based on Korean traditional cultures and explored the program's effectiveness in developing problem-solving skills. After interviewing the participating students, this study reported that 42% of the students recognized STEAM as "a process of solving a problem using convergent thinking," 34% thought it as "a knowledge reconstruction process through the exchange of opinions," and 24% perceived it as "a process of finding a solution on their own" (Table 2).

Kim and Chae (op. cit.) concluded that students who participated in the STEAM program recognized the importance of integrative knowledge in honing their problem-solving skills.

Table 2 Student's awareness of the meaning of STEAM. (From Kim and Chae (2016))

Meaning of STEAM	(N = 26)
A process of solving a problem using convergent thinking	42%
A knowledge reconstruction process through the exchange of opinions	34%
A process of finding a solution on their own	24%

7 STEAM Addresses the Limitations of Problem-Based Learning

Problem-based learning (PBL) provides students with opportunities for authentic, inquiry-based learning in which a real-world problem becomes the context for investigating what they need to know and want to know (Checkly 1997). Gallagher et al. (1995) defined three characteristics of PBL: (1) initiating learning with a problem, (2) using an ill-defined problem, and (3) having the teacher as a metacognitive coach. In PBL, students are first presented with a problem, as it might actually occur. There is rarely sufficient information to develop a solution to this problem immediately, and no obvious single right strategy, so a need to redefine the problem emerges as new information is presented (Yoon et al. 2011). Second, the problem is ill defined. This characteristic, Gallagher et al. (1995) suggests, is how problems generally present themselves in science and real life. Because the problem is unclear, students need to identify what is known, what needs to be known, and how to find a solution. The open-ended nature of the problem provides opportunity for many possible solutions (Kim et al. 2016). Finally, the teacher is simply a guide or a metacognitive coach in the PBL approach: students have the primary responsibility of determining what is learned and how learning happens. For this reason, PBL is a constructive, self-directed learning process, during which the search for meaning becomes a personal construction of the student (Yoon et al. 2011).

The PBL approach has four important limitations in terms of a broader application. First, preparing a PBL lesson takes a significant amount of time (Simons et al. 2004). For example, Simons et al. (2004)'s study showed that the teacher for a PBL effort expressed frustration at not having enough time to complete every activity and was not completely satisfied with her students' performance. Second, Ertmer and Simons (2006) also reported on limitations teachers experienced when implementing PBL, such as engaging their students and scaffolding their learning. Third, another significant limitation reported in the study was simplifying the components of the problem and content without making them superficial. Thus, Ertmer and Simons (2006) call for more studies reporting on the "how to" for teachers interested in effectively implementing PBL in the K–12 classrooms. Lastly, in PBL, the role of self-directed student is important. In general, students work in collaborative groups and engage in self-directed learning efforts to apply their new knowledge to solve the problem (Hmelo-Silver 2004).

Although PBL aims to help students develop flexible knowledge, effective problem-solving skills, self-directed learning skills, effective collaboration skills, and intrinsic motivation, the level of mastery of these skills can vary greatly among students within the given constraints of a classroom (Hmelo-Silver 2004). To this end, the aforementioned proposed framework of STEAM education facilitates, through three components, a self-directed learning process in which students enact creative design and emotional touch and contextualize a problem in a real-life situation. First, the open-endedness nature of creative design allows students time to reflect and encourages students to learn how to effectively communicate and collaborate (Park et al. 2016). "The create design also includes a provision of educational opportunities for students to experience the entire self-directed process until the final product of learning is applied in practice" (Park et al. 2016, p. 1742). Second, emotional touch addresses an aspect of education that is often overlooked, and this aspect can help the learner develop personal meaning in learning. Thus, Park et al. (2016) suggest that leveraging emotional touch that facilitates cognitive and affective development be organically connected to the process of learning. And third, students connect the contents of study to real life in a holistic manner. In summary, STEAM programs can be applied to any educational settings such as after-school program or information education and implemented to any curriculum models such as in-subject or across-subjects models.

We summarize these points as follows: STEAM education in Korea goes beyond the simple notion of integrating subject disciplines. At the core of STEAM education in Korea is the leveraging of the benefits of a self-directed and creative learning, while providing students the opportunity to solve problems in a manner that affords students a sense of accomplishment and positive emotional experience including encouragement and success. For these reasons, STEAM education continues to be promoted in the educational context of Korea.

8   Promoting Female Students in STEM/STEAM

Female students in Korea indicated significantly lower preference than male students for pursuing science and technology in terms of education and careers (MEST 2012a, b). This phenomenon could be due to institutionalized barriers that tend to favor male engineers and scientists (Hyun 2015). According to Hyun (2015), female students' interests in science and technology, and their academic performance in these related fields, are generally lower than their male counterparts. Therefore, STEM in the United States, or the United Kingdom, as well as in Korea, aims to enhance both female and male students' interest and motivation in STEM content areas. To this end, Hyun (2015) reported that implementing a STEAM program improved female students' interest and preference in science and technology, yielding greater self-efficacy in science and technology among the female students. Additionally, Jeong and Kim (2015) reported that STEAM education was effective for both male and female students in enhancing their perceptions toward STEAM, and most prominently for girls (Table 3).

Effect size in this case is a measure of the magnitude of changes in scores for pre-to-post-intervention surveys of male and female students. For example, differences in the pre-to-post survey responses for the female students were about twice as large as those for male students, for the survey topics dealing with perception of science. STEAM content knowledge of both male and female students increased, but the effect was much larger for female students. For example, female students who participated in the global climate change (GCC) activities had pre-to-post gains

Table 3 Effect size of mean scores in STEAM perceptions (Jeong and Kim 2015)

Gender	Effect size of mean scores				
	Science	Technology	Engineering	The arts	Mathematics
Male	0.28	0.06	0.16	0.13	0.32
Female	0.54	0.19	0.35	0.16	0.34

that were nearly twice as large as those of their male classmates, for perceptions of science and engineering.

Jeong and Kim (2015)'s study demonstrated that female middle school students who participated in the GCC monitoring activities had much greater pre-to-post gains than their male counterparts, with respect to their perceptions of science and engineering. Further, positive perceptions of STEAM content emerged among the participants of the GCC activities. A comparison of the gain scores in STEAM perceptions for female and male students showed that the female students' perceptions of STEAM rose dramatically and became more similar to those of males over the course of the GCC activities. In short, STEAM education effect was more pronounced for female students than for male students.

9 Effectiveness of STEAM

In 2013, the Korea Foundation for the Advancement of Science and Creativity (KOFAC) conducted a questionnaire survey on the students from 23 elementary, middle, and high schools to examine the effectiveness of STEAM education in Korea. The schools identified in the survey included 9 leader schools, 19 schools which were operating a school science teachers' study group for STEAM, 3 schools which implemented the smart class model for STEAM, and 1 science school (KOFAC 2013). Among these schools, 16 classes were implementing STEAM programs and 15 classes were in the planning stages for STEAM implementation. Nearly 1400 students were surveyed. The surveyed students included 502 elementary school students, 410 middle school students, and 461 high school students (Table 4; KOFAC 2013). Also, each of the 6 survey items showed statistically significant values ($p < 0.05$) and students who participated in STEAM programs indicated greater preferences for science (Table 5; KOFAC 2013).

In terms of "self-driven learning ability," students in STEAM programs reported a higher proficiency (Table 6). Among the other six areas, the most notable survey item was "cooperation capability," which showed highest responses by students in STEAM programs (KOFAC 2013). All of these differences were statistically significant ($p \leq 0.02$). These results can be attributed to the opportunities that students in STEAM programs have for exchanging their ideas and establishing a harmoniously functioning team through collaborative STEAM learning.

Students in STEAM programs developed greater problem-solving skills and problem cognition (Table 7), compared to students in the general class ($p = 0.000$ and $p = 0.003$, respectively). This result can be attributed to opportunities for foster-

Table 4 Demographic statistics of study participants (KOFAC 2013)

School classification	General class (male/female)	STEAM class (male/female)	Total number of students (male/female)
Elementary school	237 (123/114)	265 (114/151)	502 (237/265)
Middle school	225 (137/88)	185 (66/119)	410 (203/207)
High school	211 (103/108)	250 (108/142)	461 (211/250)
Total	673 (363/310)	700 (288/412)	1373 (651/722)

Table 5 Preferences for science (KOFAC 2013)

Dimension	Number of items	Average (standard deviation)		$t(p)$
		General class	STEAM class	
Curiosity about science	3	3.40 (1.14)	3.58 (1.07)	3.39** (0.001)
Interest in learning science	3	3.22 (1.19)	3.46 (1.14)	4.68** (0.000)
Inclusion of value on science	3	3.85 (1.00)	3.98 (0.95)	3.04** (0.002)
Belief in science	3	3.53 (1.06)	3.73 (1.00)	4.16** (0.000)
Commitment to scientific task	3	3.06 (1.13)	3.36 (1.08)	6.00** (0.000)
Choice of science-related career	3	2.80 (1.24)	3.08 (1.20)	4.61** (0.000)

**$p < 0.01$, *$p < 0.05$

Table 6 Self-driven learning ability (KOFAC 2013)

Dimension	Number of items	Average (standard deviation)		$t(p)$
		General class	STEAM class	
Curiosity about science	2	3.24 (1.05)	3.36 (1.00)	2.36** (0.018)
Interest in learning science	3	3.18 (1.04)	3.39 (1.01)	4.50** (0.000)
Inclusion of value on science	4	2.90 (1.06)	3.14 (1.04)	5.16** (0.000)
Belief in science	3	3.30 (1.01)	3.43 (0.98)	2.88** (0.004)
Commitment to scientific task	2	3.33 (1.05)	3.53 (1.00)	4.26** (0.000)
Choice of science-related career	3	3.40 (1.04)	3.56 (1.02)	3.52** (0.000)

**$p < 0.01$; *$p < 0.05$

ing students' imagination and emotional touch: Through these routes, the students can hone their scientific knowledge and creative-thinking skills (KOFAC 2015).

The results of other studies echo the findings of the KOFAC (2013)'s survey results. Notably, Kim and Chae (2016) investigated the effectiveness of STEAM implementation at the secondary level, where 26 high school students were inter-

Table 7 Creative and integrated thinking (KOFAC 2013)

Dimension	Evaluation	Average (standard deviation)		t(p)
		General class	STEAM class	
Problem cognition	Originality, applicability, validity	1.41 (1.12)	1.75 (0.936)	2.99** (0.003)
Problem-solving		1.31 (1.19)	1.76 (1.01)	3.74** (0.000)

**$p < 0.01$, *$p < 0.05$

viewed to explore students' awareness about STEAM in two categories: (1) the meaning of STEAM and (2) the necessity of STEAM education (Kim and Chae 2016). First, echoing the results of Baek et al. (2011) and Maes (2010), students in Kim and Chae (2016)'s study demonstrated their understanding of integrative STEAM knowledge in order to be successful at problem-solving. Second, students indicated that STEAM activities facilitated a procedural process, as defined by Wheeler and Jones (2008), which helped students engage in the process of "knowledge reorganization" through the mutual exchange of opinions and communications. Third, in the process of solving a problem, students indicated that their imaginations and emotional touch helped them develop their scientific knowledge and creativity as they engaged in STEAM learning. These results parallel those of studies by Kang et al. (2014), Min and Kang (2015), and Maes (2010), which demonstrated that the arts facilitated students' creativity and helped them cultivate positive attitudes as well as scientific thinking and problem-solving skills in STEAM.

10   Concluding Remarks About STEAM in Korea

In Korea, STEAM education is considered to be highly effective: STEAM programs have demonstrably impacted important aspects of science learning, such as developing integrative STEAM knowledge, problem-solving skills, etc. Especially by helping students become more successful at problem-solving, STEAM has played an important role in fostering students' emotional touch, promoting self-directed learning, and facilitating collaborations and exchange of ideas through communications (KOFAC 2015).

STEAM education in Korea has implications for teachers elsewhere. Since introducing STEAM in Korea, Korean students' interest and academic achievements in the sciences and engineering have increased; moreover, the number of students choosing to pursue their future careers in STEAM-related fields has been increasing (Lee and Lee 2014; Jeong and Kim 2015; Kim and Chae 2016). Overall, based on the research examining teachers' perception of STEAM, science teachers in Korea now are beginning to associate positive perceptions of STEAM with respect to the areas of helping students learn to communicate and develop integrative STEAM knowledge and problem-solving skills.

Continued efforts should be made in Korea to effectively leverage the advantages of STEAM education. Teachers will need to be committed to preparing and enacting effective STEAM lessons in the classrooms, which will inevitably require significant efforts. Further, to promote STEAM learning, it should be consistently implemented early, starting with elementary school children, so their interests in STEAM begin to develop. At present, STEAM education is being implemented at the college level, and this implementation will inevitably require trial-and-error adjustments. However, based on the positive results reported so far, STEAM education in Korea will continue to establish a classroom culture, which fosters students' genuine curiosity and develops their ability to successfully solve new problems that they will face in the twenty-first century.

Acknowledgments This work was supported by the Ministry of Education of the Republic of Korea, the National Research Foundation of Korea (NRF-2017S1A5A8021812), and the Korea Foundation for the Advancement of Science and Creativity (KOFAC) grant, which was funded by the Korean government (MOE) (2018EAA0002).

References

Baek, Y., Park, H., Kim, Y., Noh, S., Park, J.–. Y., Lee, J., Jeong, J.–. S., Choi, Y., & Han, H. (2011). STEAM Education in Korea. *Journal of Learner-Centered Curriculum and Instruction, 11*(4), 149–171.

Checkly, K. (1997). Problem-based learning. *ASCD Curriculum* Update, summer, 1–3:6–8.

Cho, K., & Kim, H. (2017). Effect of systems thinking based STEAM education program on climate change topics. *Journal of the Korea Contents Association, 7*(7), 113–123.

Ertmer, P. A., & Simons, K. D. (2006). Jumping the PBL implementation hurdle: Supporting the efforts of K–12 teachers. *Interdisciplinary Journal of Problem-based Learning, 1*(1), 5.

Gallagher, S., Stephien, W. J., Sher, B. T., & Workman, D. (1995). Implementing problem-based learning in science classrooms. *School Science and Mathematics, 95*(3), 136–146.

Hmelo-Silver, C. E. (2004). Problem-based learning: What and how do students learn? *Educational Psychology Review, 16*(3), 235–266.

Hyun, E. (2015). The effect of art and science technology integrated creativity education using smart devices -focusing on implementation, application, and verification of the academic achievement perception from the female high school students. *Journal of Korea Design Knowledge, 34*(1), 357–365.

Jeong, S., & Kim, H. (2015). The effect of a climate change monitoring program on students' knowledge and perceptions of STEAM education in Korea. *Eurasia Journal of Mathematics, Science and Technology Education, 11*(6), 1321–1338.

Kang, D., Yi, C., & Lee, J. (2014). A study on the content analysis of emotional touch in A-STEAM programs. *Art Education Review, 52,* 1):1–1)32.

Kim, H., & Chae, D.-H. (2016). The development and application of a STEAM program based on traditional Korean culture. *Eurasia Journal of Mathematics, Science and Technology Education, 12*(7), 1925–1936.

Kim, S. W., Chung, Y. L., Woo, A. J., & Lee, H. J. (2012). Development of a theoretical model for STEAM education. *Journal of the Korean Association for Research in Science Education, 32*(2), 388–403.

Kim, J. Y., Ju, H. Y., & Lee, K. J. (2013). The effects of STEAM program based on life science for science-related affective domain and creativity in high school students. *Biology Education, 41*(4), 531–543.

Kim, D., Yoo, M., & Woo, H. (2016). The effects of a problem-based learning program titled 'Designing safe and strong bridge' on the scientific attitudes, science career orientation, and leadership of scientifically gifted high school students. *Journal of Gifted/Talented Education, 26*(3), 449–471.

Korea Foundation for the Advancement of Science and Creative (KOFAC). (2013). *Policy research on raising scientific talented students with creativity-convergence: Focused on the analysis of the STEAM effect.* KOFAC: Seoul.

Korea Foundation for the Advancement of Science and Creativity (KOFAC). (2012). *Hand-knuckle STEAM education.* Seoul: KOFAC.

Korea Foundation for the Advancement of Science and Creativity (KOFAC). (2015). *Visible STEAM education.* Seoul: KOFAC.

Lee, H. (2013). *Understanding and application of STEM/STEAM education.* Seoul: Bookshill.

Lee, W., & Kang, S. (2014). Creative and critical thinking skills-reinforced and STEAM-oriented teaching strategy of science for students' extracurricular activities through a junior high school intervention study program. *Journal of the Research Institute of Curriculum Institution, 18*(2), 321–342.

Lee, Y., & Lee, H. (2014). The effects of engineering design and scientific inquiry based STEAM education programs on the interest, self-efficacy and career choices of middle school students. *Journal of Research in Curriculum and Instruction, Curriculum, Instruction, Education, 18*(3), 513–540.

Lee, H., & Park, K. (2010). Elementary school students' images of scientists and engineers. *The Society of Korean Practical Arts Education Research, 16*(4), 61–82.

Lee, H. N., Son, D. I., Kwon, H. S., Park, K. S., Han, I. K., Jung, H. I., Lee, S. S., Oh, H. J., Nam, J. C., Oh, Y. J., & Phang, S. H. (2012). Secondary teachers' perceptions and needs analysis on integrative STEM education. *Journal of Korean Association for Research in Science Education, 32*(1), 30–45.

Maes, B. (2010). *Stop talking about "STEM" education! "TEAMS" is way cooler.* Retrieved from http://bertmaes.wordpress.com/2010/10/21/teams/

Martin, M. O., Mullis, I. V. S., & Foy, P. (2012). *TIMSS 2011 international science report.* Boston: Boston College.

Min, I., & Kang, Y. (2015). The effect of the subject-integrated classes between art and science on learning flow and learning achievement. *Art Education Review, 55*(1), 99–128.

Ministry of Education, Science, and Technology (MEST). (2011). *The 2009 revised science curriculum.* Seoul: MEST.

Ministry of Education, Science, and Technology (MEST). (2012a). *2012 National plan for STEAM education.* Seoul: MEST.

Ministry of Education, Science, and Technology (MEST). (2012b). *A study of policy based on teacher training.* Seoul: MEST.

Mullis, I. V. S., Martin, M. O., & Foy, P. (2012). *TIMSS 2011 international mathematics report.* Boston: Boston College.

NGSS Lead States. (2013). Next generation science standards: For states, by states. Washington, DC: The National Academies Press.

Park, H. W., & Shin, Y. J. (2012). Effects of science lesson applying STEAM education on self-efficacy, interest, and attitude towards science. *Biology Education, 40*(1), 132–146.

Park, H., Byun, S. Y., Sim, J., Han, H., & Baek, Y. S. (2016). Teachers' perceptions and practices of STEAM education in South Korea. *Eurasia Journal of Mathematics, Science and Technology Education, 12*(7), 1739–1753.

Sanders, M. (2009). STEM, STEM education, STEM mania. *Technology Teacher, 68*(4), 20–26.

Shin, Y. J., & Han, S. K. (2011). A study of the elementary school teachers' perception in STEAM (Science, Technology, Engineering, Arts, Mathematics) education. *Journal of Korean Elementary Science Education, 30*(4), 514–523.

Simons, K. D., Klein, J. D., & Brush, T. R. (2004). Instructional strategies utilized during the implementation of a hypermedia, problem-based learning environment: A case study. *Journal of Interactive Learning Research, 15*(3), 213.

Wheeler, P., & Jones, D. R. (2008). The psychology of case-based reasoning: How information produced from case-matching methods facilitates the use of statistically generated information. *Journal of Information Systems, 22*(1), 1–25.

Yakman, G. (2008). STΣ@M Education: An overview of creating a model of integrative education. *Pupils Attitudes Towards Technology 2008 Annual Proceedings*. Netherlands.

Yakman, G., & Lee, H. (2012). Exploring the exemplary STEAM education in the U.S. as a practical educational framework for Korea. *Journal of the Korean Association for Science Education, 32*(6), 1072–1086.

Yoon, H., Kim, K. W., & Woo, A. J. (2011). An investigation into students' perception of problem-based learning implemented in middle school open-inquiry program. *The Korean Association for Science Education, 31*(5), 720–733.

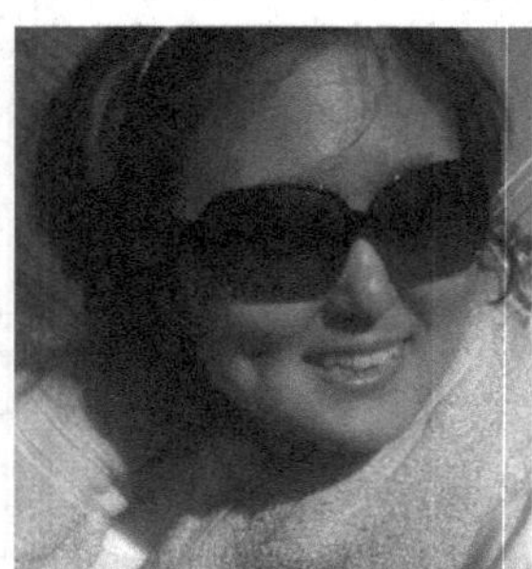

Sophia Jeong is a postdoctoral scholar in the Department of Biochemistry and Molecular Biology at the University of Georgia. Her scholarly work draws on actor-network theory to explore ontological complexities of gender and race by examining socio-material relations in science classrooms.

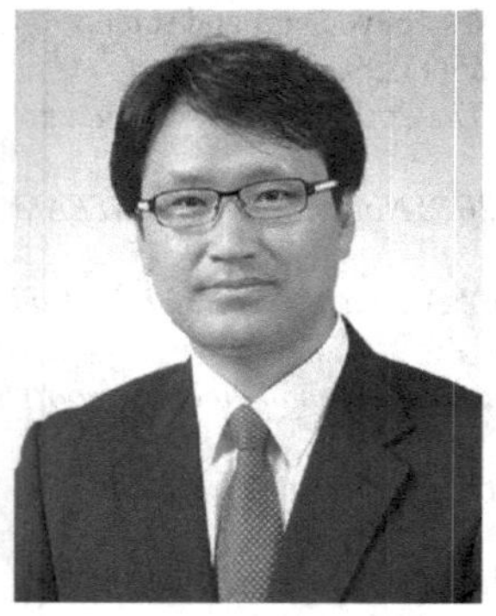

Hyounvgbum Kim obtained the B.A. degree in earth system science from Yonsei University, Korea, the M.A. degree from the School of Earth and Environmental Sciences, Seoul National University, Korea, and the Ph.D. degree in earth science education from Korea National University of Education, Korea. He had worked as postdoctoral scholar at the Science Education Research Institution of UQAM, Canada. He is currently an Assistant Professor at the Earth Science Education Department, Chungbuk National University, Korea. His research interests include science curriculum, climate change education, and STEM education (STEAM) in K–12 education.

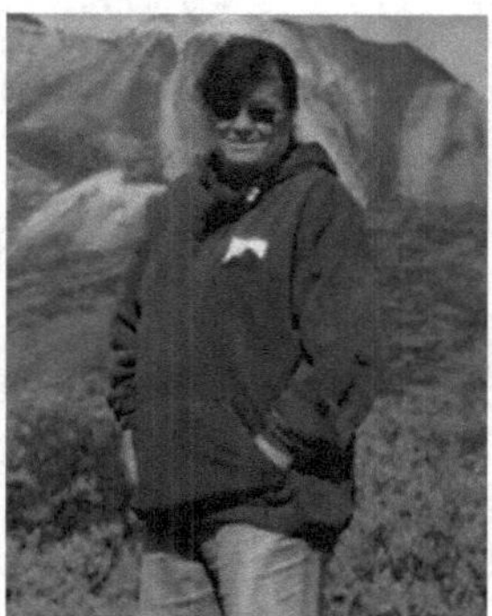

Deborah J. Tippins is currently a Professor in the Department of Mathematics and Science Education at the University of Georgia. Her scholarly work focuses on encouraging meaningful discourses around environmental justice and sociocultural issues in science education.

Emphasizing Transdisciplinary Prowess in the Evaluation of STEAM Programs

Kimberle Kelly and Erin Burr

1   Introduction

In this chapter, we consider the critical role of transdisciplinary knowledge and skills in the success of STEAM programs designed to increase the scientific and technical workforce and promote a responsible citizenry. STEAM programs by definition bridge disciplines. Thus, including the "arts" with science, technology, engineering, and mathematics programs requires the development and evaluation of both disciplinary and transdisciplinary knowledge and skills across diverse stakeholder groups. After describing the crucial role of program evaluation in the success and sustainability of STEAM programs, we use the "4 Cs" framework to describe relevant constructs of "transdisciplinary prowess." We then outline selected measurement strategies and technological tools for evaluating communication, collaboration, critical thinking, and creativity in STEAM programs as well as commenting on the need for transdisciplinary prowess in evaluators.

2   Why Program Evaluation?

The emphasis on evidence-based decision-making is prevalent in American education and government, and technological innovations have greatly expanded the landscape of available evidence. Such evidence can come from a high-quality program evaluation. As shown in Fig. 1, incorporating program evaluation before, during, and after implementing a STEM or STEAM program provides a solid foundation to support the design, development, implementation, effectiveness, dissemination,

K. Kelly (✉) · E. Burr
Oak Ridge Associated Universities, Oak Ridge, TN, USA
e-mail: Kimberle.Kelly@orau.org; Erin.Burr@orau.org

© Springer Nature Switzerland AG 2019
A. J. Stewart et al. (eds.), *Converting STEM into STEAM Programs*,
Environmental Discourses in Science Education 5,
https://doi.org/10.1007/978-3-030-25101-7_17

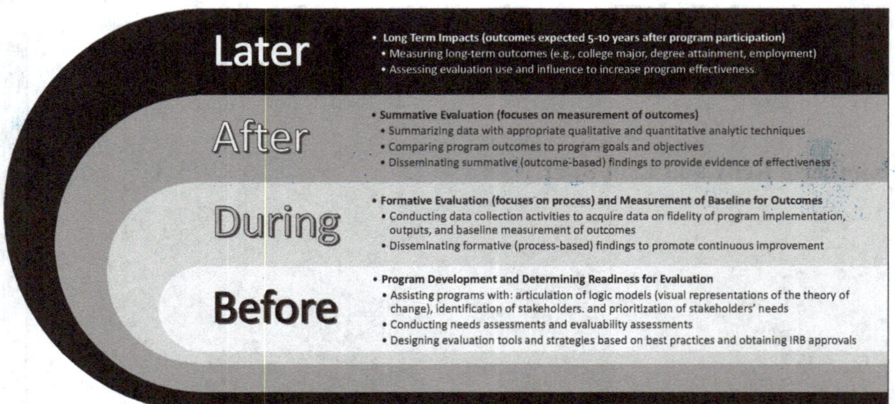

Fig. 1 Program evaluation and assessment services before, during, after, and into the future of educational programs. (Created by authors)

and sustainability of the program into the future. We further suggest that evaluation is an ethical responsibility for any program claiming to affect outcomes of the STEAM, STEM, and STEM-cognate workforce. Evaluation can help ensure that funding is spent wisely, that programs are accountable for their impact on stakeholders, and that relevant findings are communicated in the peer-reviewed academic literature.

2.1 Evaluation Use

Without evidence of effectiveness, it can be difficult to publish about, obtain funding for, or convince various audiences to participate in or support a STEAM program. Within the community of STEM and STEAM practitioners, program consumers and sponsors increasingly demand research and evaluation data of program effectiveness. For example, the National Science Foundation typically requires program evaluation as part of most grant applications for STEM and STEAM programs and offers evaluation guidance to applicants as to what constitutes credible evidence for particular research and evaluation designs (U.S. Department of Education and U.S. National Science Foundation 2013). Credible findings encourage program continuation and expansion over time and can entice desired participants during recruiting activities. As part of the academic enterprise, programs are expected to contribute what they discover to the academic research knowledge base. For this reason, highly skilled evaluators can be valuable partners in crafting such contributions, whether in the form of conference presentations, peer-reviewed journal articles, additional proposals, follow-up studies, or other academic products and services.

Research and theory regarding evaluation use considers the ways stakeholders use evaluation information and, therefore, how to best maximize the use and value of evaluation practices and findings (Patton 2008). A practical consideration is choosing an evaluation configuration that fits the program being evaluated. Typical configurations are internal evaluation, external evaluation, and metaevaluation. Different evaluation configurations have different strengths and permit different levels of access to information and stakeholders. In our experience, combinations or collaborations among these configurations provide the most useful program guidance. Further, evaluation use is highly related to successful interactions between diverse stakeholders and evaluators, without which evaluation findings are less likely to be trusted and used. The success of evaluators in engaging with diverse stakeholders relates to evaluator competencies in collaboration introduced later in this chapter.

2.2 Program Development and Continuous Improvement

STEM and STEAM education and workforce development programs typically require evaluation for two distinct purposes: (1) to inform continuous program improvement and (2) describe program effectiveness. Formative evaluation refers to the former purpose and focuses on collecting appropriate data to use in a continuous program improvement cycle. To understand the effects of a program, evaluators must first determine what the program actually delivered, compared to what was planned. The construct fidelity of implementation represents the extent to which a program was delivered as intended. This construct was developed by Jeanne Century and her colleagues into a framework (Fig. 2) and offers a suite of measurement tools (Century et al. 2010). By providing formative feedback about implementation on a regular basis, evaluators assist those who deliver a STEAM program in making data-informed decisions that can improve program delivery. We have found monthly conference calls an essential strategy for both monitoring program implementation of STEM and STEAM programs and for providing formative feedback based on evaluation data. Surveys or debriefing sessions after program activities, feedback from site visits, and regular advisory board meetings are also critical opportunities to discuss program implementation.

Formative evaluation also includes opportunities for the evaluator to engage with program stakeholders before a STEM or STEAM program is implemented. For example, evaluators can help stakeholders develop goals and objectives and inform program design by performing a needs assessment in advance of program development (Watkins et al. 2012). An evaluability assessment determines the appropriateness of program evaluation given a program's theory of change, the availability of evaluative data, and utility given the views of relevant stakeholders (Davies 2015). Figure 1 shows formative evaluation opportunities before a program begins, as well as during implementation. There are distinct benefits to engaging a team of highly qualified evaluators as early as possible in the program development process.

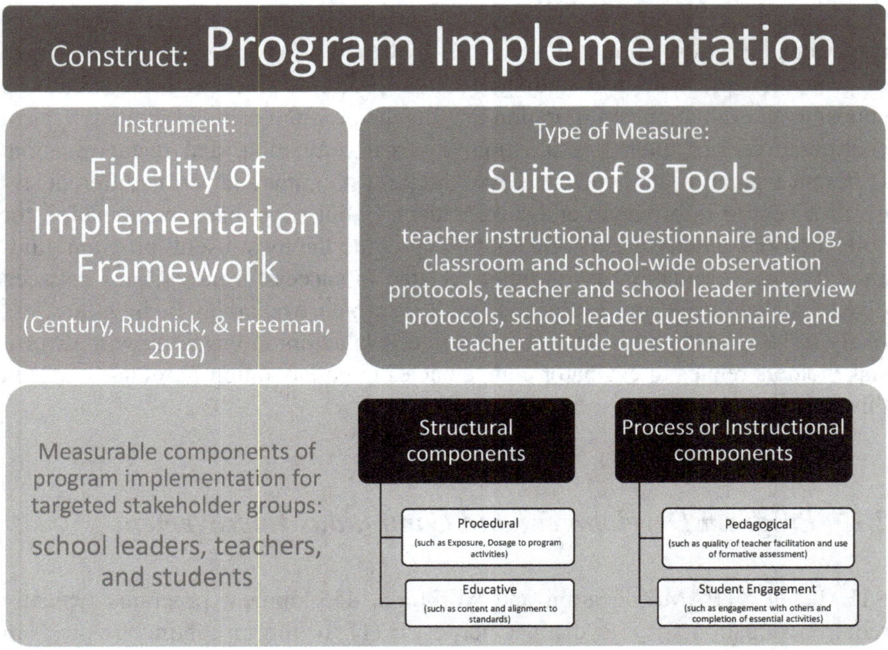

Fig. 2 Measuring program implementation as part of a formative evaluation. (Created by authors)

2.3 Program Effectiveness

In contrast to formative evaluation, a summative evaluation focuses on articulating the effectiveness or value of a program using meaningful evidence. Program success is achieved when a STEM or STEAM program does what it claims to do and can "prove it." The path to effectiveness begins with articulating the program's goals and objectives. A valuable way to frame goals and objectives is by using the attributes of SMART objectives, which are specific, measurable, achievable, relevant, and time bound (Salabarría-Peña et al. 2007). Framing these SMART goals and objectives within a logic model helps stakeholders clarify what they expect to accomplish and how they will know if in fact they accomplish it (W. K. Kellogg Foundation 2004). Figure 3 shows the components of a logic model, which traces program activities to their intended outputs and outcomes based on articulated goals and SMART objectives. Ideally, stakeholders should engage in an interactive process to generate shared understanding and commitment to a common vision. Evaluators are often critical in facilitating the articulation of program goals, objectives, and logic models.

A summative evaluation explicates the value of a program by credibly measuring outcomes specified on the logic model. What constitutes meaningful or credible evidence differs across stakeholders and contexts. For example, a headache remedy

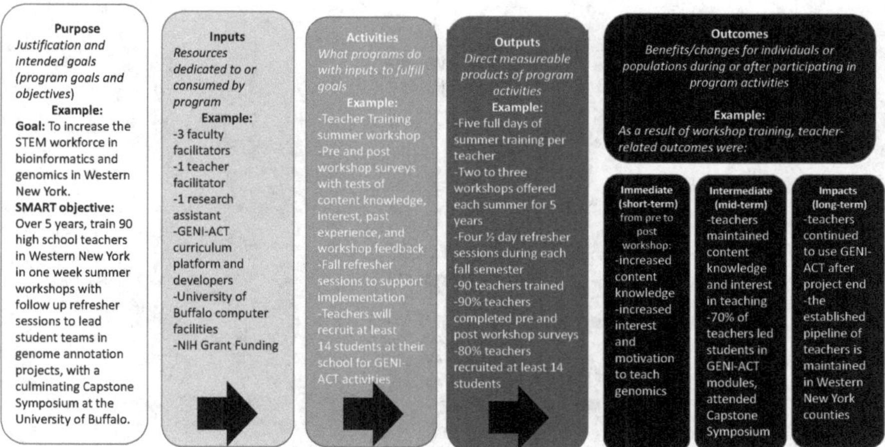

Fig. 3 Components of logic models. The example is for a project funded through the National Institutes of Health entitled *Western New York Genetics in Research and Healthcare Partnership* (Kelly et al. 2018b). (Created by authors)

that makes one feel better when they use it constitutes one type of evidence, but clinical efficacy data on headache pain established in a double-blind clinical trial is quite a different type of evidence. The *What Works Clearinghouse* is a nationally recognized resource for policy and guidance on methods requirements for credible evidence in educational research and evaluation (U.S. Department of Education 2017). While it is neither possible nor necessary to use randomized experiments to determine the effectiveness of every program, it is critical that stakeholders understand the strengths and limitations of both collected data and subsequent interpretations based on them. In sum, the value of the program evaluation depends both on the measurable articulation of the program and the credible measurement of expected outcomes across stakeholders.

Ideally, a program evaluation considers outcomes across a system of multiple stakeholders, considering programs in terms of what is delivered to whom, who provides what, and how they organize themselves to do so. This structure includes evaluators, too. STEM and STEAM programs often are conceptualized in terms of a pipeline metaphor (see Hardiman and JohnBull this volume) or, more broadly, as a learning system. Targeted participants may be anywhere along the educational continuum, from "pre-K to gray." Participants are embedded in a system or network that includes teachers in schools or faculty in universities and expands to community, government, and global partners. We have found a nested learning community model effective in portraying relationships among stakeholders (Fig. 4). Evaluation must not only measure outcomes across this diverse group of stakeholders, but must be sensitive to their unique information needs (Golder et al. 2005), a theme we will take up shortly.

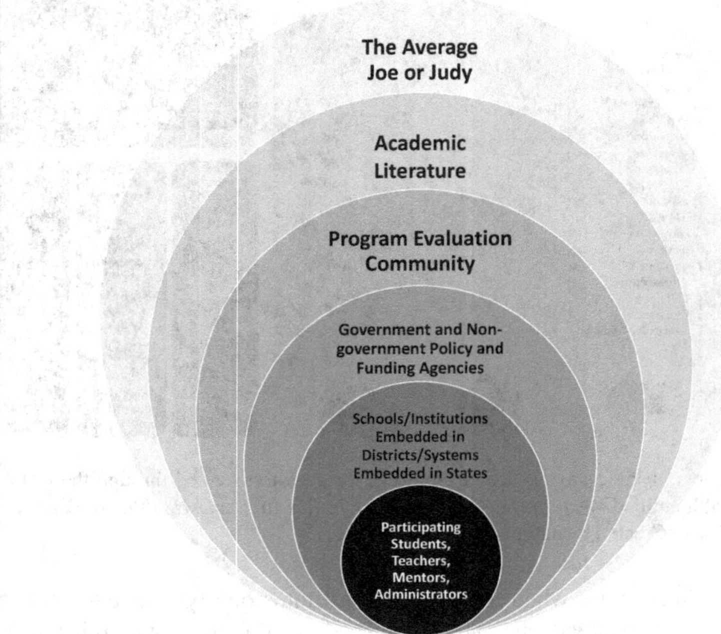

Fig. 4 A nested learning community model for STEAM programs. (Created by authors)

3 Transdisciplinary Knowledge, Skills, and Processes

Thus far, we have focused about equally on STEM and STEAM programs in terms of the value, purposes, and principles of program evaluation. While their similarities are a useful starting point, by definition, STEAM programs emphasize not only disciplinary but also transdisciplinary knowledge, skills, and processes – a range of knowledge and skills broader than traditionally considered in STEM fields. Transdisciplinary refers to knowledge and skills "beyond a specific discipline." For clarity, Fig. 5 distinguishes among multi-, inter-, and transdisciplinary models (see discussion by Evans 2014). Transdisciplinarity adds another dimension to the nested learning community model in Fig. 4 – that of intersections among disciplines within and across layers of the system. Program activities connect these networks to affect outcomes on involved stakeholders.

Tied to the need for transdisciplinary skills, STEAM program activities are often framed around projects, products, or performances that require transdisciplinary expertise. This has been dubbed project- or problem-based learning (PBL) and is a departure from traditional STEM instruction (see Gary Ubben this volume-a, this volume-b). Ghanbari (2015) explains that PBLs emphasize experiential and inquiry-based learning, whereby participants engage in authentic and relevant learning opportunities demanding transdisciplinary solutions. Further, there is an opportunity for sociocultural development as participants tackle problems with political,

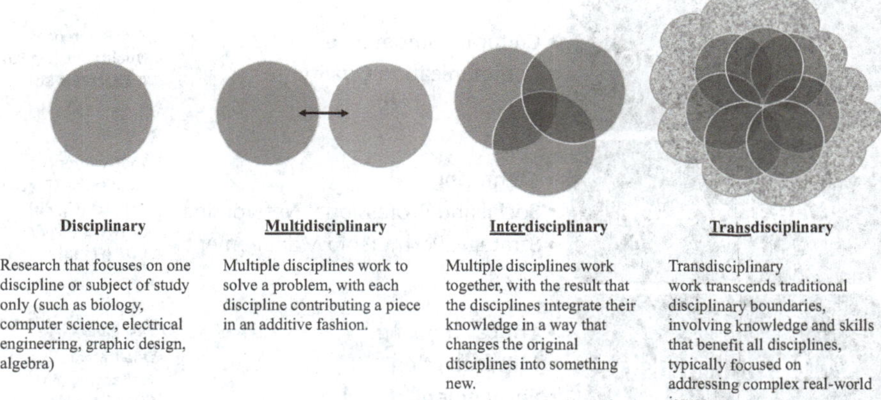

Fig. 5 Distinguishing the terms disciplinary, multidisciplinary, interdisciplinary, and transdisciplinary. (Created by authors)

economic, and social-justice implications. After all, the goal behind many STEAM efforts is to help develop both a highly trained technical workforce and a responsible citizenry, prepared to tackle the grand challenges or wicked problems of the future in a socially and morally responsible manner!

Transdisciplinary knowledge and skills are the hallmark of STEAM programs, so STEAM program evaluations must focus on their measurement. Transdisciplinary skills have been referred to in many ways, but one framework we consider particularly relevant is from the P21 STEM partnership, emphasizing what they termed twenty-first-century skills: communication, collaboration, critical thinking, and creativity (Partnership for 21st Century Learning 2007). The four Cs form core and connected domains that bridge disciplinary boundaries and represent knowledge and skills to include in the evaluation of STEAM programs (see Sanabria and Romero this volume). National accreditation of STEM programs by AdvancED includes the use of the four Cs in establishing STEM practices in every classroom (AdvancED 2018). Using the four Cs as an organizing framework, we define and demonstrate what we refer to as the measurement of "transdisciplinary prowess" among STEAM program stakeholders (Fig. 6).

4 Evidence of Transdisciplinary Prowess

In the previous section, we introduced the range of transdisciplinary knowledge and skills emphasized in STEAM programs. In this section, we introduce methods for measuring "transdisciplinary prowess." Evaluators translate a STEAM program logic model into a set of measurable variables (also called benchmarks, indicators, or metrics) that can be used in formative and summative evaluations. A classic term for this in social science research is "operationalizing" a construct. Recall, for

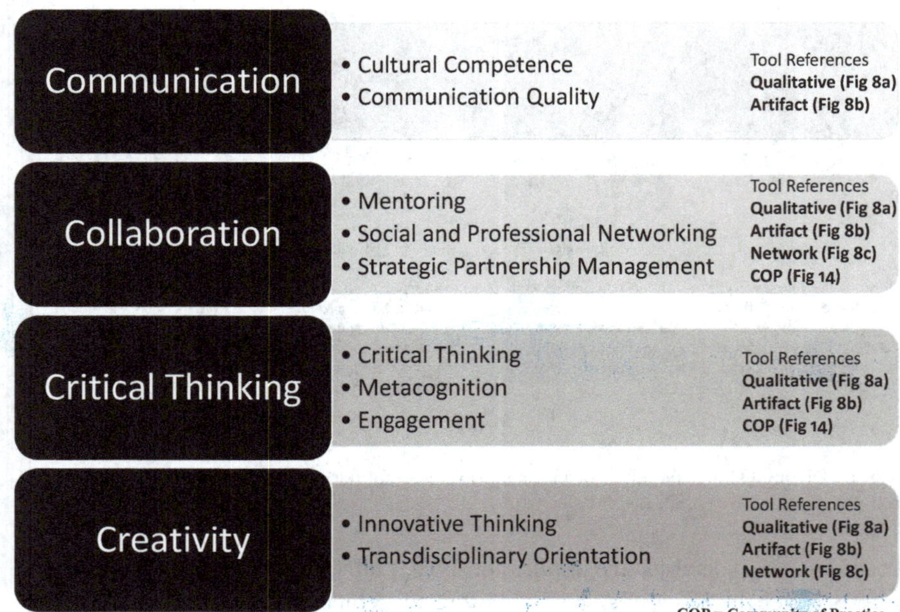

Fig. 6 Summary of four Cs and constructs of transdisciplinary prowess discussed in this chapter, matched to relevant analytic tools defined in Fig. 8 and Fig. 14. (Created by authors)

example, how implementation fidelity was operationalized by Century et al. (2010) in a framework accompanied by a suite of instruments (Fig. 2). Salabarría-Peña et al. (2007) articulated how measures based on a logic model could be developed, and Fig. 3 shows an example in which program goals were translated to specific measures in a high school bioinformatics and genomics research program (Kelly et al. 2018b), based on guidance by Brown (2005). Converging and complementary measures strengthen findings by providing both consistent and unique evidence supporting program outcomes. Metrics can be derived from multiple disciplines in addition to STEM, such as art, design, economics, and business.

For each of the four Cs, we selected representative constructs, measurement strategies, and metrics used in STEAM programs (Fig. 6). We also introduce analytical and technological tools for measuring transdisciplinary prowess in tool boxes, highlighting evolving technological abilities to measure and characterize complex systems of information and interaction. To be successful, evaluators of STEAM programs must have a certain degree of transdisciplinary prowess to optimize the value and use of evaluation evidence in STEAM programs. We conclude our discussion of each "C" by considering associated program evaluation standards set forth by the Joint Committee on Standards for Educational Evaluation, JCSEE (Yarbrough et al. 2011), as well as guiding principles for evaluators and evaluator competencies set forth by the American Evaluation Association, AEA (AEA 2018a; AEA 2018b). The JCSEE is a coalition of major professional associations in the

United States and Canada concerned with evaluation quality (see JCSEE 2018, http://www.jcsee.org/). AEA is the leading organization of evaluators in America and offers a continuing source of guidance in the field of evaluation (see AEA 2018c, https://www.eval.org/p/cm/ld/fid=4).

4.1   Communication

The first C is communication. Communication is transmission of information from a sender to a receiver. Therefore, the characteristics of both the sender and the receiver influence transmission and receipt of a communication. Many STEM and STEAM programs articulate the goal of reaching a diverse range of participants and stakeholders. For example, the National Science Foundation (2008) considers broadening participation in STEM to be a core organizational value, and this value is highly emphasized across their funding portfolio as one of two merit criteria applied to every reviewed proposal. Further, while every educational program has a range of stakeholders, STEAM programs are characterized by audiences spanning disciplinary boundaries and include parents, students, teachers, faculty, researchers, community members, government, and industry representatives. The perspectives of various stakeholders reflect complex disciplinary, cultural, and social influences and identities. Understanding and responding to the unique needs of these receivers requires **cultural competence** and **communication quality**, reflecting the ability of the sender to understand the unique needs of diverse receivers as well as employ a flexible repertoire of appropriate strategies for conveying the desired message (Fig. 6).

Measures of **cultural competence** consider the awareness and sensitivity of stakeholders to the perspectives of others with diverse disciplinary, cultural, and social identities as well as the lived experience of diverse stakeholders. Cultural competence can be assessed as a feature of persons or systems or understood by measuring dispositions, behaviors, perceptions, and lived experiences. Diller and Moule (2005) introduced five cultural competence skill areas, which can serve as a basis for developing or selecting instruments (Fig. 7). While there are examples of instruments, such as the Cross-Cultural Adaptability Inventory (CCAI; see Kelley and Meyers 1995, Fig. 7), often in-depth interviews or observations are needed to fully appreciate and articulate the experiences of underrepresented groups. For open-ended data such as these, qualitative analysis tools can help users process and make sense of complex communications, even multimedia (Fig. 8a). Measurement of cultural competence can be uncomfortable and confronting, which can produce bias in responses if stakeholders don't feel they can be honest (Byrd & Olivieri, 2014). Best practices to limit bias include interacting with stakeholders in a culturally competent manner and being explicit about the protections in place to keep respondent data private and secure.

Understanding the unique needs of diverse receivers is one part of successful communication in STEAM programs; another part is the ability to craft and deliver

Fig. 7 Measuring cultural competence. (Created by authors)

(a) Qualitative Analysis

Determines the themes or patterns present in open-ended narrative responses, multimedia, even social media. There is a wide range of analytic choices from word counts (presented in a word cloud) to detailed analytics of coded discourse using qualitative theory.

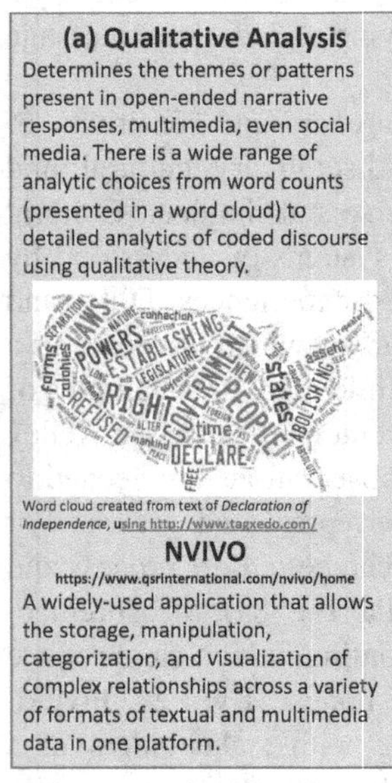

Word cloud created from text of *Declaration of Independence*, using http://www.tagxedo.com/

NVIVO
https://www.qsrinternational.com/nvivo/home
A widely-used application that allows the storage, manipulation, categorization, and visualization of complex relationships across a variety of formats of textual and multimedia data in one platform.

(b) Artifact Analysis and Validated Instruments

Constructs are operationalized as scores on customized checklists, rubrics, and rating scales. Measures may rely on participant self-report, or the judgment of raters who score responses. Similar methods can be applied to artifacts of practice or performance assessments.

Characteristics of high quality measures

Validity means that scores measure what they are supposed to. An example of validity: a score on a transdisciplinary orientation scale correlates with the number of interdisciplinary journal articles.

Reliability means that scores are repeatable across instances and raters. Extensive rater training and data gathered in pilot studies may be necessary.

(c) Network Analysis and Big Data Analytics

Data about human actions and scientific phenomena are more accessible than ever. "Big Data" analytics use large databases and computational tools to answer research questions.

Readily available tools like social network analysis can model the flow of interaction and information in a network using analysis tools like **Gephi** or **NodeXL** to visualize and analyze networks (see example below). Geospatial mapping and satellite data can track phenomena across space and time.

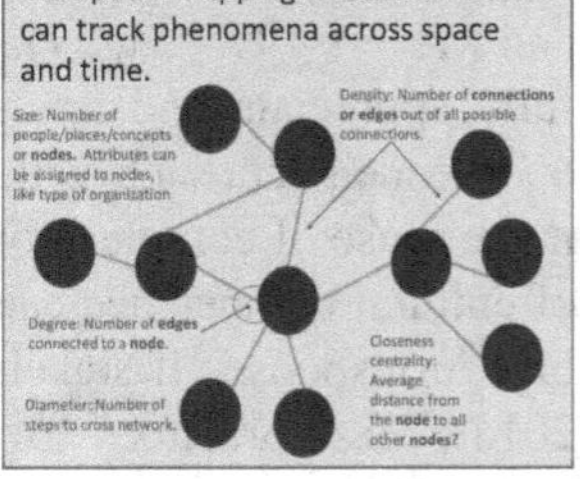

Fig. 8 Qualitative (**a**), artifact (**b**), and network analysis (**c**) tools. (Created by authors)

effective messages to these diverse receivers. **Communication quality** reflects the ability to successfully communicate ideas across audiences, media formats, and communication technologies, critical in a society where only the most effective messages are heard. The format and content of messages must be tailored to the information needs of particular stakeholders, or crafted to the context in which they are to be delivered. Artifact-based assessments operationalize communication quality through rubrics or structured rating criteria appropriate to the context and can be employed by expert(s), with peers, and even as a self-assessment. For example, Evergreen and Emery (2016) released a validated Data Visualization Checklist, useful as a guide for developing high-impact data visualizations (Fig. 9). While the use of checklists, rubrics, and structured rating scales offer a standard way of assessing artifacts of practice in educational programs, they require evidence of validity and reliability across instances and raters, which may require extensive rater training and corroborating data gathered in pilot studies (Fig. 8b).

Evaluators must consider their own cultural competence and the communication quality of their messages, because an evaluation is effective only to the extent it provides diverse stakeholders with formative and summative evaluation findings they consider useful. AEA specifically addresses the cultural competence of evaluators in its guiding principles (AEA 2018b) as well as in a public statement (AEA 2011) which highlights seven core concepts relevant to ensuring cultural competence in evaluation (Fig. 10). In regard to communication quality, Fig. 10 also identifies evaluator competencies related to how findings are communicated (AEA

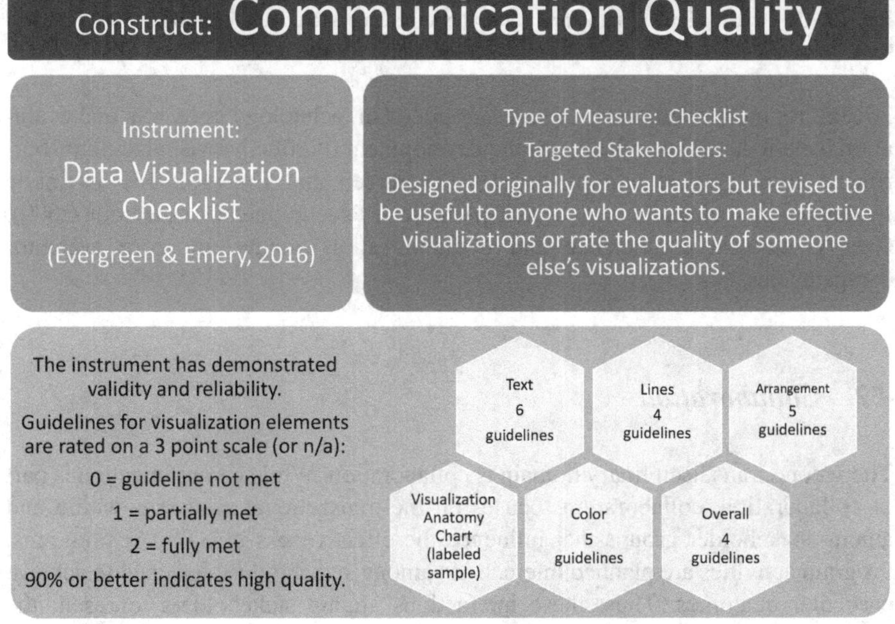

Fig. 9 Measuring communication quality. (Created by authors)

Cultural Competence (from AEA, 2011)	Communication Quality (from AEA, 2018a)
Core Concepts of Cultural Competence in Evaluation	**Core Evaluator Competencies Related to Communication Quality**
1. Cultural competence is a stance taken toward culture.	**Methodology** • 2.13: Justifies evaluation findings/results and conclusions, judging merit worth, and value as appropriate.
2. Evaluations cannot be culture free.	• 2.14: Communicates evaluation findings/results, including strengths and limitations
3. Competence in one context is no assurance of competence in another.	• 2.15: Articulates evidence-based judgments and recommendations for use.
4. Effective and ethical use of evaluation requires respecting different worldviews.	**Management** • 4.6: Documents evaluation processes and products
5. Cultural competence fosters trustworthy understanding.	• 4.9: Uses appropriate technology and other tools to support and manage the evaluation • 4.10: Communicates evaluation processes and results in appropriate, timely, and effective ways.
6. Theories are not value neutral.	**Interpersonal**
7. Cultural privilege can create and perpetuate inequities in power.	• 5.6: Communicates in meaningful ways throughout the evaluation (written, verbal, visual).

Fig. 10 Communication in evaluators: cultural competence and communication quality. (Created by authors)

2018a). As mentioned previously, recent trends in technology, research, and evaluation have focused specifically on the development of effective visualizations, borrowing from graphic and media arts (Evergreen and Emery 2016) and style guidelines in social science research (American Psychological Association 2009). Clearly, cultural competence and communication quality are core evaluator competencies.

4.2   Collaboration

The second transdisciplinary domain is collaboration. While communication is part of collaboration, collaboration focuses on the transactional processes within and among stakeholder groups that influence the effectiveness of STEAM programs. Program activities are planned interactions among stakeholders designed to achieve particular outcomes. Thus, these interactions among stakeholders represent the actual program being delivered. Success of STEAM programs, therefore, resides in effective transactional relationships. In this way, STEAM programs both cultivate and require skills in teamwork, and the need for collaboration skills is prioritized in most modern learning and work contexts. Our discussion focuses on three constructs of collaboration: **mentoring, social and professional networks,** and **strategic partnership management** (Fig. 6).

While **mentoring** is not specific to STEAM programs, this aspect of collaboration is one of the most influential relationships that students have, because advisers are gatekeepers to future success through letters of recommendation, publication opportunities, monetary support, and engagement in professional networks. Academic mentoring and advising resembles an apprenticeship model, whereby students study under senior instructors. Given that both mentors and mentees are stakeholders in a STEAM program, evaluation should focus on the impact of mentoring training and practices on both mentors and mentees. For example, we have used a mentoring scale created and later validated by Michael Fleming and colleagues (Fleming et al. 2013) that was used to study a nationwide mentoring training program (Pfund et al. 2014; Fig. 11). Mentoring can be particularly critical for students from underrepresented groups, who typically experience higher levels of isolation and barriers to success than other students. The Institute for African-American Mentoring in Computing Sciences released guidelines in 2018 for successfully mentoring Black/African-American doctoral students in these programs, which can serve as the basis for a mentoring assessment (Computing Research Association 2018, Fig. 11).

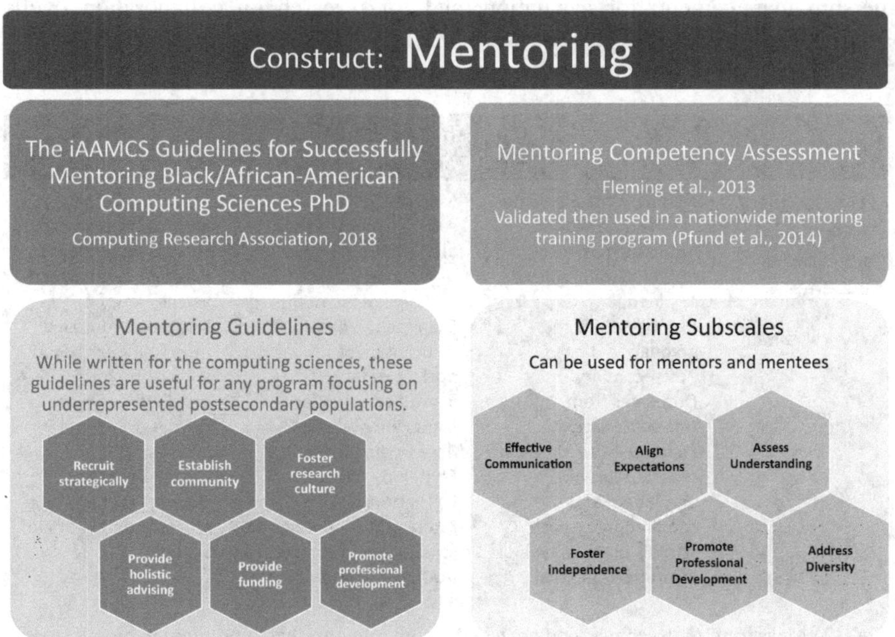

Fig. 11 Measuring mentoring. (Created by authors)

In most aspects of life, success is about both what and who you know and interact with. Particularly with the advance of technology, the study of **social and professional networking** has proven to be of increasing value in understanding transactional phenomena among entities, be they people, organizations, or nations (Taylor et al. 2014). With an emphasis on collaboration and team-based learning in STEAM programs and the often-complex system of stakeholders engaged in them, tracing the interactions among stakeholders over time and space using social network analysis (SNA) and geospatial mapping tools describes how the parts of a STEAM program work together to create the whole experience. Tools to model and depict the functioning of networks have been evolving rapidly (Fig. 8c), and Fig. 12 depicts a framework of measures based on network analysis and big data used to evaluate the postsecondary outcomes of a high school research program in bioinformatics and genomics (Kelly et al. 2018b).

Many STEAM programs are partnerships or operate as a group of diverse institutions, departments, or persons cooperating to administer and implement the planned program. Collaboration theory provides a foundation for measuring collaboration in partnerships (Gajda 2004), and data can be used to improve strategic partnership management over time. Woodland and Hutton (2012) developed the *Collaboration Evaluation and Improvement Framework* (CEIF) and associated measurement tools to characterize the level of collaboration in a partnership, organized around predictable stages of collaboration development. Figure 13 illustrates this framework, and we have used the CEIF to evaluate a multidisciplinary, cross-institutional postsecondary partnership, emphasizing the efficiency of using the collected data for partnership management, for evaluation, and for research on collaboration (Kelly et al. 2018a).

	People	**Space**	**Time**
Postsecondary Outcomes of High School Students Kelly et al., 2018b	• Number of students per teacher and school, including numbers recruited, randomized, retained • Demographics and graduation data • Postsecondary Outcomes from *National Student Clearinghouse* (date of report, dates of enrollment and completion for each school, location, type of institution, and major)	• Map the first and subsequent colleges of all students in real distance by school name and type of institution • Examine the density and distribution of postsecondary institution attendance by Region of the U.S., by Area Health Education Center, by County, and by High School. • Examine the influence of school type, distance from high school	• Map chronological pathways across high schools and colleges • Examine metrics such as number of institutions enrolled, time from high school graduation to 1st enrollment, time to 1st degree, Total time enrolled, • Changes in Major, Degree Program, Type of Institution, Name of School

Fig. 12 Measuring postsecondary outcomes of a high school research program in bioinformatics and genomics using network analysis and access to big data (Kelly et al. 2018b). (Created by authors)

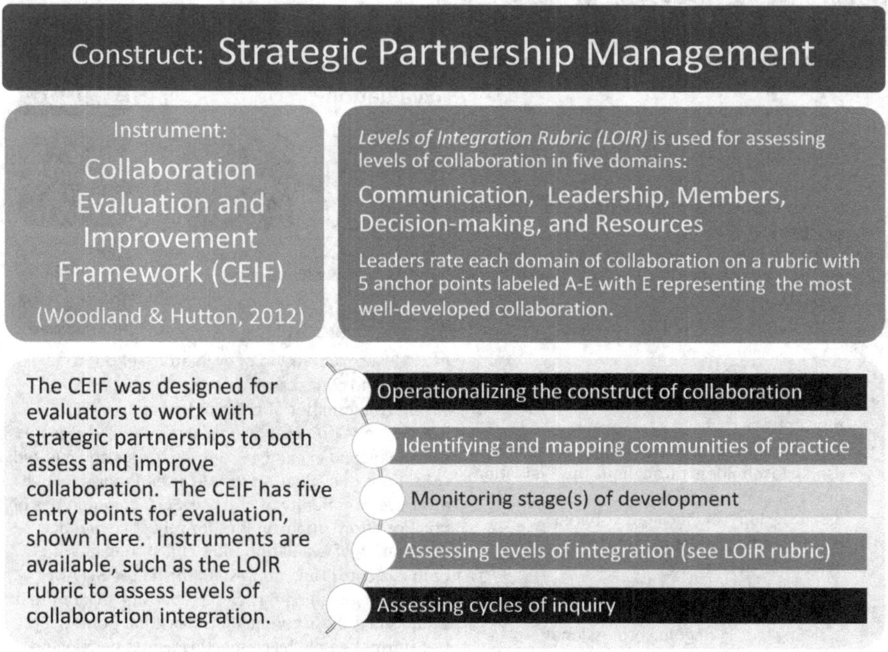

Construct: Strategic Partnership Management

Instrument:

Collaboration Evaluation and Improvement Framework (CEIF)

(Woodland & Hutton, 2012)

Levels of Integration Rubric (LOIR) is used for assessing levels of collaboration in five domains:

Communication, Leadership, Members, Decision-making, and Resources

Leaders rate each domain of collaboration on a rubric with 5 anchor points labeled A-E with E representing the most well-developed collaboration.

The CEIF was designed for evaluators to work with strategic partnerships to both assess and improve collaboration. The CEIF has five entry points for evaluation, shown here. Instruments are available, such as the LOIR rubric to assess levels of collaboration integration.

- Operationalizing the construct of collaboration
- Identifying and mapping communities of practice
- Monitoring stage(s) of development
- Assessing levels of integration (see LOIR rubric)
- Assessing cycles of inquiry

Fig. 13 Measuring strategic partnership management. (Created by authors)

Evaluators must have the knowledge and skills to measure but also to engage successfully in collaboration with program staff and other stakeholders to support the delivery of a program in STEAM communities of practice. Examples of evaluators engaging collaboratively with stakeholders have been mentioned throughout this chapter, such as facilitating logic model development, providing formative feedback, and establishing rapport and trust to encourage evaluation use. There are evaluator competencies related to successful collaboration with stakeholders, across five domains of competence – professional practice, methodology, context, planning and management, and interpersonal. Figure 14 itemizes these competencies (AEA 2018a). To improve professional practice, it is important that evaluators engage regularly with the greater STEAM evaluation community of practice as learners and leaders (Fig. 14). There is a clear benefit to collaboration skills in evaluators to facilitate partnership functioning as well as their own professional learning and growth.

4.3 Critical Thinking

Critical thinking, the third C, reflects the use of higher-level intellectual skills, including problem-solving, logic and reasoning, data analysis and interpretation, the development of expertise, and the transfer of skills across domains. The latter is

Core Evaluator Competencies Related to Collaboration (from AEA, 2018a)

Interpersonal
- 5.2: Values and fosters constructive interpersonal relations foundational for professional practice and evaluation use.
- 5.3: Uses appropriate social skills to build trust and enhance interaction for evaluation practice.
- 5.8: Collaborates and engages in teamwork.
- 5.9: Negotiates decisions for evaluation practice.
- 5.10: Addresses conflicts constructively.

Context
- 3.3: Addresses systems and complexity with the context
- 3.4: Identifies and engages diverse users/stakeholders throughout the evaluation process.
- 3.8: Develops and facilitates shared understanding of the program and its evaluation in context.

Professional Practice
- 1.7: Engages in ongoing professional development to extend personal learning, growth, and connections.

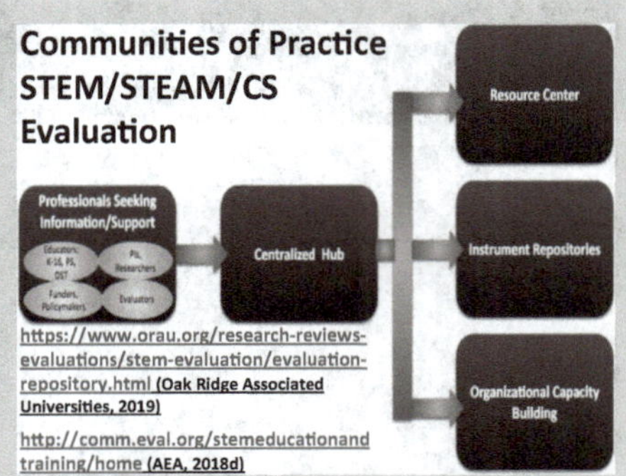

https://www.orau.org/research-reviews-evaluations/stem-evaluation/evaluation-repository.html (Oak Ridge Associated Universities, 2019)

http://comm.eval.org/stemeducationandtraining/home (AEA, 2018d)

In 2012, a collaboration of evaluators set out to develop and connect online resource centers and virtual communities of practice focused on STEM, STEAM, and Computer Science evaluation. The goal is to connect and expand existing capacities and curated collections of evaluation resources and tools through an active community of practice. Such communities of practice afford opportunities for reflective practice and sharing of evaluation tools and strategies with both evaluators and non-evaluators in the STEAM pipeline. In pursuing this work, it became evident that establishing these connections has great potential to transform the field by connecting across people and organizations that otherwise might not have learned of each other's very similar work.

Fig. 14 Collaboration competencies and communities of practice for evaluators. (Created by authors)

conceptually related to generating transdisciplinary solutions. A familiar way to think about the range of intellectual skills is to consider Bloom's Taxonomy, which organizes intellectual skills into a hierarchy from low to high levels of cognitive challenge (Bloom et al. 1956). The words in the description of critical thinking in the first sentence of this paragraph map to more complex, cognitively challenging intellectual skills in the taxonomy (Fig. 15). Research on learning also suggests that **metacognitive thinking** facilitates learning, and we employ metacognitive strategies and measurements in program evaluations whenever possible. Lastly, we will examine the gatekeeper to all thinking – **engagement** in a learning activity (Fig. 6).

Measuring **critical thinking** (CT) requires making the invisible process of thinking both observable and measurable. CT can be measured as performance in the context of a specific STEAM activity or more as an aptitude (designed tests of reasoning and analysis). The emphasis on PBLs and therefore on contextually appropriate and relevant artifacts of performance favors the former measurement strategy. The tools discussed earlier in this chapter are applicable for measuring CT in context, including qualitative analysis of open-ended narrative (Fig. 8a) and artifact analysis (Fig. 8b). Models of thinking such as mind maps or other expressions of conceptual knowledge can be used to access or represent thinking, for which public

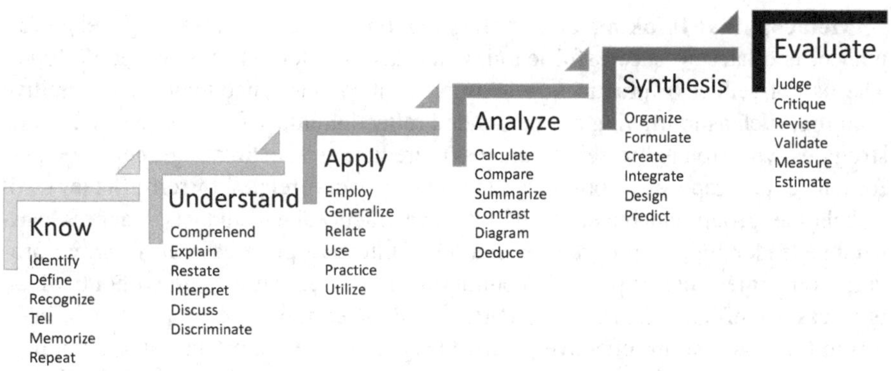

Fig. 15 Power or thinking verbs for levels of Bloom's taxonomy. (Created by authors)

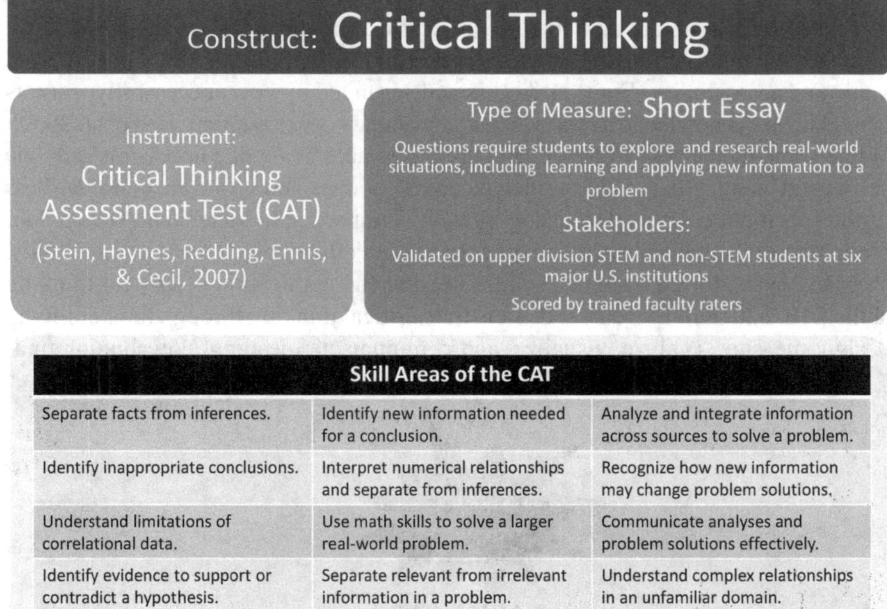

Skill Areas of the CAT		
Separate facts from inferences.	Identify new information needed for a conclusion.	Analyze and integrate information across sources to solve a problem.
Identify inappropriate conclusions.	Interpret numerical relationships and separate from inferences.	Recognize how new information may change problem solutions.
Understand limitations of correlational data.	Use math skills to solve a larger real-world problem.	Communicate analyses and problem solutions effectively.
Identify evidence to support or contradict a hypothesis.	Separate relevant from irrelevant information in a problem.	Understand complex relationships in an unfamiliar domain.

Fig. 16 Measuring critical thinking. (Created by authors)

tools are available as for network analysis (Fig. 8c). Figure 16 provides an example of an instrument that measures critical thinking using short-answer questions in response to real-world situations, as well as solving a complex real-world problem through learning and applying new information (Stein et al. 2007). An important conclusion is that the measurement of critical thinking involves collecting expressions of critical thinking in sufficient detail to render a reliable judgment of the intellectual demands of a task. This requires time, resources, and expertise of the evaluator, as well as time and intellectual demands on participants.

Metacognitive thinking, or the ability to monitor one's own learning and performance, is related to successful learning and task mastery (Bransford et al. 2000). The use of reflective practices in learning contexts can encourage metacognitive thinking, such as journaling or open-ended reflection, identifying concepts that one struggles with (related to self-awareness), predicting or estimating one's own performance, and reporting confidence in one's answers (related to self-efficacy). All stakeholder groups can benefit from reflective practice, be it students, teachers, partnership leadership, or program evaluators. Reflective practices can improve program outcomes and support a continuous improvement cycle, without which progress is inefficient and unclear. Participating in communities of practice offers a forum for engaging in reflective practice (Fig. 14). The important point is not how to measure metacognition, but rather the importance of engaging all stakeholders in reflective practice to improve program outcomes.

In today's world, a vast amount of information competes for stakeholders' attention, and the gatekeeper to success for any STEAM program is engagement of the stakeholders in program activities. Using a mind-mapping approach (Fig. 8c), we note that the main influences on engagement and successful learning are perceived value and accessibility (Fig. 17). Access is determined by how successfully learners can relate new knowledge to knowledge they already possess (Bransford et al. 2000). While some mediators of engagement are observable, many are accessible only via self-report using rating scales, written feedback, or interviews (Fig. 8a, b). Regardless of the specific mediator or measures used, it is important to understand stakeholder feelings about value and access in order to interpret their engagement in the activity.

Many evaluator competencies are related to the application of critical thinking skills (AEA 2018a). Evaluators demonstrate critical thinking through their ability to design, measure, analyze, visualize, and communicate meaningful evaluation find-

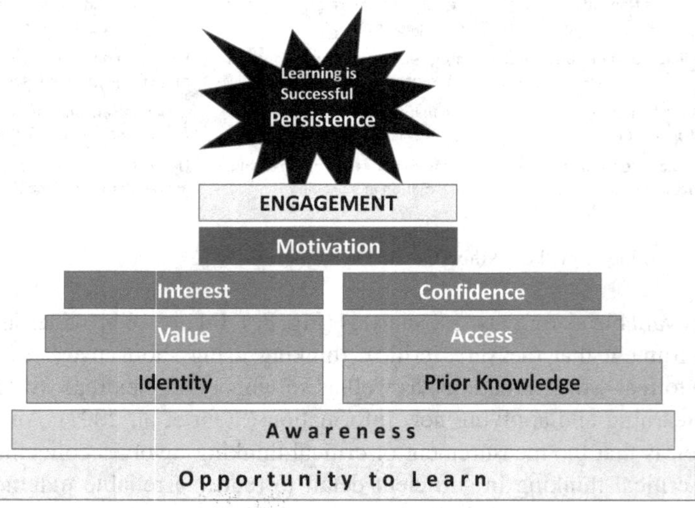

Fig. 17 A house of constructs related to successful engagement and positive learning outcomes. (Created by authors)

Core Evaluator Competencies Related to Critical Thinking
(from AEA, 2018a)

Professional Practice

- 1.2 Knows and applies evaluation foundations that ground and guide evaluative thinking and professional practice.
- 1.3 Knows and applies multiple evaluation approaches and theories.
- 1.4 Uses evidence and logic in making evaluative judgments.
- 1.6 Articulates personal evaluator competence and perspectives, areas for growth, and implications for professional practice.

Methodology

- 2.2 Understands the range of methodologies that ground inquiry in evaluation practice.
- 2.6 Frames evaluation questions.
- 2.7 Designs sound, credible, and feasible studies that address evaluation purposes and questions.
- 2.8 Determines appropriate methods, including quantitative, qualitative, and mixed methods.
- 2.9 Identifies data sources and samples.
- 2.10 Collects data using sound, credible, and feasible procedures.
- 2.11 Analyzes data using sound, credible, and feasible procedures.
- 2.12 Interprets findings/results and draws conclusions by identifying possible meanings in context.
- 2.13 Justifies evaluation findings/results and conclusions, judging merit, worth, and value as appropriate.
- 2.16 Conducts informal and formal meta-evaluations of studies, identifying their strengths and limitations.

Fig. 18 Critical thinking competencies of evaluators. (Created by authors)

ings using appropriate, valid, and reliable strategies (Fig. 18). Evaluators can both engage in and encourage reflective practice, and we suggest that evaluators not only measure the occurrence and impact of metacognitive thinking in a STEAM program but also facilitate and encourage its use with all stakeholder groups as part of evaluation activities. Reflective practice can drive a continuous improvement cycle based on evaluation evidence. While facilitating reflective practice as part of evaluation activities is a value-added service that a skilled evaluator can provide to stakeholders, evaluators must also be willing to apply metacognitive thinking to their own evaluation practice for best results.

4.4   Creativity

The fourth C reflects the domain of creativity. Creativity is associated with a complex range of knowledge and skills; consider constructs such as aesthetic appeal and expressive behavior traditionally associated with art. Personal attributes associated with creative thinking include independent attitudes, divergent thinking, problem-finding, incubation, diverse interests, rejection of external controls, high confidence levels, risk-taking inclinations, and an ability to engage in exploratory work (Gomez 2007). Evaluators can measure creativity by the way an audience behaves in response to a creation, by the features of the creation, or by the abilities of the maker to create (Candy and Bilda 2009). We will discuss constructs related to the latter two of these perspectives. Creative abilities of a maker include not only the opportunity and capability for self-expression but also the ability to generate **innovative thinking**, leading to unique and surprising creations. Features of a creation are amenable to appropriate artifact analysis (Fig. 8b). In the context of STEAM programs, we can also measure attitudes and aptitudes like **transdisciplinary orientation** as well as the ability to generate transdisciplinary solutions. In a world where the technologies and jobs of the near future are unknown today, creativity becomes a very important intellectual commodity (Fig. 6).

Most pertinent to the present discussion is that creativity is related to constructs such as innovation, ingenuity, flexibility, adaptability, cleverness, thinking outside the box, and transdisciplinary – necessary competencies in keeping up with technological innovation and tackling the grand challenges of our age. The concept of **innovative thinking** can be applied to a person, a team, or an organization. While attitudes and aptitudes related to innovative thinking can be measured using established instruments, success metrics hailing from art, business, economics, and engineering can be used to demonstrate and measure creativity as the "best" solution appropriate for the context and relevant for the stakeholders. In this way, measuring creativity is similar to measuring critical thinking, because they both interact in generating transdisciplinary solutions to problems. An example of a STEAM assignment in an elementary science classroom is writing a tanka poem about a celestial body in the solar system. A **tanka poem** is similar to a haiku but has two additional lines. A tanka consists of 5 lines and 31 syllables, with 5 or 7 syllables per line. The tanka poem assignment contains components of both critical thinking and creativity. These poems would be appropriate for an artifact analysis.

As the need for transdisciplinary knowledge and skills increases, so does the need to measure stakeholder aptitude and orientation toward work and associated scholarship that spans disciplinary boundaries, which is inherently creative and cognitively challenging. Misra et al. (2015) developed a Transdisciplinary Orientation Scale (TDO) to use with researchers involved in work that spans disciplinary boundaries. We have used the TDO with faculty and student stakeholders to measure changes in dispositions, perceptions, and behaviors related to interdisciplinary work over the course of a graduate training program (Fig. 19). Other metrics to consider are related to the extent to which scholarly activity spans disciplinary boundaries, such as publications in interdisciplinary journals, joint faculty appointments, and participation in an interdisciplinary research center (Institute of Medicine 2005). Increased access to reference information in today's technological age can facilitate the measurement of scholarly productivity beyond the curriculum vitae, such as citation analysis (Fig. 8c).

Just as the STEAM workforce must possess creative talent, so must evaluators be adaptive designers, employ innovative strategies, and embrace emergent technolo-

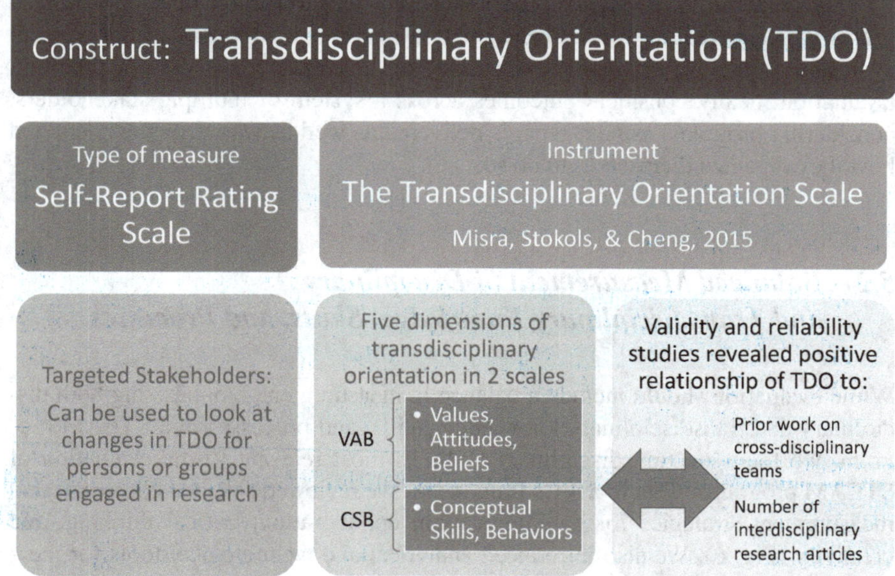

Fig. 19 Measuring transdisciplinary orientation. (Created by authors)

gies. Above all, evaluators must be responsive to the diverse contexts of STEAM programs, able to demonstrate flexibility and adaptability when needed. Some examples from our own work include developing unique measures of transdisciplinary skills and embedding program evaluation within program delivery to encourage data collection efficiencies and reflective practice. It is also important to remain abreast of technological innovations and analytic tools for improving the evidence base. Other examples from our work specifically designed to facilitate evaluation use include modifying program evaluation strategies on an ongoing basis in response to stakeholder feedback and refining report structures and data visualizations to promote evaluation use, program improvement, and dissemination across diverse STEAM audiences.

5 In Conclusion

5.1 The Value-Added of Program Evaluation for STEM and STEAM Programs

Program evaluation is a vital component of STEM and STEAM programs: it provides guidance for evidenced-based decisions before, during, after a STEAM program. Without evidence of effectiveness, it becomes difficult to publish about, obtain funding for, or convince various audiences to participate in or support a

STEAM program. STEM and STEAM education and workforce development programs typically require formative evaluation to foster a continuous improvement cycle and summative evaluation to measure the value or effectiveness of a program. Evaluation ideally considers outcomes across a system of multiple stakeholders, considering programs as both what is delivered as well as who provides them and how they organize themselves to do so.

5.2 Balanced Measurement of Disciplinary and Transdisciplinary Knowledge, Skills, and Processes

While evaluation should include a balance in measurement, emphasizing both disciplinary and transdisciplinary knowledge, skills, and processes in a STEAM program, we focus on transdisciplinary skills as most relevant to the evaluation of STEAM programs. Cast in a 4 Cs framework, we reviewed selected constructs and measurement strategies for communication, collaboration, critical thinking, and creativity (Fig. 6). We also introduced analytical and technological tools for measuring transdisciplinary prowess in Fig. 8a–8c and Fig. 14, highlighting evolving technological capabilities for measuring and characterizing complex systems of information and interaction.

5.3 The Transdisciplinary Prowess of the Evaluator

Evaluators must possess the transdisciplinary prowess to optimize the value and use of evaluation evidence across multiple stakeholders in STEAM programs. An evaluation is effective when it provides diverse stakeholders with findings they consider useful, requiring both cultural competence and communication quality from the evaluator(s). Evaluators must collaborate successfully across STEAM communities of practice, from working with program staff to support the delivery of a program to the greater STEAM evaluation community of practice. Evaluators demonstrate critical thinking through their ability to select and apply evaluation strategies that are appropriate, valid, informative, reflective, and useful to stakeholders. Incorporating reflective practices with all stakeholders around evaluation findings encourages professional learning, promotes evaluation use, and can help drive a continuous improvement cycle. Just as the STEAM workforce must have creative talent to thrive in an unknown future, so must evaluators be adaptive designers, employ innovative strategies, and embrace emergent technologies.

We conclude that the secret to STEAM success lies in the "transdisciplinary prowess" of its stakeholders, including participants, program staff, and evaluators.

References

AdvancED. (2018). *STEM certification: Meeting the demands of the future.* Retrieved from https://www.advanc-ed.org/services/stem-certification

American Evaluation Association. (2011). *American evaluation association public statement on cultural competence in evaluation.* Retrieved from https://www.eval.org/p/cm/ld/fid=92

American Evaluation Association. (2018a). *The AEA Evaluator Competencies.* Retrieved from https://www.eval.org/d/do/4382

American Evaluation Association. (2018b). *Guiding Principles for Evaluators.* Retrieved from https://www.eval.org/p/cm/ld/fid=51

American Evaluation Association. (2018c). *About AEA.* Retrieved from https://www.eval.org/p/cm/ld/fid=4

American Evaluation Association. (2018d). *Welcome to the STEM Education and Training TIG.* Retrieved from http://comm.eval.org/stemeducationandtraining/home

American Psychological Association. (2009). *Publication manual of the American Psychological Association.* Washington, DC: American Psychological Association.

Bloom, B. S., Engelhart, M. D., Furst, E. J., Hill, W. H., & Krathwohl, D. R. (1956). *Taxonomy of educational objectives, handbook I: The cognitive domain.* New York: David McKay Co Inc.

Bransford, J. D., Brown, A. L., & Cocking, R. R. (2000). *How people learn: Brain, mind, experience, and school.* Washington, DC: National Academy Press.

Brown, C. (2005). *Teaching American history: Creating an evaluation framework.* Center for Evaluation & Education Policy, http://ceep.indiana.edu/index.html

Byrd, M., & Olivieri, B. S. (2014). Measuring teacher cultural competence. *International Review of Social Sciences and Humanities, 8*(1), 55–67. www.irssh.com, ISSN 2248-9010 (Online), ISSN 2250-0715 (Print).

Candy, L., & Bilda, Z. (2009).*Understanding and evaluating creativity: A tutorial.* Tutorial presented at ACM Creativity & Cognition 2009, Berkeley, CA. Retrieved from http://www.eecs.qmul.ac.uk/~nickbk/creativityandcognition09/tutorials/t1-candy.pdf.

Century, J., Rudnick, M., & Freeman, C. (2010). A framework for measuring fidelity of implementation: A foundation for shared language and accumulation of knowledge. *American Journal of Evaluation, 31*(2), 199–218. https://doi.org/10.1177/1098214010366173.

Computing Research Association. (2018, September 18). iAAMCS releases guidelines for successfully mentoring Black/African-American computing sciences doctoral students. *CRA Bulletin.* Retrieved from https://cra.org/iaamcs-releases-guidelines-for-successfully-mentoring-black-african-american-computing-sciences-doctoral-students/?utm_source=CRA+Board%2c+Committee+Participants%2c+Member+Contacts+and+Staff&utm_medium=email&utm_campaign=cffe64dddf-RSS_Bulletin_Mailchimp&utm_term=0_0f664bcc92-cffe64dddf-82482937

Davies, R. (2015). *Evaluability assessment.* BetterEvaluation. Retrieved from http://betterevaluation.org/themes/evaluability_assessment

Diller, J., & Moule, J. (2005). *Cultural competence: A primer for educators.* Belmont: Wadsworth.

Evans, J. (2014, July 29). *What is transdisciplinarity?* Retrieved from https://polytechnic.purdue.edu/blog/what-transdisciplinarity

Evergreen, S., & Emery, A. K. (2016). *Data visualization checklist.* Retrieved from http://stephanieevergreen.com/wp-content/uploads/2016/10/DataVizChecklist_May2016.pdf

Fleming, M., House, S., Hanson, V. S., Garbutt, J., McGee, R., Kroenke, K., & Runio, D. M. (2013). The mentoring competency assessment: Validation of a new instrument to evaluate skills of research mentors. *Academic Medical Journal, 88*(7), 1002–1008. https://doi.org/10.1097/ACM.0b013e318295e298.

Gajda, R. (2004). Utilizing collaboration theory to evaluate strategic alliances. *American Journal of Evaluation, 25*(1), 65–77. https://doi.org/10.1177/109821400402500105.

Ghanbari, S. (2015). Learning across disciplines: A collective case study of two university programs that integrate the arts with STEM. *International Journal of Education and the Arts, 16*(7), 1–21.

Golder, B., WWF-US, & Gawler, G. (2005). *Cross-cutting tool: Stakeholder analysis*. Retrieved from https://intranet.panda.org/documents/folder.cfm?uFolderID=60976

Gomez, J. G. (2007). What do we know about creativity? *The Journal of Effective Teaching, 7*(1), 31–43.

Hardiman, M. M., & JohnBull, R. M. (this volume). From STEM to STEAM: How can educators meet the challenge? In *Converting STEM into STEAM programs*. Cham: Springer.

Institute of Medicine. (2005). *Facilitating interdisciplinary research*. Washington, DC: The National Academies Press. https://doi.org/10.17226/11153.

Joint Committee on Standards for Educational Evaluation. (2018). *About JCSEE*. Retrieved from http://www.jcsee.org/

Kelley, C., & Meyers, J. (1995). *CCAI: Cross-cultural adaptability inventory*. Arlington: Vangent, Inc.

Kelly, K. A., Murphrey, T. P., Koswatta, T. J., & Burr, E. M. (2018a, November). *Application of a collaboration framework to facilitate and evaluate the development of collaboration in a post-secondary academic alliance*. Poster presented at evaluation 2018: Speaking truth to power, Cleveland, OH.

Kelly, K. A., Slawsky, E., Koury, S. T., & Carlin-Mentor, S. (2018b, November). *A model for geospatial analysis and longer-term tracking of outcomes in STEM pipeline building*. Poster presented at evaluation 2018: Speaking truth to power, Cleveland, OH.

Misra, S., Stokols, D., & Cheng, L. (2015). The transdisciplinary orientation scale: Factor structure and relation to the integrative quality and scope of scientific publications. *Journal of Collaborative Healthcare and Translational Medicine, 3*(2): 1042. http://www.jscimedcentral.com/TranslationalMedicine/translationalmedicine-3-1042.pdf

Oak Ridge Associated Universities. (2019). *Evaluation repository*. Retrieved from: https://www.orau.org/research-reviews-evaluations/stem-evaluation/evaluation-repository.html

Partnership for 21st Century Learning. (2007). *The intellectual policy foundations of the 21st century skills framework*. Retrieved from http://www.p21.org/route21/images/stories/epapers/skills_foundations_final.pdf

Patton, M. Q. (2008). *Utilization-focused evaluation* (4th ed.). Thousand Oaks: Sage.

Pfund, C., House, S. C., Asquith, P., Fleming, M. F., Buhr, K. A., Burnham, E. L., et al. (2014). Training mentors of clinical and translational research scholars: A randomized controlled trial. *Academic Medical Journal, 89*(5), 774–782. https://doi.org/10.1097/ACM.0000000000000218.

Salabarría-Peña, Y, Apt, B.S., Walsh, C.M. (2007). Practical use of program evaluation among sexually transmitted disease (STD) programs. Centers for Disease Control and Prevention: Atlanta.

Sanabria, & Romero. (this volume). Emerging scenarios to enhance creativity in smart cities through STEAM education and the gradual immersion method. In *Converting STEM into STEAM programs*. Cham: Springer.

Stein, B., Haynes, A., Redding, M., Ennis, T., & Cecil, M. (2007). Assessing critical thinking in STEM and beyond. In M. Iskander (Ed.), *Innovations in E-learning, instruction technology, assessment, and engineering education* (pp. 79–82). Dordrecht: Springer.

Taylor, M., Plastrik, P., Coffman, J., & Whatley, A. (2014). *Evaluating networks for social change: A casebook*. Retrieved from http://www.evaluationinnovation.org/sites/default/files/NetworkEvalGuidePt2_Casebook.pdf.

U.S. Department of Education, Institute of Education Sciences, National Center for Education Evaluation and Regional Assistance, What Works Clearinghouse. (2017). *WWC version 4.0 standards handbook*. Retrieved from https://ies.ed.gov/ncee/wwc/Docs/referenceresources/wwc_standards_handbook_v4.pdf

U.S. Institute of Education Sciences, & U.S. National Science Foundation. (2013). *Common guidelines for education research and development*. Retrieved from https://www.nsf.gov/pubs/2013/nsf13126/nsf13126.pdf.

U.S. National Science Foundation. (2008). *Broadening Participation at the National Science Foundation: A Framework for Action*. Retrieved from https://www.nsf.gov/od/broadeningparticipation/nsf_frameworkforaction_0808.pdf

Ubben, G. C. (this volume, Chapter 6). Using project-based learning to teach STEAM. In *Converting STEM into STEAM programs*. Cham: Springer.

Ubben, G. C. (this volume, Chapter 7). How to structure project-based learning to meet STEAM objectives. In *Converting STEM into STEAM programs*. Cham: Springer.

W. K. Kellogg Foundation. (2004). *W.K. Kellogg Foundation logic model development guide*. Battle Creek: W.K. Kellogg Foundation. Retrieved from https://www.wkkf.org/resource-directory/resource/2006/02/wk-kellogg-foundation-logic-model-development-guide

Watkins, R., West Meiers, M., & Visser, Y. L. (2012). A *guide to assessing needs: Essential tools for collecting information, making decisions, and achieving development results*. World Bank. © World Bank. https://openknowledge.worldbank.org/handle/10986/2231

Woodland, R. H., & Hutton, M. S. (2012). Evaluating organizational collaborations: Suggested entry points and strategies. *American Journal of Evaluation, 33*(3), 366–383. https://doi.org/10.1177/1098214012440028.

Yarbrough, D. B., Shulha, L. M., Hopson, R. K., & Caruthers, F. A. (2011). *The program evaluation standards: A guide for evaluators and evaluation users* (3rd ed.). Thousand Oaks: Sage Publications.

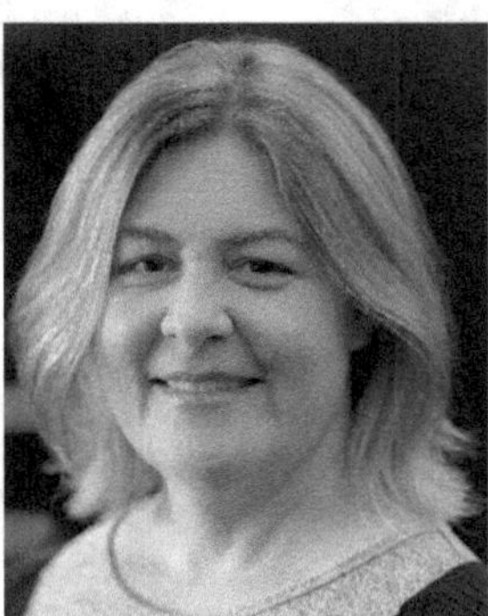

Kimberle Kelly is a senior evaluation specialist within the Scientific Assessment and Workforce Development Program at Oak Ridge Associated Universities. She holds a BS in Psychology, a PhD in Psychology/Cognitive Brain Sciences, and completed a postdoctoral traineeship in psychological research on schizophrenic conditions. Kim's professional interests include STEM education and workforce development evaluation from cradle to career, mixed methods research and evaluation, clinical and cognitive neuroscience, and underrepresented groups in STEM.

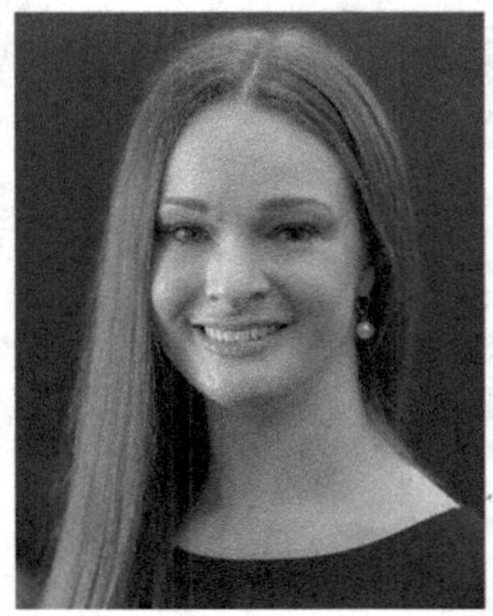

Erin Burr is a senior evaluation specialist and section manager for assessment and evaluation within the Scientific Assessment and Workforce Development Program at Oak Ridge Associated Universities (ORAU). She holds BS and MS degrees in Psychology, a PhD in Educational Psychology and Research with a concentration in Evaluation and Assessment and completed a postdoctoral fellowship at ORAU with the Oak Ridge Institute for Science and Education (ORISE). Erin's professional interests include STEM workforce development and education evaluation in K–20 settings, strategic needs assessment in national laboratories and laboratory systems, impact evaluations/long-term follow-up studies, measurement of evaluation use and influence, and applied experimental psychology.

STEM Education and the Theft of Futures of Our Youth: Some Questions and Challenges for Educators

Sophia Jeong, Deborah J. Tippins, Kimberly Haverkos, Mel Kutner, Shakhnoza Kayumova, and Stacey Britton

1 Introduction

The headlines of a recent issue of the National Science Teacher Association's "NSTA Reports" highlight a middle school STEAM project in the article "Off to the STEM races." In this project, the students use interdisciplinary knowledge of physics, engineering, and mathematics to construct miniature racecars propelled by a carbon dioxide cartridge. In the broadest sense, discourse surrounding the meaning of STEM/ STEAM education, both in theory and practice, suggests interdisciplinary dialogue or interaction between two or more disciplines, as the above example illustrates. While STEAM education typically emphasizes the forging of connections between science, mathematics, engineering, technology, and the arts, connections clearly can occur across any discipline, in the intersections between disciplines or even by transcending

---

The original version of this chapter was revised: The chapter author's, "Sophia (Sun Kyung) Jeong", first and last names has been corrected now. The correction to this chapter is available at https://doi.org/10.1007/978-3-030-25101-7_19

S. Jeong · D. J. Tippins (✉) · M. Kutner
University of Georgia, Athens, GA, USA
e-mail: sj33678@uga.edu; dtippins@uga.edu; mrk19412@uga.edu

K. Haverkos
Thomas More University, Crestview Hills, KY, USA
e-mail: haverkk@thomasmore.edu

S. Kayumova
University of Massachusetts, Dartmouth, MA, USA
e-mail: skayumova@umassd.edu

S. Britton
University of West Georgia, Carrollton, GA, USA
e-mail: sbritton@westga.edu

what are traditionally perceived as disciplinary boundaries. Thus, it comes as no surprise when scholars such as Bennington (1999) and Moran (2010) discuss the ambiguity surrounding the term "interdisciplinary," which has led to the emergence of other terms such as "post-disciplinary," "anti-disciplinary," and "transdisciplinary," to name a few. Other scholars have argued that if we are going to teach in interdisciplinary ways, then we must develop interdisciplinary assessments. While an in-depth theoretical examination of the relationship between disciplines is beyond the scope of this chapter, we recognize that, as authors, our own disciplinary histories (i.e., biology, geology, psychology, anthropology, literacy, genetics) influence what we perceive as significant questions and challenges for STEM/STEAM education. We ask readers to keep this in mind in our attempt to generate productive conversations around the questions and challenges we pose to extend thinking about STEM education. We begin the conversation by first posing some questions at the heart of the STEM education for all metaphor prevalent in the recent STEM/STEAM discourse. We ask:

1.1   Question # 1: What Are the Challenges of STEM Education for All Within Contemporary Reform Efforts?

STEM education and participating in STEM knowledge production is highly important for *all*, and especially for low-income and racially diverse communities traditionally minoritized from participating in science education, since STEM knowledge might offer a unique position from which to understand, combat, and transform inequities and sociopolitical circumstances (Weinstein 2008). However, studies in science and mathematics education have shown that while policy and education documents promoting STEM education claim prosperity for all, in reality, "only those with the most wealth have received any financial benefit from these endeavors" (Basile and Lopez 2015, p. 521). Similar disparities in STEM education in the United States have been documented for decades. The overwhelming evidence demonstrates persistent inequalities faced by underrepresented groups from traditionally minoritized communities (e.g., girls, students from low-income backgrounds, students of color, Hispanics, African-Americans, and English language learners) from the outset of their STEM educational experiences to choosing and staying in their career paths (Basil and Murray 2015).

It is a grave problem in the US context given that most of the "spatial patterns in environmental injustice and inequality" are also concentrated in areas where "low-income nonwhite young children" live and go to school (Clark et al. 2014). Thus, the underrepresentation of people and youth from these communities in STEM education perpetuates injustices not only in economic terms but also by constraining minoritized communities from the involvement in critical issues directly impacting their lives, work, health, and social and emotional conditions. As Basil and Murray (2015) argue,

> Especially in our segregated society, it can be difficult for a collective of white middle-class scientists to understand the lives, experiences, needs, and desires of peoples of color. …that this inability to see might in turn lead to scientific knowledge or applications of that scien-

tific knowledge that are unjust … when we do not include diverse voices in the project of generating scientific knowledge and deciding how we ought to put that knowledge to use in our society. It is difficult to imagine, for example, that low-income and Latino communities of central and southern California, plagued by birth defects, would be home to the only toxic waste sites in the state if scientists originating from those communities weren't involved in the scientific community. (p. 257)

To this end, Dana Zeidler (2014) argues that before jumping on the bandwagon of the current STEM education movement, it is important to pose contextualized questions like "What does an educative experience embedded in STEM entail for the daily lives of citizens?" "How does that experience play out throughout life-long development?" (p. 12). We find Zeidler's (2014) questions and remarks to be relevant also to the scholarship that centers on issues of equity and considers sociopolitical problems within science education as embedded and interconnected to the STEM education reform documents. This perspective also allows us to acknowledge that science- and STEM-related knowledge and skills are recognized as important, both in terms of keeping up and progressing in a knowledge economy and making informed decisions related to STEM. Arguably, however, mere access to knowledge and skills is not enough to tackle the pernicious sociopolitical and environmental problems which disproportionally impact many communities marginalized by environmental injustices, socioeconomic disparity, access to affordable healthcare, and all the ways science and technological changes impact daily lives, work conditions, health, and psychology of people.

In the recent water crisis in Flint, Michigan, for example, over a hundred thousand people were exposed to lead in drinking water. Many of those that were exposed were children, and the majority of those exposed were people of color living in low-socioeconomic areas (Butler et al. 2016). Indeed, recent research has shown that the social disproportionality often observed in economically depressed areas coincides with pollution disproportionalities: the very highest polluters tend to be located in the very poorest areas (Collins et al. 2016). These events once again remind us of the importance of place-based science education and empowering citizen scientists who can observe, identify, and act on local environmental issues. For instance, some science educators have used ecojustice pedagogy, an emerging framework, which helps teachers and students to approach environmental issues with place-orientated consciousness, emphasizing a culturally situated knowledge of ecology and thus helping students to position themselves as "stewards" of the natural world (Bricker et al. 2015, p. 85). This is one example of an expansive view of science education that extends existing techno-narratives of science, existing solely in an economic context, to one that seeks to understand science principles in a more systems-based approach.

In light of our fundamental concerns regarding the positioning of low-income and racially diverse youth within STEM initiatives, we reflect more specifically on the extent to which STEM/STEAM efforts may continue to reify traditional gender roles. We situate our discussion of this question within Kim's story of her experiences with a STEM camp for middle and high school girls as we ponder:

1.2 Question # 2: In What Ways Does STEM/STEAM Continue to Reify Traditional Gender Roles?

For the past three years, Haverkos and several colleagues joined together to run a week-long STEM camp for middle school and early high school girls. We have a passion to share our love of STEM with these young women and hopefully show them how our love of the processes and ideas and nature of STEM can lead to a life (semi) fulfilled. We care about their interests and their futures and push to both be role models and help craft them into role models—we encourage returning to camp as a mentor in the following years and several girls have done so. We are not all STEM professionals. While two of us are in the field of STEM as an ecologist and a geneticist, the other two come at STEM from different perspectives—one as a philosopher and one as a teacher-educator. Our varied and diverse backgrounds help us share different possibilities and different avenues of moving into and through STEM. We bring varied adults in to speak with the girls about their experiences in STEM and love of their field.

Each day of this camp ties into one of the STEM ideas—science, technology, engineering, and math—and the week wraps up with the girls presenting what they've learned to an audience of parents, siblings, grandparents, and professors. We attempt to incorporate a little of the arts by engaging in imaginative and creative activities that link to what we are learning about (origami, designing and screen-printing T-shirts, and journaling). But woven into the week also is conversation around what it means to be a woman in STEM, and a girl interested in STEM. Discussions around stereotypes and messaging in STEM and the girls' broader worlds allow them to share their own stories around issues of gender[1] and how it plays out in their love for STEM and, at this point, schooling. For all the strides forward that some claim STEM provides and the consistent push for more STEM engagement, we, the adults running the camp, often find ourselves frustrated at how little progress has been made and how far we have yet to go. But we will continue—to provide a space for girls who love STEM, to provide a space to explore their identity in relation to STEM, and to provide a supportive and caring environment as these girls make decisions about their futures and STEM.

Culturally, awareness for bringing girls, and women more broadly, into the STEM fields has been a bright spot. Success for the movie *Hidden Figures*, the GoldieBlox toys (Loudenback 2016), and more gender-neutral or science-themed clothing for girls specifically, and children generally (Bologna 2015), has brought STEM and women and girls to the cultural forefront. Women now make up almost half of all incoming medical students (a drop from the early 2000s when women made up more than half of all applicants, AAMC 2016), and shows like *The Big Bang Theory* provide *some* cultural awareness around women in STEM. Twitter

---

[1]The authors recognize the complex and thorough work being done around broader issues of gender, both in and out of the STEM fields, but space prohibits the depth of conversation necessary to do justice to this broader conversation.

hashtags #ILookLikeAnEngineer and #distractinglysexy add to that momentum with real-world representations of women in STEM fields. But the backlash for the "nontraditional" roles of women in STEM fields and studies has agitated us culturally. We see examples such as "trolling" online to death threats for engaging against misogyny in computer science (see https://www.elephantinthevalley.com/ for stories of misogyny in the tech world), Malala's fight for the right to an education, and rampant anti-intellectualism (found in such STEM areas as anti-vaccine movements, climate deniers, and evolution deniers). Society is working hard to reconcile women in STEM (and perhaps educated women more generally) with their traditional cultural roles. In *Geek Chic*, Lorna Jowett (2007) states, "Both real and fictional female scientists continue to struggle against ingrained ideas about what women are and what science is" (p. 45). Her study of female scientists on television shows just how ingrained those images are and how women portrayed as scientists often have to give up part of their identity as female to be considered a successful scientist. Caroline Merchant (1980) reminds us that this tension was built into the very fabric of the scientific revolution and enlightenment (i.e., STEM) thinking:

> …sexual politics helped to structure the nature of the empirical method that would produce a new form of knowledge and a new ideology of objectivity seemingly devoid of cultural and political assumptions. (p. 172)

The role of women in society, and the ways in which they live out their connections to the fields within STEM, is tightly bound to traditional expectations of caregiving, motherhood, and femininity—traits that are considered in opposition to scientific thinking, the base of STEM work, objectivity, and reason. While gains have been made in many STEM fields, statistics suggest that we still have a long way to go toward equity in the STEM fields. The National Girls Collaborative Project's tracking of these issues shows that while women make up half the total of US college educated workforce, within science and engineering, that number is 29% (NGC Project 2017). Looking through a lens of intersectionality,[2] that number is even more discouraging, even though movement has been made to increase diverse representation in STEM fields (NGC Project 2017). While the STEM fields represent huge and diverse fields of work, the National Science Board's 2016 report suggests that engineering has the most to make up, with women only constituting 15% of the engineering workforce (National Science Board 2016). In a society that is calling for more engagement in STEM fields, engineering in particular, work around equity has a lot of ground to make up.

If we look more closely at the fields where equity is being chased, a theme arises—caring. Research suggests that the idea of caring is a tool for getting girls (and hence women) interested in STEM fields (Capobianco and Yu 2014). By exploiting their "traditional nature" for caring, we can increase the number of women in STEM fields. In the description for our summer STEM camp described earlier, we thoroughly engage that theme. The title of our camp is *STEM Girls: Transforming the World*, and we seek out daily connections to real-world community-

---

[2] Again, the authors recognize the work being done in intersectionality, but space prohibits an in-depth discussion here.

based problems and solutions *that the girls might care about*. Together we brainstorm issues that the girls are concerned about and have deep discussions about how STEM can work to solve them. The girls engage in discussions of identity and oppression alongside water quality and sustainability—but all within the framework of caring for others, finding solutions for others that they care for. This is not to suggest that caring is a "bad" thing, but rather to understand that in pulling on this thread, we walk a fine line between reifying traditional gender roles and challenging the status quo in STEM fields. In our own work for the camp, we walk that line. Looking within the medical fields, we can see this played out through time. The culturally expected roles of nurses (i.e., caring female) versus doctors (i.e., rational male) and the gender breakdown are keys to this point (Anthony, 2004), but perhaps more telling is the recent research that looked at patient success and how women doctors had higher success rates with their patients (Tsugawa et al. 2017). On the surface, this sounds like a fabulous result; scratch more deeply, and we can see that it also represents a pushback against women in STEM. While the authors suggest they don't know why female physicians have lower mortality rates, previous research suggests some possibilities: the (slightly) longer amount of time that female physicians spent with their patients and better communication (Reese 2011). However, the institutionalized system of medicine works against a structure that would allow physicians to spend that time with their patients and actually penalizes them for that time. In a world of insurance-driven constraints, time with patient is reduced to maximize profit (Dugdale et al. 1999; Rabin 2014). Physicians who spend more time with their patients lose status within the institution—they are reimbursed at a lower rate and are not seen as effective as their counterparts.

It is vitally important for women's voices, ideas, and knowledges to be a part of STEM conversations, careers, and future directions. But it is also vitally important that women's ideas, voices, and knowledges be legitimated. There has been an uptake in the number of women pursuing STEM careers in college. However, the number of women who stay in STEM positions as long-term careers slips drastically, with more than half leaving their careers after 10 years (Hill et al. 2010, p.19). Biology, where one of the biggest increases can be found, is also the most explicitly interrelated—a space where caring is often culturally apparent. When embedded in a "caring" science—such as biology, medicine, or ecology—this works to keep women in marginalized positions in relation to the larger scope of science and STEM. Traits deemed necessary to be successful in STEM are often linked to masculine performances (i.e., rational, objective, reasoned). For performances of caring to be linked to feminine performances, women and their fields are distanced from STEM; their knowledge and fields are delegitimized. By removing scientific legitimacy from caring and placing it in the realm of traditional women's work, an activity culturally normalized as female, women are not allowed into science through caring because these performances of and traits of caring are considered unnatural science performances. So, the activities that are culturally normalized for women, caring activities, are the same activities that are made unnatural in science. This feminization of performances of caring also works to keep issues around community and relationships marginalized in larger societal discourses. Climate change deniers aren't simply denying the science—they are reducing care of the environment to a feminized space that is delegitimated scientifically because it is women's work. Women are doing caring work because commonsense tells them to. It is "natural" for women to care; it is a "natural" part of feminine nurturing and cultural expectations. This allows for the legitimacy of science to be upheld at the expense of both women and caring. The women do not challenge science's meaning-making or rituals or norms because they cannot access the science through their performances of caring. A recent article on men and service in academia suggests white men need to learn to say yes to service rather than women and academics of color learning to say no (Portillo 2017). Do we need to think of caring in STEM this way, as well? Do we need men in STEM to see the importance of caring? Or does this feed into the anti-intellectualism that is plaguing the country? Does this move STEM discourses into a dangerous area of disbelief by the public, by feminizing STEM knowledge? Perhaps the focus may need to shift toward another question: Not just why is there a lack of women and girls in STEM, but rather, are women and girls rejecting the community that STEM has created and normalized in society as "natural" just because it continues to reify traditional gender roles and expectations and delegitimize their experiences and knowledges?

Many scientists, philosophers, and educators are collaborating at the borders of biology, social sciences, and the humanities. However, the techno-narratives we noted earlier may be constraining not only the participation of marginalized communities and girls in STEM but also opportunities for all youth to make knowledgeable decisions and act on issues with relevance to local place-based issues. Biologist Donna Haraway moved into cultural studies and history of science because of her concern for the driving force of techno-science and its claims to neutrality and objectivity. Haraway (1997) argues for repositioning science as a situated form of knowledge. She emphasizes the importance of recognizing that science, while it may have the appearance of isolated disciplinary silos, is always part of place-based narratives and knowledges. Our concern is that many of the place-based narratives we might draw on to encourage the participation of all youth in STEM are missing from the current STEM rhetoric, leading us to ask:

1.3 Question # 3: How Can We Conceptualize STEM/STEAM While Maintaining a Focus on Community, Local Knowledge, and a Connection to Place?

The underlying components of human nature are to consider self and immediate needs—such as safety, space, and nourishment. In his review of literature, Krug (2012) poses a challenge that preservice teachers be developed in ways that allow a sustainable mindset to be exemplified in their future classrooms. However, most STEM/STEAM initiatives rarely integrate issues of sustainability; therein lies a disfunction within the current drive for STEM/STEAM education policy. With current initiatives in STEM/STEAM focusing more on societal advancement and personal economic growth, how can we alter this approach in such a way as to promote a central consideration for justice while understanding the need for progress within place? More importantly, how can we, as educators, change the focus from progress for anthropocentric gain to progress that promotes equity for all? What does that STEM/STEAM look like?

1.3.1 Critical Pedagogy

According to Krzesni (2014), one presumed goal of STEM/STEAM educators should be to function within a critical pedagogy—what you teach and how you perceive/portray value in that instruction is essential for developing learners who are better able to encounter the world and act justly. While the ultimate argument made by Zeidler (2016) is for STEM to be addressed through socioscientific issues, his point that learning should be contextualized and include embedded cultural components is critically important when considering how STEM might be framed. Zeidler also suggests that STEM alone, as an educational construct, is incomplete. Rather, pedagogical approaches to STEM/STEAM must include a focus on developing global citizens who are intrinsically aware of their space and the role of and value inherent within *all* players. Commensurate with Aikenhead's (2006) humanistic approach to science education, learners must be encouraged to seek out opportunities to encounter the natural world in ways that help them make connections to what they learn within a context of the community and place in which they live.

1.3.2 Transdisciplinary Instruction

Instruction in STEM/STEAM could be considered an aspect of social practice, meaning that communication should be encouraged through collaborative decision-making and dialogue related to scientific praxis. Guyotte et al. (2014) suggest that the typical STEM focus on engineering and more technical components of learning should integrate the arts as a way to encourage social practice and awareness of these connections. They emphasize the possibility of transdisciplinarity through the researchers' mindful approach of having three different disciplines grouped. Groups of students from arts, landscape design, and engineering worked together in solving problems using content knowledge unique to their discipline and collaborative dialogue with group members who varied in perspective and background. This approach suggests that STEAM as a social practice could be especially beneficial for science educators: multiple players working together involve developing and communicating knowledge that encourages equity and a consideration for place and communities. Instruction might include real-world challenges that may not be immediately solvable but encourage students to make meaning of their experiences through the application of multiple disciplines. This approach to instruction might support community-driven science that aligns with place-based instruction and quite possibly encourage a deeper connection to the environment and awareness for social issues that are affected by the decisions of the group. Quigley and Herro (2016) conducted research with 21 middle school classroom teachers using a STEAM-based approach, with a strong emphasis on problem-based instruction. Their research revealed that one of the main challenges to transdisciplinary instruction was a question of how to conduct assessments that aligned with teaching approaches rather than evaluations that directly correlated to the state-mandated multiple-choice assessments. The teachers in that study felt driven to use traditional approaches to testing, even though they realized these were a direct contradiction to the collaborative approaches to learning. Unfortunately, the challenge of state testing is not new and is not one that is easily solvable. However, a transdisciplinary approach allows students to engage in multiple content areas while developing a greater level of social awareness and personal responsibility for learning.

1.3.3 Student Ownership

In their recent STEM research, Quigley and Herro (2016) also found that student choice and relevance were key components to student interest and success. With technology serving as an essential component of instruction in the middle school classrooms where they conducted their research, the teachers in the study moved beyond self-developed PowerPoints®, allowing for student-created blogs and websites that fostered greater ownership and possibly greater content knowledge. Teachers' discussion regarding this work led to the suggestion that others who plan similar types of work should consider the scope, content authenticity, and the students' sense of ownership and responsibility to learn what is needed in order to be successful. With this in mind, STEAM has the potential to encourage collaboration and problem-solving skills that are necessary to function as part of a community. Students have opportunities to develop connections to place when they experience teaching that encourages problem-solving as it relates to issues which are familiar, either as a result of locality or basic interest. While there are many other approaches for considering place, local knowledge, and community in science instruction, these aspects are essential for a sustainable future that promotes health and well-being for all. We must move past an anthropocentric lifestyle to a more ecocentric/concentric purpose that improves the planet, rather than simply strengthening our economy or defense.

This broader conceptualization of the purpose of STEM brings to the forefront a question of the role of ethics and morals within STEM education. With the recognition that hidden forces of race, class, and gender shape an individual's educational experiences, we believe that it is extremely important for educators to ask:

1.4 Question # 4: What Role Should Ethics and Morals Have in STEM Education? The Missing "E" and "M" Pieces in STEM Education

Judith A. Ramaley, the former director of the National Science Foundation's Education and Human Resources Division, coined the term STEM (an acronym for science, technology, engineering, and mathematics) in 2001. Since then, among the different stakeholders such as various agencies in government, education, and research, as well as within different contexts, the definitions of STEM education have varied. Generally, conceptualizations of STEM and what STEM entails fall into one of two domains: education or occupation (Koonce et al. 2011). According to Koonce et al. (2011), from an occupational perspective, jobs related to the four disciplines in STEM are important because they exert more influence on economic growth and development than jobs categorized under non-STEM fields. An important priority of science education reform, and by proxy STEM education, is to promote scientific literacy for all (Fowler et al. 2009). Though different in the conceptualizations of STEM, various stakeholders seem to agree that STEM education is important in producing scientists, engineers, and workers within technically advanced fields. This outcome helps maintain the economic competitiveness of the United States in the global market.

There is a long history of efforts to improve science and mathematics education in the United States documented in reports such as the following: (1) *Science for All Americans* (American Association for the Advancement of Science 1989), (2) *No Child Left Behind* (U.S. Department of Education 2001), (3) *Answering the Challenge of a Changing World* (U.S. Department of Education 2006), (4) *Rising Above the Gathering Storm* (Committee on Science 2007), and (5) *Race to the Top* (U.S. Department of Education 2015). The discourse characterizing these reports reflects support for neoliberal ideology with respect to STEM education. For example, the National Academies' report on *Rising Above the Gathering Storm* emphasizes the importance of focusing on developing STEM skills to ensure the "future prosperity of the United States." Similarly, educational researchers and scholars have claimed that mastery of science and mathematics subjects is key in college success and retention, economic growth and development, national security and innovation, as well as the US competitiveness in the global market (Breiner et al. 2012).

Notably among these reports, *Answering the Challenge of a Changing World* underscores that education is critical to innovation and improving "America's financial security" as well as national security (U.S. Department of Education 2006). This report, along with others, emphasizes the importance of developing scientific literacy, because improving science and mathematics education for all students is key to sustaining American quality and way of life. However, none of these reports mention the importance of ethics and morals and the role of values with respect to STEM education. Tuana (2007) states that "the absence of moral literacy is a glaring omission from our national efforts to strengthen education" and that to fully answer "the challenge of a changing world, we cannot ignore the essential role of moral literacy in our children's education" (p. 365). In this section, we explore the role of ethics, morals, and values in STEM education and we propose that ethical and moral dimensions should be considered an important aspect of STEM education for our youth in the twenty-first century.

The call to encompass ethical and moral dimensions as part of fostering scientific literacy for our youth is not new. However, the extent of the discourse regarding the potential role of ethics, morals, and values in STEM education is underwhelming when compared to the broader scope of discourse in STEM/STEAM/STREAM/STEMM education. Thus, the topic of ethical and moral dimensions with respect to STEM education is an important one to revisit. We suggest that scientific literacy encompasses the ability to "negotiate and make decisions regarding complex social issues with theoretical and/or conceptual links to science" (Fowler et al. 2009, p. 279). Along the same vein, to prepare for the challenges of a changing world in the twenty-first century, scholars have argued that students should learn important skills such as "adaptability, complex communication, social skills, nonroutine problem-solving, self-management, and systems thinking to compete in the modern economy" (Bybee 2010, p. 996). Alongside this growing challenge, socioscientific issues have begun to garner the attention of science education researchers and teachers, as "socioscientific issues provide situations where teachers and students need to analyze complex issues associated with ethical, political, and social dilemmas" (Mueller and Zeidler 2010, p. 105). Science is a social process, and scientific knowledge is formulated within particular social contexts, so compartmentalizing science as separate from ethics, morals, and values would be impossible in the context of socioscientific issues. Thus, an argument follows that "socioscientific decision-making occupies a seminal space in scientific literacy and attention to morality and ethics must be included in the science curriculum" (Sadler 2004, p. 39). Sadler (2004) further argues that one's inability to "successfully utilize any of these three aptitudes will significantly hamper one's ability to make judgments regarding socioscientific issues and by extension limit scientific literacy" (p. 42). Similarly, Fowler et al. (2009) explicitly address moral aspects of issues using socioscientific issues-based science teaching. In this approach, they found that regular exposure to socioscientific issues promoted students' development of moral sensitivity that was context dependent. Thus, Fowler et al. (2009) drew similar conclusions: that moral considerations are critical to the negotiation of socioscientific issues. Tuana (2007) conceptualized basic components of moral literacy to include ethics sensitivity, ethical reasoning skills, and moral imagination. Tuana makes a similar argument—that moral literacy should be developed under the guidance of an experienced teacher who is well versed in moral subject matter. Tuana posits that only an integrated approach to developing all three components can lead to moral literacy.

We agree with the scholars who call for emphasizing the importance of ethical and moral dimensions in today's landscape of promoting STEM education and fostering scientific literacy for our youth. We also agree that our youth must learn to negotiate and make decisions when faced with complex, social issues that are intricately linked to science. However, we take a moment to take up a critical stance on the means by which we might achieve a goal of heightening various aspects of ethical and moral dimensions in STEM education (i.e., moral sensitivity, moral literacy).

As noted previously, conceptualizations of STEM education and scientific literacy differ in different contexts for various stakeholders. Even though the long-standing goal of science education is to promote scientific literacy, the meanings of the term remain ambiguous (Sadler 2004). If one were to follow the recommendations of Tuana (2007), there are challenges to the development of moral literacy which integrate all of the three components. First, secondary science teachers lack suitable material for including ethical and moral aspects of science in their teaching (Reiss 1999). Second, materials suitable for training and preparing teachers do not adequately address issues of morality and ethics in science (Sadler 2004). In this regard, we echo certain aspects of Reiss' (1999) arguments against teaching ethics as part of the domain of science: that is, science and ethics occupy separate spheres of knowledge. "Science teachers are generally educated in science, not in moral philosophy. It is therefore unrealistic and unfair to expect them to teach ethics" (Reiss 1999, p. 119). To explicitly teach, develop, and foster moral literacy as part of scientific literacy adds to the conundrum of the ambiguity of scientific literacy by further broadening the scope and meanings of the term. Further, to teach ethics and promote moral sensitivity, as well as develop moral literacy, scholars argue that the teacher must be trained and well versed in moral subject matter. This is an additional challenge that science teachers today face in light of the notion that STEM education is an integrated curriculum of four (or more) disciplines. In the context of STEM education, science teachers are often expected to teach the separate disciplines of science, technology, engineering, and mathematics as one cohesive unit, which requires deep understanding of all four subjects (Breiner et al. 2012). How the next generation of science teachers can be prepared to become experts in four disciplines or even more in order to effectively teach STEM as a cohesive entity remains to be seen and is a continuing challenge for science teachers.

Despite these challenges, science cannot be divorced from values, and thus, we critically evaluate the aims of highlighting ethical and moral considerations in science education. Reiss (1999) provides four recommendations:

> First, teaching ethics in science might heighten the ethical *sensitivity* of participants.
> Second, teaching ethics in science might increase the ethical *knowledge* of students.
> Third, teaching ethics in science might improve the ethical *judgments* of students.
> Fourth, teaching ethics in science might make students *better people* in the sense of making them more virtuous or otherwise more likely to implement normatively right choices.
>   (pp. 123–124)

These goals are important, especially in today's landscape of STEM education propelled by neoliberal ideology. However, these aims cannot be all equally met by a single teacher, or by a single teaching method. These goals would certainly be another challenge that teachers must overcome in terms of developing their own familiarity and expertise in moral subject matter. Coupled with these challenges, our next question is precisely "the question of how, if ethics is to be taught in science, it should be" (Reiss 1999, p. 117). To answer this question, we want to return to one of Fowler et al.'s (2009) findings that students' development of moral sensitivity was context dependent, which helped them understand the direct impact of socioscientific issues on humans and society. This finding supports similar findings by other scholars, such as Zeidler et al. (2005), where science became more meaningful to students when placed in a context where students could understand how science affected technology and how technology, in turn, directed society.

In light of these accounts, instead of creating another knowledge domain such as moral literacy that teachers must master in order to teach students ethics, we suggest that science educators and teachers create teachable moments to hold dialogues about ethical and moral considerations in science, as well as guide students to develop their own value judgments with respect to science. An account of such dialogue is offered in Jeong et al.'s (2017) *A story of chicks, science fairs and the ethics of students' biomedical research*. The case narrative in Jeong et al.'s (2017) chapter demonstrates a meaningful conversation between a science educator and a high school student about the student's science fair experiment conducted on prematurely born chicks to test the effect of alcohol and the projected implications on human pregnancy. The intention of the science educator was to encourage the student to explore and engage in independent questioning of her own science experiment with respect to ethics, morals, and values. The student in this case was encouraged by her teacher to conduct the experiment on chicks; the student, who was intelligent and articulate, showed initiative and diligence in carrying out the experiment to yield significant results. As one can see from this example, fostering value judgments in students can be challenging; however, it is feasible to achieve. First, as supported from the findings of other research, developing value judgments needs to occur in a context that is meaningful for the student. When the student (Jenna) in this case was guided to think about the direct impact with respect to human babies, she came to a realization on her own about the potential ethical considerations of dosing chicks with alcohol. Second, asking a student non-leading questions so that he/she can independently think about the issues of ethics, morals, and values is important. In this case, for example, the science teacher educator never explicitly mentioned or used the terms ethics, morals, or values. Yet, she facilitated dialogue with the student by asking questions that allowed Jenna to be able to think for herself about the implications of the experiment that neither she nor her teacher had considered, which were ethical and moral considerations of science. As seen in the conduct of Jenna's high school teacher who suggested the original experiment characterized by the quest for "right answers," it is easy to suggest an answer to what was supposed to be an inquiry-oriented science fair project. What is perhaps more important about eliciting these critical questions in conversations with students is that it is equally as easy to allude to a solution or an answer to an ethical or moral issue in science during these discussions. However, teachers can avoid these pitfalls by encouraging independent thinking on issues involving ethics and morals, without judgment.

We argue that ethical and moral considerations should be brought to the forefront of the STEM education discourse. However, we propose a way to do so without putting additional burden on science teachers to have specialized knowledge or understanding of moral subject matter. We recommend having meaningful conversations with students within the appropriate context that matters to them and their lives and facilitating critical questions that will help them come to their independent value judgments and conclusions about issues of ethics and morals in science.

From a postmodern perspective, the questions we ask are often more important than the products or answers. The questions we pose in this chapter are intended to be divergent rather than convergent—they do not require right or wrong answers. Nevertheless, the landscape of STEM education reflects an underlying neoliberal ideology. As Foucault reminds us, there is no field of knowledge that is not also imbricated in (4) forces of power, but also that knowledge/power constructions are not in and of themselves morally charged—it's how they are used. Including arts into STEM curricular programs and philosophies should be considered by examining what types of knowledges and ways of being it explicitly and implicitly legitimizes. "Understanding STEM through a discursive lens points toward the play of power in the current education policy environment and the possibility that what would appear on the surface to be an opening for the promotion of scientific, mathematical, and technological knowledge may, in fact, constitute a form of closure" (Ellison and Allen 2016). In the following question, we explore the ways that STEM is imbricated in neoliberal discourses, and to what extent the shift to STEAM education represents a continuation or departure from neoliberalism by raising the following question:

1.5 Question # 5: What Is Neoliberalism and Why Is It Relevant to STEM/STEAM Education?

The *liberalism* of neoliberalism refers to the economic liberalism that holds it is the role of the government to protect individual citizens' right to engage free market choices. *Neo*liberalism refers to political and social perspectives that have become more expansive since the 1980s when we saw economic ideals increasingly encroach political and social spheres. One example is the argument that government should withdraw from funding public projects that promote the public welfare and instead reduce taxes so individuals have the freedom to individually purchase services from private corporations. Another defining aspect of neoliberalism is idealizing the market values of cost-benefit analyses and efficiency across our private and social lives, usually working hand in hand with technological advances. The focus of individual efficiency is usually presented hand in hand with the government withdrawal from providing social services. For example, in a time where public and private higher education costs are skyrocketing because of reduction in tax revenue and aversion to regulatory practices, we see a rise in "flexible" online learning opportunities so that people can work and go to school at the same time. Mobile applications that collect and track data on our habits and health that can be used to make recommendations on how to change our behavior are another example of neoliberalism at work. Again, we can see correlates in education in data-driven instruction and scripted curriculum programs.

The STEM movement in education has been critiqued as a neoliberal education initiative that emphasizes the production of future capitalist producers and consumers rather than critical thinkers prepared to engage in civil democratic institutions. Bencze et al. (2018) reviewed "websites, curriculum policy and instructional materials, STEM education brochures, publications about revelation research, etc." and found *neoliberal capitalism* was "perhaps *the* – driving force determining much of STEM education" (pp. 70–71). The rationale for arguing for STEM education is founded in the idea that American students are falling behind global peers in proficiency in science, technology, engineering, and mathematics, which is disastrous because it would put the United States at a competitive disadvantage in the global marketplace. Or, as Zeidler (2016) puts it, there is "laser-like focus on technocratic issues embedded in the language of STEM initiatives" (p. 14). Granted, the technological and scientific breakthroughs that are the goals of STEM education are not overtly, purely profit driven. Advances in health, medicine and energy technology are embedded in humanitarian as well as economic ideals. Moreover, STEM proponents argue that they are taking a pragmatic, even social-justice-oriented approach to ensuring that the next generation is well prepared for global realities.

However, a deeper investigation of these arguments for STEM education has found that they are largely spurious. The claim that there is a skills gap for American students (versus specific differences in scores on standardized tests) came from a group of chief executive officers in US companies, in the *Closing America's Skills Gap* report; this report also blamed the K–12 education system for the gap (Sharma 2016). However, research indicates that "this crisis exists only on the pages of reports and position papers emanating from groups and organizations connected with big business" (Capelli 2015 and Salzman et al. 2013, as cited by Sharma 2016, p. 44). Further, Ellison and Allen (2016) reviewed extensive labor and wage reports and projects that show a decrease in value of STEM jobs due to a decreased need of workers, along with increased output of those with STEM training and the number of overqualified workers who have to take jobs outside of their fields. A critical examination of STEM initiatives through this lens could contend that the benefactors of this shift are corporations that can choose to pay lower wages and can capitalize on the results of pushing an educational system specifically tailored to create the most cost-effective worker.

Another general critique of STEM education is that the economic instrumentalism of STEM risks erasing the particular social and cultural contexts of science, technology, engineering, and mathematics individually. "On moving from science to STEM, technical fields are linked into a mutually reinforcing and referential sphere, and critically, science is unlinked from issues of history, sociology, and ethics" (Weinstein 2016, p. 238). The idea of STEAM itself can be considered as a response to STEM's intellectual flattening of disciplines. As a proponent of socioscientific practices, Zeidler (2016) argued that "a focus on STEM sans 'Arts' necessarily excludes important areas that inform and contextualize science by grounding them in sociocultural contexts" where "arts" are defined broadly by social sciences and humanities (p. 17). For educators who value the studies of humanities and the liberal arts, the expansion of STEM to STEAM could be heralded as a move that could rescue STEM. The potentiality of this scenario rests on what is being understood by "arts."

1.5.1 What Is STEAM Art?

At face value, including the "A" with S-T-E-M appears to indicate that the arts as a content area is going to be held at the same level with and become a part of the study of the interaction between science, technology, engineering, and math. However, a review of STEAM-focused educational resources reveals that art does not refer to the subject matter or content of what is taught, but rather how it is taught. A cornerstone of STEAM education is project-based learning and arts integration that allows students to express their knowledge of STEM topics through visual, dance, or performative arts. The Kennedy Center—an organization one would think would put primacy on the study and lifting up of art for understanding and appreciating art for its own sake—has an extensive arts-integration program which defines arts integration as "an approach to teaching in which students construct and demonstrate understanding through an art form" (https://artsedge.kennedy-center.org/educators/how-to/series/arts-integration/arts-integration). To the extent that art content is incorporated, its focus is on engaging students in STEM content or spurring on creative forces. For example, the website steamtostem.org states that "In this climate of economic uncertainty, America is once again turning to innovation as the way to ensure a prosperous future" and that "Art + Design are poised to transform our economy in the 21st century just as science and technology did in the last century." We can see the same strains of neoliberalism tying together the rationale of STEAM education as that underpinning STEM education, without any signs of fraying. If anything, it appears that a new "problem" of creativity and innovation has been constructed for public education to solve. Moreover, the reduction of "art" to how it can be used to create and design products elides how the intellectual engagement of social sciences and humanities can be a critical part of public discourse. It appears to us that neoliberalism is interested in art only insofar as it can be commodified.

1.5.2 STEAM

Neoliberalism is a concept that we can see enacted out in policy, ideology, and Foucauldian governmentality, the way that social and cultural discourses are embodied and integrated into individuals' decision-making processes (Larner 2000). Including "arts" into STEM may be offered as a surface-level response to the neoliberal critiques of STEM education; but it ignores the depth and complexities of each discipline. A more in-depth analysis of how "arts" is being understood in STEAM indicates that the linguistic move to add an "A" is not indicative of liberal arts or humanities. To the extent that art is studied as a subject in STEAM, it appears to primarily be in service of understanding product design and marketing, in other words, a "neoliberal appropriation of the arts in STEAM transform art into a corporatist practice of techno-rational 'design'" (Weinstein 2016, p. 239).

While student engagement and culturally responsive STEM instruction can be enriched with arts-based instruction, this is not the equivalency of reintroducing social sciences, liberal arts and humanities on their own merits. Some proponents of STEAM education point to the ways that high-stakes testing has squeezed the opportunities for creative exploration out of schools, but as noted, neoliberalism is only interested in art insofar as it can be commodified. Those that are invested in the sociocultural contexts of STEM disciplines should not uncritically or unconditionally embrace STEAM education initiatives.

1.5.3 Some Concluding Thoughts

The questions we have posed in this chapter, and the tensions they reflect, lead us to conclude that there is an urgent need for an inclusive view of STEM education. This view extends existing techno-narratives of science, existing solely in an economic context, to one that seeks to understand science principles in a more systems-based approach. We recognize that the questions we pose defy easy answers or solutions. They are questions rooted in an awareness of our own histories, theories, methodologies, and subject matter preparation, in particular disciplines. While it may seem that the long-standing division between science and the humanities remains a resilient obstacle to STEM-STEAM reforms, we are encouraged by the everyday examples of productive STEM spaces. For example, noted oceanographer and researcher Samantha Joye has looked to art to better understand the impact of the Deepwater Horizon oil spill on biodiversity. Likewise, we are not so naïve as to think that the questions we have posed in this chapter are the only or most important ones. We concur with Kincheloe et al. (1999) when they state that "knowledge is intimately tied with asking, not answering, and learning becomes insight into what to ask and what asking means" (p. 176).

References

Aikenhead, G. S. (2006). *Science education for everyday life: Evidence-based practice.* New York: Teachers College Press.

American Association for the Advancement of Science. (1989). *Science for all Americans.* Washington, DC: American Association for the Advancement of Science.

Anthony, A. S. (2004). Gender bias and discrimination in nursing education: Can we change it? *Nursing Educator, 29*(3), 121–125.

Association of American Medical Colleges. (2016). *Current trends in medical education.* Retrieved from http://www.aamcdiversityfactsandfigures2016.org/report-section/section-3/#figure-1http://www.edweek.org/ew/articles/2011/12/01/13steam_ep.h31.html

Basil, V., & Murray, K. (2015). Uncovering the need for diversity among K–12 STEM educators. *Teacher Education and Practice, 28*(3), 255–268.

Basile, V., & Lopez, E. (2015). And still I see no changes: Enduring views of students of color in science and mathematics education policy reports. *Science Education, 99*(3), 519–548.

Bencze, L., Reiss, M., Sharma, A., & Weinstein, M. (2018). STEM education as 'Trojan horse': Deconstructed and reinvented for all. In L. Bryan & K. Tobin (Eds.), *Thirteen questions in science education* (pp. 69–87). New York: Peter Lang.

Bennington, G. (1999). Inter. In M. McQuillan, G. MacDonald, R. Purves, & S. Thomson (Eds.), *Post-theory: New directions in criticism* (pp. 103–119). Edinburgh: Edinburgh University Press.

Bologna, C. (2015, April 7). *12 Brilliant kids' clothing lines that say no to gender stereotypes.* Retrieved from http://www.huffingtonpost.com

Breiner, J. M., Harkness, S. S., Johnson, C. C., & Koehler, C. M. (2012). What is STEM? A discussion about conceptions of STEM in education and partnerships. *School Science and Mathematics, 112*(1), 3–11.

Bricker, P., Jackson, E., & Binkley, R. (2015). Building teacher leaders and sustaining local communities through a collaborative farm to school education project—What EcoJustice work can PreService teachers do? In M. P. Mueller & D. J. Tippins (Eds.), *EcoJustice, citizen science and youth activism* (pp. 83–97). Geneva: Springer.

Butler, L., Scammell, M., & Benson, E. (2016). The Flint, Michigan, water crisis: A case study in regulatory failure and environmental injustice. *Environmental Justice, 9*, 93–97. https://doi.org/10.1089/env.2016.0014.

Bybee, R. W. (2010). What is STEM education? *Science, 329*(5995), 996. https://doi.org/10.1126/science.1194998.

Capelli, P. H. (2015). Skill gaps, skill shortages, and skill mismatches: Evidence and arguments for the United States. *ILR Review*, 0019793914564961.

Capobianco, B., & Yu, J. (2014). Using the construct of care to frame engineering as a caring profession toward promoting young girls' participation. *Journal of Women and Minorities in Science and Engineering, 20*(1), 21–33.

Clark, L. P., Millet, D. B., & Marshall, J. D. (2014). National patterns in environmental injustice and inequality: outdoor NO2 air pollution in the United States. *PloS One, 9*(4), e94431.

Collins, M., Munoz, I., & JaJa, J. (2016). Linking 'toxic outliers' to environmental justice communities. *Environmental Research Letters, 11*, 015004. https://doi.org/10.1088/1748-9326/11/1/015004.

Committee on Science, Engineering, and Public Policy. (2007). *Rising above the gathering storm: Energizing and empowering America for Brighter economic future.* Washington, DC: The National Academies Press.

Dugdale, D., Epstein, R., & Pantilat, S. (1999). Time and the patient-physician relationship. *Journal of General Internal Medicine, 14*(1), 34–40.

Ellison, S., & Allen, B. (2016). Disruptive innovation, labor markets, and big Valley STEM School: network analysis in STEM education. *Cultural Studies of Science Education.* https://doi.org/10.1007/s/11422-016-9786-9.

Fowler, S. R., Zeidler, D. L., & Sadler, T. D. (2009). Moral sensitivity in the context of socioscientific issues in high school science students. *International Journal of Science Education, 31*(2), 279–296. https://doi.org/10.1080/09500690701787909.

Guyotte, K. W., Sochacka, N. W., Costantino, T. E., Walther, J., & Kellam, N. N. (2014). STEAM as social practice: Cultivating creativity in transdisciplinary spaces. *Art Education, 67*(6), 12–19.

Haraway, D. J. (1997). *Feminism and technoscience.* New York: Routledge.

Hill, C., Corbett, C., & St. Rose, A. (2010). *Why so few? Women in science, technology, engineering, and mathematics.* Washington, DC: AAUW Research Report.

Jeong, S., Tippins, D. J., & Kayumova, S. (2017). A story of chicks, science fairs and the ethics of students' biomedical research. In M. P. Mueller, D. J. Tippins, & A. J. Stewart (Eds.), *Animals and science education: Ethics, curriculum and pedagogy* (Vol. 2, pp. 99–121). Cham: Springer.

Jowett, L. (2007). Lab coats and lipstick: Smart women reshape science on television. In S. Inness (Ed.), *Geek Chic: Smart women in popular culture.* New York: Palgrave Macmillan.

Kincheloe, J., Steinberg, S., & Tippins, D. (1999). *The stigma of genius: Einstein, consciousness and education*. New York: Peter Lang.

Koonce, D. A., Zhou, J., Anderson, C. D., Hening, D. A., & Conley, V. M. (2011). *What is STEM?* Paper presented at the American Society for Engineering Education annual conference proceedings.

Krug. (2012). *STEM Education and Sustainability in Canada and the United States*. Paper presented at the 2nd international STEM in Education conference. Retrieved from http://stem2012.bnu.edu.cn/data/long%20paper/stem2012_87.pdf

Krzesni, D. (2014). *Pedagogy for restoration: Addressing social and ecological degradation through education*. New York: Peter Lang Publishing.

Larner, W. (2000). Neo-liberalism: Policy, ideology, governmentality. *Studies in Political Economy, 63*, 5–26.

Loudenback, T. (2016, June 15). This company's toys are helping mold in a new generation of engineers. *Business Insider*. Retrieved from http://www.businessinsider.com

Merchant, C. (1980). *The death of nature: Women, ecology and the scientific revolution*. San Francisco: Harper & Row.

Moran, J. (2010). *Interdisciplinarity*. New York: Routledge.

Mueller, M. P., & Zeidler, D. L. (2010). Moral–ethical character and science Education: EcoJustice ethics through socioscientific issues (SSI). In D. J. Tippins, M. P. Mueller, M. van Eijck, & J. D. Adams (Eds.), *Cultural studies and environmentalism: The confluence of EcoJustice, place-based (science) education, and indigenous knowledge systems* (pp. 105–128). Dordrecht: Springer.

National Girls Collaborative Project. (2017). *Statistics: State of girls and women in STEM*. Retrieved from https://ngcproject.org/statistics

National Science Board. (2016). *Science & engineering indicators*. National Science Foundation. Retrieved from https://www.nsf.gov/statistics/2016/nsb20161/uploads/1/nsb20161.pdf

Portillo, S. (2017, August 23). White men must learn to say yes. *Inside Higher Ed*. Retrieved from https://www.insidehighered.com

Quigley, C. F., & Herro, D. (2016). "Finding the joy in the unknown": Implementation of STEAM teaching practices in middle school science and math classrooms. *Journal of Science Education and Technology, 25*(3), 410–426.

Rabin, R. C. (2014, April 21). 15-Minute visits take a toll on the doctor-patient relationship. *Kaiser Health News*. Retrieved from http://khn.org

Reese, S. (2011, June 23). Women MDs spend more time with patients: Does it matter? *Medscape*. Retrieved from http://www.medscape.com

Reiss, M. J. (1999). Teaching ethics in science. *Studies in Science Education, 34*, 115–140. https://doi.org/10.1080/03057269908560151.

Sadler, T. D. (2004). Moral and ethical dimensions of socioscientific decision-making as integral components of scientific literacy. *Science Educator, 13*(1), 39–48.

Salzman, H., Kuehn, D., & Lowell, B. L. (2013). *Guestworkers in the high-skill U.S. labor market*. Retrieved from http://www.epi.org/publication/bp359-guestworkers-high-skill-labor-market-analysis/

Sharma, A. (2016). The STEM-ification of education: The Zombie reform strikes again. *Journal for Activist Science & Technology Education, 7*(1), 43–51.

Tsugawa, Y., Jena, A., Figueroa, J., Orav, J., Blumenthal, D., & Jha, A. (2017). Comparison of hospital mortality and readmission rates for Medicare patients treated by male vs female physicians. *JAMA Internal Medicine, 177*(2), 206–213.

Tuana, N. (2007). Conceptualizing moral literacy. *Journal of Educational Administration, 45*(4), 364–378. https://doi.org/10.1108/09578230710762409.

U.S. Department of Education. (2001). *No child left behind*. Retrieved from http://files.eric.ed.gov/fulltext/ED447608.pdf

U.S. Department of Education. (2006). *Answering the challenge of a changing world: Strenghtening education for the 21st century*. Retrieved from http://files.eric.ed.gov/fulltext/ED493422.pdf

U.S. Department of Education. (2015). *Fundamental change: Innovation in America's schools under Race to the Top*. Retrieved from https://www2.ed.gov/programs/racetothetop/rttfinalrpt-full.pdf

Weinstein, M. (2008). Finding science in the school body: Reflections on transgressing the boundaries of science education and the social studies of science. *Science Education, 92*(3), 389–403.
Weinstein, M. (2016). Imagining science education through ethnographies of neoliberal resistance. *Mind, Culture, and Activity, 23*(3), 237–246.
Zeidler, D. L. (2014). Socioscientific issues as a curriculum epmphasis: Theory, research and practice. In N. G. Ledermand & S. K. Abell (Eds.), *Handbook of research on science education* (Vol. 11, pp. 697–726). New York: Routledge.
Zeidler, D. L. (2016). STEM education: A deficit framework for the twenty first century? A sociocultural socioscientific response. *Cultural Studies of Science Education, 11*(1), 11–26.
Zeidler, D. L., Sadler, T. D., Simmons, M. L., & Howes, E. V. (2005). Beyond STS: A research-based framework for socioscientific issues education. *Science Education, 89*(3), 357–377. https://doi.org/10.1002/sce.20048.

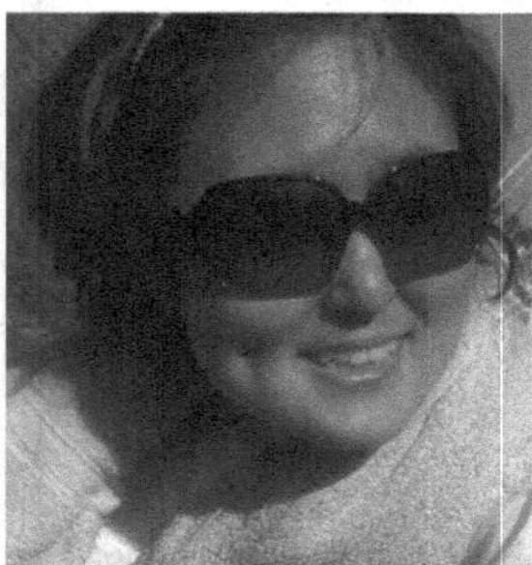

Sophia Jeong is a postdoctoral scholar in the Department of Biochemistry and Molecular Biology at the University of Georgia. Her scholarly work draws on actor-network theory to explore ontological complexities of gender and race by examining sociomaterial relations in science classrooms.

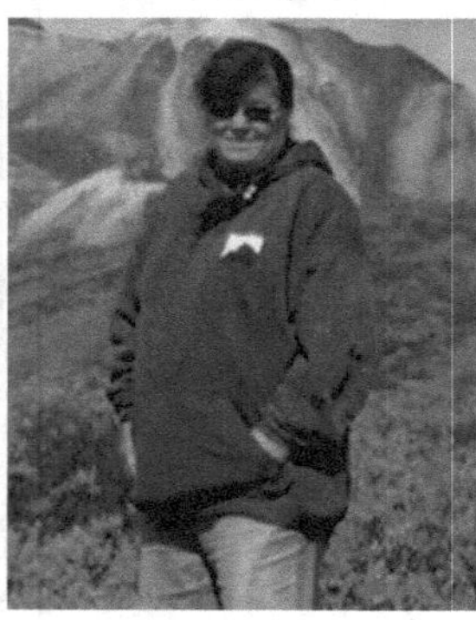

Deborah J. Tippins is currently a professor in the Department of Mathematics and Science Education at the University of Georgia. Her scholarly work focuses on encouraging meaningful discourses around environmental justice and sociocultural issues in science education.

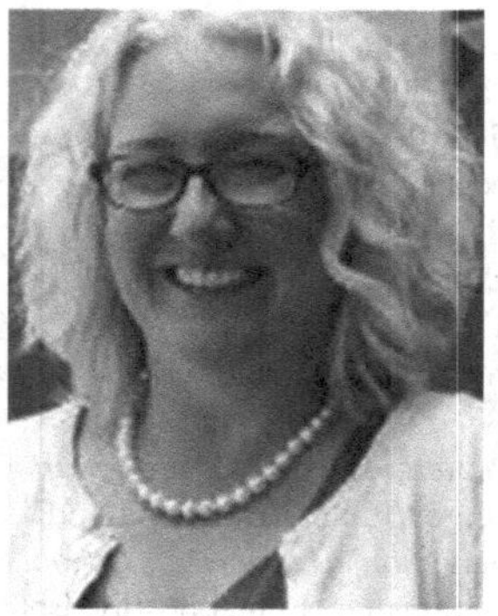

Kimberly Haverkos is founding dean of the College of Education and Health Sciences at Thomas More University. She earned a Bachelor's degree from Xavier University in Biology and a Master of Education in Science Education from the University of Cincinnati. She joined Thomas More's Education Department in the fall of 2012 after receiving her doctorate from Miami University, where her dissertation work explored girls' attitudes and behaviors around "going green" and science, technology, engineering, and math (STEM). Her areas of research and teaching continue to focus on STEM education and cultural studies. Haverkos also teaches in the graduate and undergraduate programs in the Education Department.

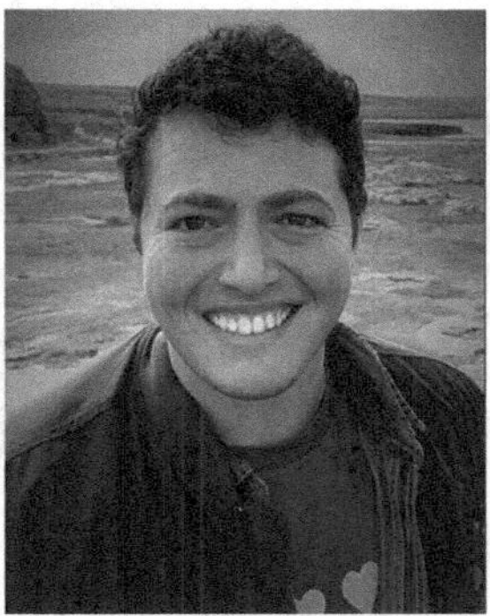

Mel Kutner is currently a doctoral student in Educational Theory and Practice at the University of Georgia. Mel has a Master of Science degree in Conflict Analysis and Resolution, and their research interests include using pre-personal affect theory and queer theory to understand comedy and conflicts in education. Pronouns I use: they/them/theirs.

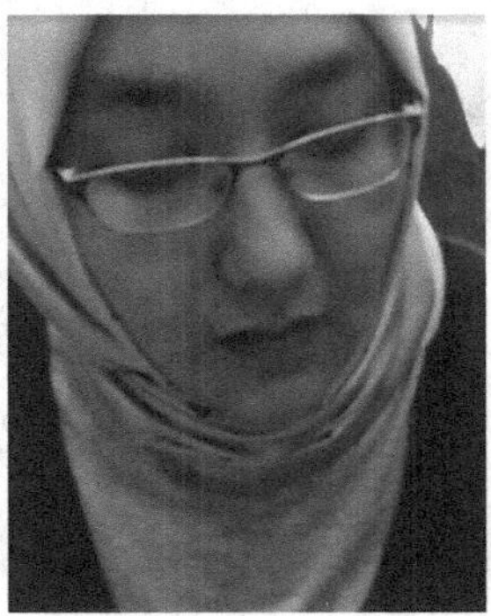

Shakhnoza Kayumova is currently an assistant professor in the STEM Education and Teacher Development Department at the University of Massachusetts, Dartmouth. Her interdisciplinary work focuses on the empowerment of bilingual/multilingual students and girls using equitable STEM practices.

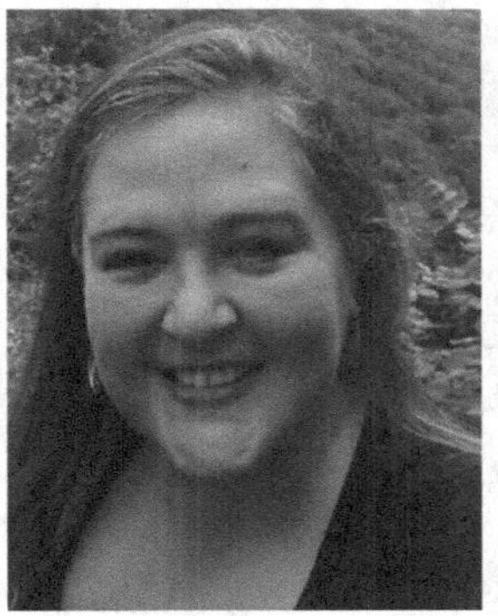

Stacey Britton is an assistant professor of Science Education at the University of West Georgia and hosts a monthly science café in the local community. Her research seeks to connect Ecojustice Theory within the framework of STEM education in ways that increase access and understanding of the larger community. When Stacey is not working on "school-related" items, she enjoys quilting and spending time in the outdoors.

Correction to: Converting STEM into STEAM Programs

Arthur J. Stewart, Michael P. Mueller, and Deborah J. Tippins

Correction to:
A. J. Stewart et al. (eds.), *Converting STEM into STEAM Programs*, Environmental Discourses in Science Education 5,
https://doi.org/10.1007/978-3-030-25101-7

In the original version of the book, the following correction has been incorporated: The chapter author's, "Sophia (Sun Kyung) Jeong", first and last names have been corrected. The correction chapter and book have been updated with the change.

The updated version of these chapters can be found at
https://doi.org/10.1007/978-3-030-25101-7_16
https://doi.org/10.1007/978-3-030-25101-7_18